The Agronomy of the Major Tropical Crops

The Agronomy of the Major Tropical Crops

C. N. WILLIAMS

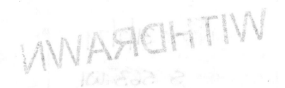

KUALA LUMPUR
OXFORD UNIVERSITY PRESS
LONDON   NEW YORK   MELBOURNE
1975

Oxford University Press

OXFORD LONDON GLASGOW NEW YORK
TORONTO MELBOURNE WELLINGTON CAPE TOWN
DELHI BOMBAY CALCUTTA MADRAS KARACHI DACCA
KUALA LUMPUR SINGAPORE JAKARTA HONG KONG TOKYO
NAIROBI DAR ES SALAAM LUSAKA ADDIS ABABA
IBADAN ZARIA ACCRA BEIRUT

Typeset by Art Printing Works, Kuala Lumpur
Printed litho by Dainippon Tien Wah Printing (Pte) Ltd. Singapore
Published by Oxford University Press, Bangunan Loke Yew, Kuala Lumpur

CONTENTS

TABLES

FIGURES

Acknowledgements

I am grateful to Dr. K.T. Joseph for reading the manuscript, and to Miss Loh Im Hoe for much editorial assistance.

Note

Since the book reviews diverse material both metric and other units are used.

1 INTRODUCTION

THIS book is written as a guide for students and research workers interested in tropical crops. It gives a general account of the botany and the agronomy of the crops considered including the research work carried out on these crops up to about the early 1970s.

The major tropical crops have been chosen on the basis of their importance in the economies of tropical countries. Primarily we are concerned with the equatorial regions of the world which support largely a forest climax under natural conditions and have been used for plantation agriculture with perennial crops; and with the extensions of this environment outside the tropics on the seaboards of the southern continents of Australia, Africa and South America, and to the north in Indo-China. The broad areas under consideration are indicated in Fig. 1.1 below, and the climates of these areas may range from ever-humid, in which all months are humid, to the monsoonal climate in which a sub-humid or dry period of several months may occur. In all areas, however, a perennial, essentially non-deciduous type of vegetation can be supported, and low temperatures, except in the case of high elevation sites and the extension of tropical crop growing outside the equatorial zone, are seldom a limiting factor.

Fig. 1.1 Distribution of humid tropical zones

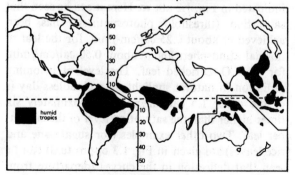

Source. Adapted from Blummenstock and Thornthwaite, 1941.

Certain areas within the equatorial zone, notably the East African savannahs away from the coast and the great lakes, and southern Brazil/northern Argentina, are more arid and more suitable for arable cropping and grazing. Some of the agriculture systems and crops of these regions are much closer to those of the sub-tropical and temperate regions and are adequately dealt with in a number of excellent texts.

The way in which the environment is exploited in carrying on agricultural practices is of particular importance in the tropics because of the monocropping necessary with perennial crop species, the absence of fallow periods necessitated by seasonal climate variations, and also because of the tendency to exploit sometimes the environment with a relatively short-term view. The tropical agriculturist should always have in mind the question of conservation of resources by correct farming procedures, the preservation of natural watersheds, areas of unstable soils, zones of natural forest and so on, and should have as an overall aim, the evolution of stable systems of agriculture to be inherited in good order by the generations hereafter. The ultimate agricultural systems of the near future will have to be based on closed cycling or near closed cycling of plant nutrients and the careful utilization of water resources.

Tropical agriculture, at the present time, presents a devious and confusing polyglot of agricultural practices, sometimes generally backward but sometimes also surprisingly good, as will be seen in the following chapters. Before considering the individual crops, however, some general considerations on three aspects of crop production which merit special attention need to be dealt with. The first concerns the assimilation apparatus of the crop, the second, propagation and nursery techniques used in many tropical crops, and the third, herbicide use and the associated development of minimal and zero cultivation techniques. Questions of soil fertility and fertilizer use are not considered here because of the large number of books dealing with these subjects.

THE ASSIMILATION APPARATUS

Traditionally, agronomy has been much concerned with the soil, since this is one phase of the environment which lends itself to manipulation. Much of the early work on crop production was concerned with the overcoming of nutrient deficiencies which have often constituted the major bottle-neck to production. In recent years, however, widespread fertilizer use has greatly reduced losses due to soil factors and interest has become focused on other factors controlling yield. Central among these is the crop canopy and its properties of assimilation; the way in which light is used to carry out photosynthesis which constitutes, by far, the greatest part of crop yield. Of course, it is not as easy to manipulate the aerial environment of a crop as it is with the root environment (by irrigation, fertilizer use etc.), but it is possible to manipulate by genetic and other means the crop canopy or assimilation apparatus to make better use of the available raw materials of carbon dioxide and light. Probably the greatest yield increases have been achieved by these procedures and examples of extremely high-yielding tropical crop varieties are seen in maize (see Tanner and Jones, 1965), soybeans (Anderson *et al.* 1965), rice (Chapter 5), tea (Chapter 8) and tapioca (Chapter 10). Such varieties probably occur in all crop species, and the potentials for improving yield in most crops undoubtedly exist. A general feature of all these high-yielding canopy varieties is that they tend to have smaller and more vertically-orientated leaves, and furthermore, they require higher fertilizer levels, particularly nitrogen, and closer spacing for full yield expression to occur. Given these circumstances, such varieties do not always show up well in variety trials conducted at a standard spacing and fertilizer treatment.

The reasons for the superiority of these varieties is connected with the utilization of light in the crop canopy and is discussed briefly below; further discussions occur in some of the following chapters on individual crops.

Leaf and canopy characteristics in assimilation

Photosynthesis involves the photolytic splitting of water followed by the reduction of CO_2 from the atmosphere. Consequently, the availability of CO_2 affects the rate of photosynthesis and the uti-

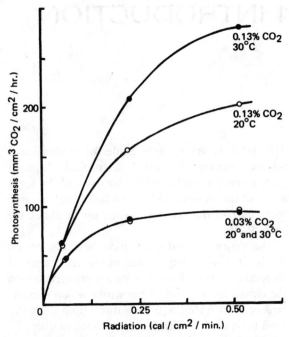

Fig. 1.2 Photosynthesis of cucumber leaves in relation to light intensity, in normal air and in CO_2-enriched air at 30°C (near optimal) and 20°C (below optimal)

Source. After Gaastra, 1963.

lization of incoming light energy which must be used more or less instantaneously as it is received. The average level of CO_2 in the lower atmosphere is about 0.03 per cent and in fact this does constitute a severe limiting factor in the utilization of light when considering the exposed leaf surface. If the level of CO_2 in the atmosphere can be increased, then the level of photosynthesis increases and the utilization of more light occurs. In general, however, the quantities of light available for photosynthesis in natural sunlight are greater (sometimes far greater) than that which can be utilized by the leaf surface receiving only atmospheric CO_2. This is indicated in Fig. 1.2 above. Here it can be seen that saturation (threshold photosynthesis) has been achieved at about 0.25 cal/cm²/min. for the leaf in normal atmosphere and at about 0.50 cal/cm²/min. for the CO_2-enriched leaf. The level of incoming radiation in natural sunlight on a cloudless day is of the order 1.5 cal/cm²/min., approximately six times higher than the saturation level of the cucumber leaf. Two other examples (for sugar-cane and rice leaves) are given in Fig. 1.3 opposite. It can be seen that deflection in the curves (departure from a linear relationship between radiation and photosynthesis) occurred at about 0.2 cal/cm²/min. in rice

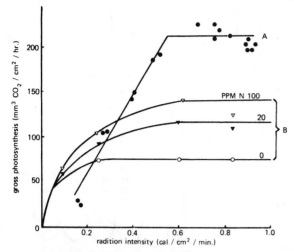

Fig. 1.3 Relationship between photosynthesis and radiation in sugar-cane and rice leaves

Sources. A. After Waldron *et al.* 1967.
   B. Adapted from Navasero and Tanaka, 1966.

Notes. A. Mutual shading was prevented by arranging the leaves on a frame so that each was fully exposed to the incoming light.
   B. The radiation figures were obtained by converting the light measurements of the authors (in lux) to energy units, and this is an approximation only.

leaves, and about 0.6 cal/cm²/min. in sugar-cane leaves. This brings out the interesting point that the sugar-cane leaf was able to utilize light of much higher levels (up to half of natural sunlight); and this may be attributed to basic differences in the photosynthetic apparatus and pathways (see Hatch, 1971). Plant species of this type, represented by sugar-cane, have a more complex photosynthetic apparatus which allows more efficient use of light particularly at high light levels. In these species chloroplasts occur in two zones; an outer layer (in the mesophyll cells) is thought to fix first of all CO_2 in the C-4 carboxyl of C-4 acids (malate, aspartate and oxalacetate). This CO_2 is then released to an inner zone of chloroplasts located in cells around the vascular bundles where it is fixed into 3-phosphoglycerate and thence to the photosynthetic products of sucrose and starch.

It will be apparent from the above that a crop canopy, as opposed to a single leaf surface, which absorbs less light in the upper layers (since saturation occurs well below full sunlight) and transmits more to lower layers will be more efficient in utilizing sunlight in photosynthesis. The way in which light is cut off as it penetrates into a crop canopy is approximately exponential, when considering light from any particular direction. Fig. 1.4 illus-

trates a hypothetical canopy in which each layer of leaves absorbs 50 per cent of the light reaching it; this is in fact generally too much for the efficient utilization of natural sunlight.

The factors which control utilization of light are also dependent on the stage of development of the crop (i.e. whether leaf area is small as in the seedling stage, or large as in a mature crop). This has been discussed by Verhagen *et al.* (1963). The factors controlling light penetration are primarily leaf orientation relative to the direction of incoming light, leaf size (many small leaves as opposed to few large ones), vertical distribution above the ground and optical properties of the foliage. Some of these factors are illustrated diagrammatically in Fig. 1.5 (see p. 4).

When considering the photosynthesis of a crop canopy it is also necessary to take into account the movement of the sun. The depth of penetration of light into the canopy will vary with canopy type and sun elevation and Fig. 1.6 shows leaf area index (LAI, which is the area of leaf relative to area of ground) illuminated in hypothetical canopies of different foliage orientation, assuming both an incoming radiation (direct) as shown, and a compensation point of 0.02 cal/cm²/min. This figure is based on model studies but it conforms fairly well to practical observation in that broad-leaved spe-

Fig. 1.4 Absorption of light by successive layers of leaves of a hypothetical crop in which each layer absorbs 50 per cent of the light reaching it

Source. After Williams and Joseph, 1970.

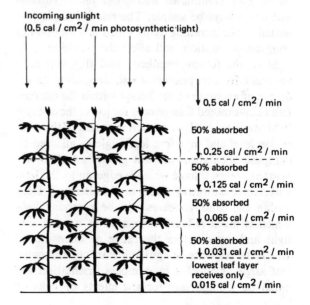

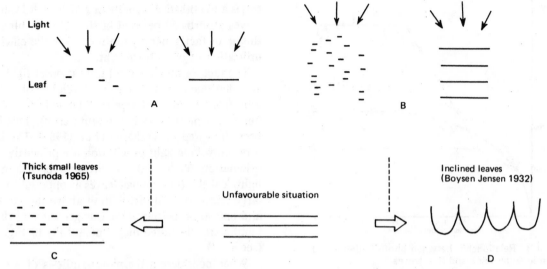

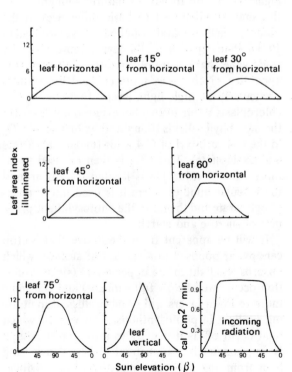

Fig. 1.5   Factors which control light penetration
Source.   After Tsunoda, 1965.
Notes.   A. When leaf quantity per land area is small, large
leaves have advantages in light utilization.
B. When leaf mass per land area is large, small
leaves are advantageous.
C and D. Two ways of maintaining a large leaf
mass or area in well-organized situation.

cies can support a leaf area index of about 4 to 5
while vertical-leaved species such as some of the
higher-yielding rice and maize varieties can support
a much higher and more productive canopy with
LAI values of 12 to 15. Because the relationship
between light intensity and photosynthesis is not
linear over the whole range of natural light intensities (cf. Figs. 1.2 and 1.3), the relationship between LAI illuminated and gross photosynthesis
will not always be simple. The relative angle presented to the incoming light, which is a function of
compass orientation, will affect the partitioning of
light to the foliage members, and this will have
various effects on photosynthesis depending on the
degree of exposure of the foliage within the canopy.
In a fully-exposed foliage near the top of the canopy,
light levels are likely to be well in excess of the saturation level and therefore leaf angle as affected by
compass orientation will not greatly affect photosynthesis (even foliage which is edge on to the light
will receive saturating values of illumination). Deep
within the canopy, leaves will be receiving light
which is much reduced compared to natural sunlight and more within the range of intensities in
which the response to light is linear; consequently,
leaf angle as affected by compass orientation will
again be unimportant for the effects of light on

photosynthesis of each leaf surface will be additive.
(In other words, photosynthesis will not be affected if some leaf surfaces receive a relatively higher
proportion of light than others, for in the linear
range the total photosynthesis will be independent
of light distribution.) Foliage in intermediate layers,

Fig. 1.6   Daily changes in LAI illuminated in hypothetical
foliage types
Note.   The pattern of incoming radiation used to compute
the curves is shown.

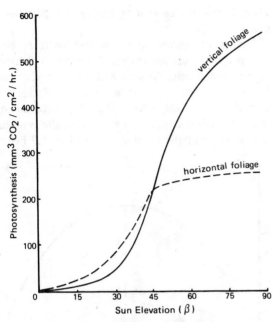

Fig. 1.7 Changing gross photosynthesis in model horizontal and vertical canopies with sun elevation
Source. After Williams and Joseph, 1970.

in which the light level is near the deflection of the light/photosynthesis response curve, will show a complex relationship depending on the range of angles presented as a function of compass orientation. Photosynthesis in this intermediate zone can only be determined by numerical integration of the photosynthesis possible from leaves over the range of angles presented. Estimations of gross photosynthesis from foliage which is orientated vertically and

Fig. 1.8 Diurnal change of apparent assimilation rate 80 days after transplanting of 'Peta' rice population planted at 30×30 cm spacing
Source. After Tanaka *et al.* 1966.
Note. Light intensity at 12 noon was about 90 to 95 k. lux.

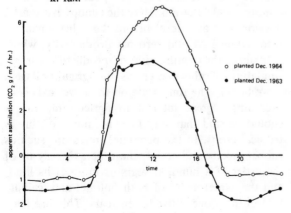

horizontally, using the photosynthesis/light curve of Gaastra (Fig. 1.2) and the energy and sunlight conditions of Fig. 1.6, are shown in Fig. 1.7. Quite marked differences appear between vertical and horizontal foliage which are related to LAI illuminated, as shown in Fig. 1.6. Horizontal foliage shows a higher rate of photosynthesis at lower sun elevations (β), but photosynthesis of vertical foliage greatly exceeds this at higher sun elevations. Studies on the rice crop (Tanaka *et al.* 1966) in which foliage would be predominantly vertical show that diurnal changes in net photosynthesis conform to the pattern which would be expected from the above considerations (Fig. 1.8). It will be obvious from the above that varieties permitting a high degree of foliage illumination will be responsive to close spacing and high levels of fertilization (particularly with nitrogen) which will promote canopy development. Examples of this are seen in the following chapters on individual crops.

There is one other important factor concerned with light penetration into the canopy which is particularly related to climbing plants and species producing runners. This is the vertical distribution of the foliage which also affects light penetration.

A very clear example of yield increases as a result of experimentally altering leaf distribution has been reported by Chapman and Cowling (1965) in sweet potato varieties (Table 1.1). The yield of tubers from one variety used in this investigation was usually of the order of two tons per acre at maturity. By training the vines to grow up wire-netting frames, however, yields were increased to 6–8 tons/acre. These research workers suggest that their findings may be of use to plant breeders in the selection of high yield characteristics. It will be noticed from Table 1.1 that nitrogen (which increased leaf area), actually depressed yield in the absence of the wire frame but when light penetration was improved, yield was increased.

Similar results have been obtained with other crops. Campbell and Gooding (1962) report 50–100 per cent yield increases in yams by staking at a height of six feet. Studies of this nature are likely to yield interesting results in all plants with a climbing habit.

There is yet another way in which light interception is affected by the canopy structure and that is in the amount of solar radiation which is reflected. This differs with the nature of the crop surface. For a 'rough' crop, a higher proportion of reflected radiation is trapped between the leaves than with a fine crop. The amount of reflection also varies with the angle of the sun. Montieth (1965) lists measurements made on reflected radiation with different crops. For smooth-surface crops such as pasture grasses, wheat and so on, reflection may range from about 20 per cent when the sun is at median elevations to 30 per cent with the sun at low elevations. With rough crops such as pineapple, reflection reaches only about 15 per cent at its highest values. These values represent considerable losses in incoming radiation, particularly at a time when the sun is at low elevations and limitations to photosynthesis imposed by moisture stress will be minimal. In rubber cultivation there is evidence that 'hedge planting', which breaks up the canopy into a series of folds that can be seen clearly when viewed from the air, may give a better yield, and it has been suggested that this could have resulted from the absorption of a higher proportion of light by the canopy, particularly in the lower ranks of leaves (see Rubber, Chapter 9).

TABLE 1.1   Effect of leaf spacing on yield of tuber of sweet potato.

	NITROGEN		
WIRE	Absent	Present	Mean
Absent	2.107	1.539	1.823
Present	5.128	7.316	6.222
			±0.336
Mean	3.618 ±0.366	4.428	4.023

Source. After Chapman and Cowling, 1965.

Shade requirements of leaves

The question of light saturation in individual leaves is tied in with canopy structure, in that the leaves of some crop species are adversely affected by excessive sunlight. Evidence for cocoa (Chapter 7) and tea (Chapter 8) suggests that this reaction may be more related to temperature build-up in the leaf than to actual light values, and in this regard vertically orientated foliage will obviously be superior. The whole question of shade in perennial crops is considered in more detail in the following chapters on individual crops.

Development of the canopy and growth patterns in crops

Studies on numerous crops and pastures have revealed that a certain pattern of growth is common to all plant communities which are not restricted by moisture stress or other limiting factors. If growth rate (of the whole crop) is plotted against time or against canopy development, the relationship may be of the form shown in Fig. 1.9 below.

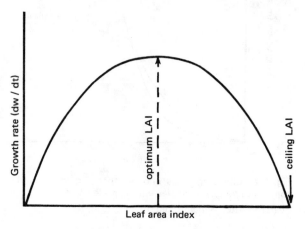

Fig. 1.9   Changing crop growth rate (dw/dt) with increasing leaf area (canopy) development or with time
Note.   w=weight per unit area and t=time.

From this it is clear that a condition of the canopy exists in which the array of foliage is optimal for productivity. If the crop is perennial, it may, if left unharvested, continue development to the stage of zero growth or zero net productivity. It will be obvious that the highest agricultural return can be achieved by maintaining the canopy in the state of optimal LAI. A simple case would be in the clipping of pastures, and the plucking of tea (Chapter 8). It is probably not often that this simple situation applies however, and Fig. 1.10 possibly represents a more usual situation. Here the canopy has ceased development at a level before the achievement of true ceiling LAI and zero net productivity, which still allows the continuation of production (or assimilation) of other plant parts (the agricultural yield —we hope). The time scale on the lower axis is of exceedingly great interest in agriculture, and it should be said that very little is known of this in tropical crops. In the perennial tree crops such as rubber, cocoa and coffee, the time axis may be of the order of a hundred years or more at its limit, and the achievement of both optimal and 'ceiling' LAI of the order of five to ten years. This has been

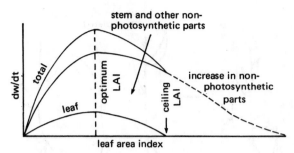

Fig. 1.10 Trends in crop growth rate (dw/dt) of leaf and other plant parts
Source. After Donald, 1961.

discussed to some extent in Rubber—Chapter 9, but very little is known of the situation in other crops; in coffee, tea and citrus, these quantities are also affected by pruning and in a sense this procedure conforms to the idea of maintaining the canopy closer to optimal LAI.

In short (or shorter) term crops (e.g. rice and tapioca) the check on the vegetative proliferation of the canopy to the point of zero net production is obviously linked with the diversion of assimilates to the developing grain and tubers, and this also seems likely to a certain extent in sugar-cane and pineapples. In bananas, the pruning of old shoots also tends to maintain the canopy in a more productive condition.

The palms (coconut and oil palm, Chapters 11 and 12) present an interesting case, for here the proliferation of the canopy is restricted by the single growing point; consequently these palms are particularly sensitive to spacing.

Climate and leaf area index

If it is accepted that both optimal and ceiling LAI are achieved as a result of progressive utilization of the light space by the growing canopy, followed by an increase beyond this to a condition in which the canopy can just maintain the crop in a static condition, then it can be readily appreciated that an overall change in the quantity or intensity of incoming light will allow the development of different values of these leaf-area quantities. Radiation used in photosynthesis is a function of both total incoming light and also of moisture conditions, because moisture stress limits the period of effective photosynthesis. Thus useful radiation is dependent on broader aspects of the climate and soil than simply the total radiation. A climate optimal for

growth would be one in which soil moisture is not limiting, evaporative conditions are not excessively high and there is plenty of sunlight and air movement—a rare combination indeed and one which is probably never achieved without supplementary irrigation.

Sink activity of the crop and assimilation

Agricultural crops are of many different types and the harvestable part varies from virtually the whole plant as in a fodder crop to a relatively small part of the crop such as grains, the yield or the latex from a rubber tree, fruits and so on. Unless plant organs and structures are present to absorb photosynthate from the leaves, the process could become limited by the accumulation of assimilate. The ability to grow the part of the plant which is of agricultural interest is an intrinsic plant factor with a strong genetic component. The development of plant organs often depends on special environmental stimuli (particular temperatures, photoperiods, etc.) but superimposed on this are more fundamental requirements. These are the presence of the necessary shoot or bud primordia or growing tissues, and the necessary system for translocating photosynthate and nutrients. The activity of these tissues has been termed 'sink-strength' and it can be thought that different parts of the plant compete for photosynthate (and other nutrients), the final distribution being determined by the relative sink-strengths and translocation pathways of the different organs. Plant breeders are often concerned, either consciously or unconsciously, with selecting and breeding crop varieties with a high sink-strength for the desirable plant parts. A good example is seen in tapioca (Chapter 10) in which large tubers appear to affect the assimilation rate of the canopy.

The growth patterns of plants and the yield of crops appear to be strongly influenced by such a hierarchy of 'sinks' which determine the translocation and accumulation of assimilates. The quantity of assimilates available for growth is determined by light and photosynthesis, but the distribution between the different plant parts is affected by the additional environmental factors as indicated in Fig. 1.11. Evans *et al.* (1964) have suggested that the activity of sinks is largely determined by endogenous growth substance levels which are affected by environmental factors. Photoperiod, for example, is an environmental factor which produces morphological and other changes which suggest

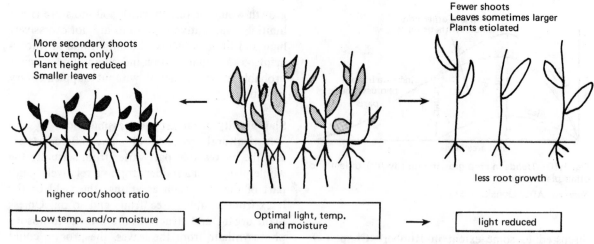

Fewer shoots
Leaves sometimes larger
Plants etiolated

More secondary shoots
(Low temp. only)
Plant height reduced
Smaller leaves

less root growth

higher root/shoot ratio

| Low temp. and/or moisture | ← | Optimal light, temp. and moisture | → | light reduced |

Fig. 1.11 Diagrammatic presentation of the effects of low light and temperature/moisture stress on the growth of a hypothetical crop
Source. After Williams and Joseph, 1970.

major changes in the activity of growth regulators including also possibly growth inhibitors. In cases where the sink consists of expanding organs it is likely that auxins will be involved in their growth. For example, Arnold *et al.* (1967) have shown that both auxin (naphthalene acetic acid) and gibberellic acid stimulate leaf growth (in a pasture grass *Phalaris*) and in this case it is the leaf—the sink—which is providing the agricultural yield. Auxin(s) are also central in the establishment and maintenance of apical dominance. In both these facets of plant growth, the primary role of auxins appears to be in cell growth, more particularly in cell expansion.

Respiration losses in crops

The net rate of photosynthesis or assimilation depends not only on CO_2 reduction but also on the rate at which the products of photosynthesis are being used up in respiration. Respiration supplies energy for the complex living processes of growth, translocation, cell division, the establishment and maintenance of osmotic gradients, the reduction of nitrogen and so on.

Measurements of respiration rates made in crop plants vary considerably, and evidence suggests that some species have relatively high respiration rates while others are much more efficient in their use of respiratory substrates. Estimation of respiration in alfalfa, for example, indicates that respiration accounts for about 33 per cent of gross photosynthesis (Thomas and Hill, 1949). On the other hand, the respiration levels of other crops, such as sugar-

cane and maize, are much lower (Moss, 1962). Respiratory efficiency, particularly of the foliage, also affects photosynthesis by influencing the diffusion gradient of CO_2 into the leaf. Species with a high leaf respiration have a reduced CO_2 gradient between the atmosphere and the leaf and hence have a lower potential rate of photosynthesis. In this respect maize and sugar-cane are regarded as efficient, and others (for example tobacco and tomato) as inefficient. This is discussed in some detail by Zelitch (1967). It is not entirely a question of monocot species versus dicot species, although it is conceivable that the foliage of the latter should have a relatively high respiration rate because of the location of the meristems, while monocots (with a basal intercalary meristem) may be expected to have low leaf respiration rates.

The actual need for respiration to take place is high in young and actively-growing tissues. In mature tissues, such as a mature leaf canopy at optimal LAI, respiration needs should theoretically be very low. Even so, relatively high respiration rates are often observed (rates far in excess of those required to complete the actual work done in growth and differentiation). This actual work involves only relatively small quantities of energy for the establishment of osmostic potentials, the formation of the architectural structure of the crop and so on. Relatively high rates of respiration (even in so-called efficient species) can be attributed to the extraordinarily low efficiency of many of these biological processes. Considerable amounts of energy are simply expended as heat. These aspects of plant respiration have been excellently reviewed by Wohl and James (1942).

As far as the quantities of energy required are

concerned, considerable differences occur as a result of the nature of the elaborated materials formed by the plant. If glucose is regarded as the primary product of photosynthesis, the condensation of this to sucrose, starch, cellulose and other carbohydrates uses only small amounts of energy. The heat of formation of one mole of cellulose, for example, is only 4.2 calories. Carbohydrate-producing species (e.g. sugar-cane and maize) have a relatively low absolute demand for energy. On the other hand, protein-forming species (such as legumes and fodder grasses) require considerable expenditure of energy for nitrate reduction and protein formation. Nitrogen fixation in legumes consumes approximately 60 cal/mole, and nitrate reduction has a similar energy requirement. These differences may contribute to the observed high respiration rates of some species such as alfalfa, tomatoes, tobacco, etc., and the low respiration rates of other species such as maize and sugar-cane.

Interaction with the root environment

The functioning of the assimilation apparatus depends largely on a supply of water *via* the root environment of the crop. In most types of agriculture, the root environment consists of soil of one kind or another which may be treated in various ways to improve or decrease its suitability for the establishment, growth and harvesting of crops. From the point of view of water supply, the suitability of a soil for crop production is very closely linked with the climatic factors of rainfall and rainfall distribution and saturation deficit of the air. For example, a very poor soil in the general agricultural sense (a coarse sand for example) may yield excellent crops if adequate watering and fertilization are carried out. The example of the very high yields obtained in sand culture may be cited in support of this. In non-irrigated cultivation, a climate with an adequate evenly-distributed rainfall will allow the use of poorer soils for agricultural purposes. Conversely, in geographical situations where a marked dry season occurs, the nature of the soil, particularly its moisture-holding and storing properties, becomes a central factor in agricultural production. To grow perennial crops successfully in such situations, the soil must be deep and capable of holding sufficient water to last through the adverse season.

As far as plant or genetic factors are concerned, characteristics of the genotype which make for economical water use (deep-root penetration and moisture-conserving mechanisms) become important in determining the varieties most successful under conditions of moisture stress, whether these be due to irregular or infrequent rains or to poor moisture properties of the soil. On the other hand, when moisture stress is low (as on a good moisture-holding soil, or when the rainfall is adequate and so on), other characteristics of the genotype will favour high yields. The fact should be stressed that an experiment, such as a variety trial, which does not take into account these factors that affect the performance of different varieties, is likely to be of little use except in very restricted situations. Many tropical crop species, and most of those under consideration in this book, are particularly sensitive to moisture stress and are not generally well adapted to survive drought. Tropical crop species are mainly affected by intermediate conditions of moisture stress, as indicated in the following chapters, and functioning of the assimilation apparatus may be affected at soil tensions close to field capacity under highly evaporative conditions. Stomatal opening is one of the most sensitive processes affected by moisture stress, and mid-day stomatal closure is probably a common occurrence under tropical conditions. Furthermore, in the case of tree species, declining yields with age may also be related to elevation of the canopy which imposes a tension of approximately one atmosphere for each thirty feet of trunk height. This may well be the case with coconuts and oil palms (Chapters 11 and 12).

The factors in drought resistance and plant adaptations to survive moisture stress have been considered in some detail in recent books by Salter and Goode (1967) and Williams and Joseph (1970).

PROPAGATION AND NURSERY TECHNIQUES

It will be apparent from the following chapters on individual crops that a nursery phase is a common feature of tropical crop production. Many tropical crop plants are propagated vegetatively, and the raising of the propagules and their care before planting out forms a major aspect of their husbandry. Even with others which are propagated by seed, an important phase of planting might involve the raising of the seedlings at a special site or nursery to a stage when they are growing strongly and can be transplanted to the field. Another reason for the widespread use of nurseries is connected with the fact that the species are often trees or large-sized plants which are planted at relatively low densities. Also, the standard of field and seed-bed preparation (which sometimes involves simply the cutting and burning of jungle) and of weed control is lower and often more difficult with such crops. (The problem of weed control centres on the fact that for a large part of the early life of crops in which the individual plants are spaced at a relatively low density, there is an extensive ground area between the plants and consequently care of the individual plant and the suppression of weeds over large areas become difficult. Seedling protection from pests and selection at the time of planting out are also facilitated by growing plants in nurseries. Even in rice cultivation, the traditional methods of planting involve the raising of seedlings in a nursery. (The advantages and disadvantages of transplanting in padi cultivation are considered in Chapter 5.) For the present, we may summarize the reasons for maintaining nurseries in the planting of tropical crops as follows:

1. It is generally considered cheaper to maintain seedlings in a nursery up to the size at which they are planted out, and more attention or better husbandry (weeding, fertilization, watering etc.) can be applied to each seedling under nursery conditions for less cost.

2. Protection from insects, rodents and other animal pests, and from fungal, bacterial, and other diseases can also be more easily carried out under nursery conditions.

3. The selection and rejection of seedlings for field planting can be conveniently done at the nursery stage. This is very important when considering the expected mature yield of a long-term crop.

4. Where vegetative propagules are used, it is often essential to cultivate these in special nurseries until they have struck root or budded etc.

Although it is undoubtedly cheaper to control weeds, diseases and pests under nursery conditions, this saving must be offset against the increased cost of planting seedlings in a nursery and then transporting and transplanting these, often when they are quite large, to the field site. With improved weed-control chemicals developed in recent years and pest and disease-control methods, it is possible that the raising of seedlings directly in the field may in fact be the more economical operation for some crops. The question has seldom (if ever) been critically examined and in any case it must also depend on local conditions of terrain, labour availability and so on. In coffee growing for example, it is a general practice in Brazil to plant the seeds directly in the field (see Chapter 6), but in Java, direct-planted coffee was found to require a great deal of attention and expense (deLight, 1937). However, in those days, chemical weed control was practically unknown. The technique of direct field planting in Brazil is apparently acceptable for conditions there.

In nursery practices, tradition often plays a major role, and individuals who have become used, to one method or another often resist change, even if it is of an economical nature. In fact one of the problems of evaluating nursery techniques is that planters tend to become very efficient at their own particular methods, even if they are cumbersome, and consequently when attempting a new technique this may tend to come off as second best. However, one modern innovation which has been universally accepted in plantation agriculture is the polythene bag. It is not intended to attempt an evaluation of nursery methods here, but simply to consider some agronomic aspects of nursery practice and vegetative propagation.

Nursery practice

Choice of nursery soil. Since nursery practice aims to provide a high standard of husbandry to the young developing plant, it is usual to select a good agricultural soil for the location of nurseries or for the filling of bags or other containers if the nursery plants are raised in this way. Frequently a free-draining, fertile soil is chosen, and fertilizers where necessary are used correctly to maintain good nutrient conditions for the plants. With most plants, a free-draining soil will provide a better medium for growth but there are also two other factors which are important here. The first is that a free-draining soil also dries out rapidly so that some provision for watering is usually necessary under most conditions. The second consideration is that when seedlings are grown in the ground (as opposed to growing them in bags or other containers), the raising of the seedling at transplanting time becomes a problem; consequently it is sometimes necessary to use a heavier soil than is ideal for seedling growth to facilitate the extraction of the roots in a block of earth.

Watering. With many nurseries, particularly those in which the seedlings are established in separate bags or containers of soil, some supplementary water will be needed. The need for water and the quantity will of course depend on rainfall and crop evaporation rates. As a general rule, in the equatorial tropical areas, a water delivery system capable of supplying about 5 mm per day will be satisfactory. Whether this is supplied mainly by a watering-can, or by a modern sprinkler is a matter for local consideration in each case, but a source of water must be close at hand.

Shade and spacing. With young plants which have recently germinated and have not yet developed a large leaf area, it is usual that the amount of light received by the unit leaf surface will be too high, and leaf growth or leaf expansion may improve with some degree of shade (see following chapters on individual crops). Shade is recommended in the young stages for coffee, cocoa, tea, oil palm, rubber and many other crop nurseries. In coconut nurseries, however, shade is not generally recommended except in very hot and dry conditions—in fact shade is found to reduce rates of growth in most instances with this crop. The desirability and amount of shade depend to a great extent on the climatic conditions. Under hot drying conditions some shade will reduce the loss of water and promote leaf expansion. This has, for example, been shown to be the case for oil palm seedlings by Rees (1963) and is also known to be a general phenomenon in plants. A second reason for shading nursery plants and nursery material occurs as a result of the operations involved in vegetative propagation. With cuttings, for example, the material will be very sensitive to moisture stress before rooting has occurred, for loss of water at the foliage surfaces may occur much more readily than the uptake of water by the cut stem base. This is one of the reasons why leaf surfaces are usually severely pruned in cuttings (e.g. in cocoa—Chapter 7).

In regard to shade therefore, it is often desirable to provide this for young seedlings before they become very leafy, and for cuttings and other vegetatively propagated material. After some time however, young plants will grow larger and inter and intraplant shading will occur so that it usually becomes necessary to remove the shade, and at a later stage to reduce the density of the seedlings by placing them further apart. If this is not done, crowding will result in the etiolation and weakening of seedlings. Quite often nurseries involve two stages: a first in which seedlings are kept closely planted and shaded (while they are very small), and a second in which the plants are more widely spaced and unshaded, and kept until they are ready for field planting.

Another reason for removing shade at later nursery stages is to allow the plants to harden before they are planted out to the less reliable conditions of the field. In a nursery, hardening can be done in stages by progressively removing the shade.

Germination of seeds. The seeds of most tropical crops germinate readily when fresh but they lose their viability on storage. This is true of tea and coffee, cocoa, rubber and coconut. (Although with the coconut, the embryo is usually dormant for up to six weeks after plucking.) With one important tropical crop however, namely the oil palm, seed germination is very low when the seeds are fresh. Oil palm seeds require a period of heat treatment to raise germination levels (see Chapter 12).

A comparison of ground nurseries and bag nurseries

With all crops, nurseries can be established by planting the seedlings in the ground (usually in specially prepared beds), or in bags or individual containers. Polythene bags are now widely used for this latter type of nursery, but other containers, notably various types of woven baskets in which the seedlings can actually be planted are also used. Polythene bags must however be punched with holes to provide drainage. Which type of nursery (in the ground or in containers) will be most satisfactory is largely dependent on local conditions, for there are advantages and disadvantages in both types. The following comparison of both bag and ground nurseries indicates the main features of each type (see next page).

The chief advantages of the bag type of nursery are the elimination of transplanting shock and the ease of transport. The advantages must be weighed against the cost of filling the bags or containers with soil. When transport from the nursery site is difficult because of distance or because of poor roads or paths (as is often the case) then it is likely that the advantages of a bag nursery will outweigh those of a ground nursery.

Two-phase nurseries

With some crops, the nursery actually consists of two phases, one of which is permanent and the other which is continuously changing. This occurs

BAG NURSERY	*SOIL NURSERY*
1. A good quality friable soil can be used.	A heavier clay soil is usually required to raise seedlings at transplanting time.
2. Supplementary watering is usually essential.	In some climates groundwater will be sufficient and supplementary watering will not be necessary.
3. The cost of filling bags with soil is often considerable.	The cost of ground preparation is relatively low.
4. Bag seedlings are easy to handle and transport and can usually take a moderate amount of rough handling.	Lifting field seedlings is often a tricky task and seedling losses can be quite high. When seedlings are lifted in a block of earth, special care is needed in transport to avoid shattering of the block.
5. Seedlings, except in very large bags, are usually ready to plant sooner and therefore the length of field life before maturation is longer and cost of field maintenance may be higher.	Seedlings can often be raised to a large size in ground nurseries thereby reducing the length of the premature field life.
6. Transplanting shock to the seedlings is minimized.	Transplanting shock to seedlings lifted from the ground is often severe.
7. No special treatment is required at or before transplanting time.	Root pruning of seedlings to confine the roots in the region of the block of soil is often necessary before lifting (see Fig. 1.13), and leaf pruning to reduce transpiration losses in the post transplanting phase is sometimes necessary.

when material for vegetative propagation is raised in the nursery. (In rubber planting for example, the permanent phase of the nursery consists of areas of closely planted clonal material which is planted in the ground and kept in a specially pruned condition to produce branches which are taken for bud wood.) The second phase of the nursery consists of the raising of cuttings in beds or bags etc., or the grafting of buds onto special stock seedlings grown for the purpose. As with other materials, the latter can either be raised in bags or in the ground as discussed.

Transplanting of seedlings

If it is assumed that an economy is achieved by retaining plants in a nursery during the early phase of their lifespan, then it is obvious that the longer the plants are maintained in the nursery in a productive state, the greater will be the economic saving. There are, however, two factors which limit the period during which plants can be held in a nursery:

1. With the closer than field spacing of nursery plants, competition for light and root space will check growth when the plants become large.

2. The larger the plants become in the nursery, the more difficult it will be to handle and transplant them.

Consequently, a compromise is reached about the length of nursery life, and transplanting is usually carried out at a stage when the seedlings are still easy to handle and can easily be carried by hand. Quite often, climatic factors affect the length of the nursery period, for in a monsoonal climate there is sometimes only one season in which planting out can be done. In other climatic regions there are two moist seasons in which planting out can be done. The use of rainfall confidence limits is useful in deciding on a safe planting out season (see Manning, 1956).

To understand more clearly the question of correct timing in a nursery and the consequence of planting out late, we may refer to Fig. 1.12 which shows the course of development of a plant in the exponential phase of growth that would be appropriate for the nursery and early field stages of planting. If nursery and planting out operations are well carried out, there should both be a minimal disturbance to the course of development as expressed by this curve, and a minimal departure of the actual growth curve from this theoretically optimal one. In practice of course, timing is not the only factor which can cause a loss of development; another is transplanting shock. In this respect, polythene bag or container nurseries, as explained, are superior for planting shock is minimized. A well-timed polythene bag nursery will give therefore the best possible rate of development. Other methods of lifting the seedlings can cause considerable setback to the seedlings themselves. There are basically two methods of lifting the seedlings from nurseries; one is the bare-root method in which the plant is pulled from the ground, thus exposing the roots and at the same time breaking most or all of the fine root ends and the root hairs. The other method is to lift a block or core of the soil surrounding the plant roots. This also causes considerable damage to the root system but much less than with the bare-root method. For a comparison of the two methods, see Chapter 6 on coffee. In transplanting both bare-root and soil-block seedlings, it is often a good idea to prune off some of the leaf surface to reduce transpiration losses during the period before the new roots are established.

Root pruning is sometimes carried out at some stage prior to lifting the soil blocks. This is done in oil palm and in other crops. The effect of root pruning in confining the roots to the soil block is illustrated in Figure 1.13. Usually vertical cuts are made a few weeks before lifting and not all the roots are pruned on the same occasion.

Previously it was said that for the satisfactory lifting of soil-block seedlings, a heavy type of clay soil gives the best results. This in fact is generally true when soil blocks are cut with no special equipment, for the clay blocks will stay together much better than a more friable soil. Various types of core lifters, however, are used by planters to enable the lifting in more friable soils. One of these was

Fig. 1.13 Illustration of the effect of root pruning before lifting a soil block

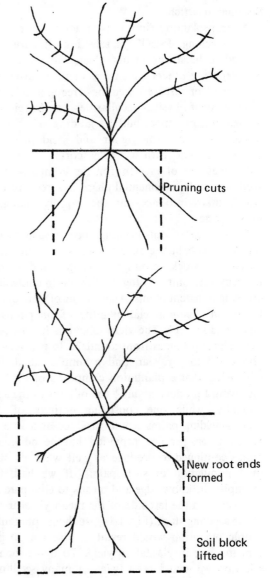

Fig. 1.12 Development curves for a correctly-timed nursery and for a delayed nursery

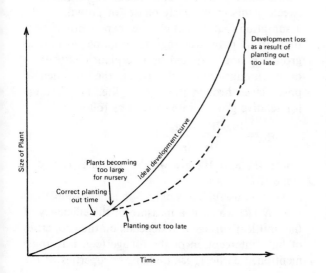

developed for lifting rubber seedlings at the Rubber Research Institute in Malaysia (see Chapter 9).

Transporting seedlings from the nursery to the planting site often presents some difficulty in new plantations because of rough terrain and poor roads, and some thought should be given therefore to ease of transport. In this regard polythene bag seedlings are superior to others. The size of bags must also be taken into account and convenient sizes are such that one or a number of them can be carried by each worker. Tall thin bags are superior to short fat ones because more of them can be stacked on a trailer or a lorry. These considerations are very important in keeping down planting costs.

Seedling selection

At transplanting time it is usual to carry out some screening of seedlings, selecting what are considered to be strong plants which will grow well, and rejecting others which are weak or diseased. Abnormal or diseased seedlings,. recognizable because of their physical peculiarities, can be rejected at planting out time, but the grading of normal-looking seedlings for vigour and yield potential cannot be done with a single scoring. The only valid measure of growth for the comparison of seedlings in the exponential phase is a relative one which takes into account the rapidly changing growth rate.

In the case of out-breeding species such as the oil palm, coconut, certain varieties of coffee and cocoa, the stock seedlings in rubber and so on, progeny variability is often high, even in controlled crossing of parent plants. An example of yield distribution in a commercial planting of oil palms is given in Fig. 1.14. The yield of palms in this particular group ranges from about 25 lb per year to about 875 lb per year, and a simple calculation will reveal that a plantation composed of the former would produce about ½ ton of fruit to the acre annually, while one consisting entirely of the higher-yielding palms would produce at a rate of about 20 tons per acre, provided that the individual palms continued to produce as well when planted in a population of such palms. If we had, for example, the knowledge and means to eliminate all the palms on the left side of the mean yield arrow at the seedling stage (Fig. 1.14), such an apparently simple operation would result in about a 30 per cent increase in plantation yield. To make such a selection we would need twice as many seedlings

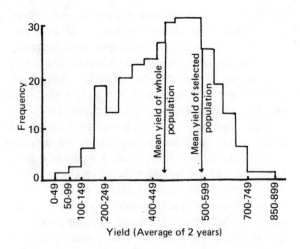

Fig. 1.14 Yield distribution (average two years) of individual oil palms in a field planting

to choose from, and a reliable index of seedling growth which is correlated with the mature yield of the palms. The latter requirement is where the problem lies. There are many factors affecting yield including the functioning and efficiency of the photosynthesis apparatus and the sink activity of the harvested organs; the situation is even more complex because of environmental interactions. As far as the relationship between seedlings and the mature crops are concerned, it might be assumed that a vigorous fast-growing seedling will produce a vigorous high-yielding mature plant, and the problem here would be to find a valid index of seedling growth.

Formulae have been derived for growth rates which allow accurate comparisons to be made between plants in the early phases of growth on the assumption that weight changes exponentially with time. In oil palm seedlings, for example, as long as growth is not restricted by interplant competition or by limitations to the roots, the exponential phase lasts through the nursery life. The formula for relative growth rate (R_w), is as follows:

$$R_w = \frac{\log_e W_2 - \log_e W_1}{t_2 - t_1} \tag{1}$$

where W_2 and W_1 are weights at times t_2 and t_1 respectively.

Another useful index of growth is net assimilation rate (NAR) which is a measure of the efficiency of the unit leaf surface and depends on the properties of light interception of the foliage (e.g. leaf angle mentioned earlier), gas exchange properties of the leaf, characteristics of drought resistance and water use, and many other factors. The validity of this growth index is based on the fact that assimilation of dry matter takes place at the leaves, and therefore leaf efficiency affects and gives a measure of the growing ability of the plant. A usual formula for NAR is shown below:

$$NAR = \frac{W_2 - W_1}{t_2 - t_1} \times \frac{\log_e A_2 - \log_e A_1}{A_2 - A_1} \qquad (2)$$

The first term represents the change of weight and the second is the reciprocal of the mean leaf area between times t_1 and t_2 (A_1 and A_2 being areas at respectively times t_1 and t_2).

Leaf area ratio can also be a useful index of yield ability and it is almost a direct measure of foliage growth. In the cultivation of oil palm under humid conditions leaf area ratio may be of considerable significance in regard to yield because a strong correlation between yield and leafiness has been demonstrated (see Hardon *et al.* 1969; Chapter 12).

Leaf area ratio $\left(\dfrac{A}{W}\right)$ or mean area over mean weight between times t_1 and t_2 is obtained as follows:

$$\frac{A}{W} = \frac{A_2 - A_1}{\log_e A_2 - \log_e A_1} \times \frac{\log_e W_2 - \log_e W_1}{W_2 - W_1}$$

Quite clearly since growth primarily depends on photosynthesis of the leaf surfaces, it becomes a function of the product of leaf area ratio $\left(\dfrac{A}{W}\right)$ and the efficiency of the leaf surface (NAR). It can be seen that relative growth rate R_w is in fact equal to net assimilation rate multiplied by leaf area ratio, thus:

$$R_w = \frac{W_2 - W_1}{t_2 - t_1} \times \frac{\log_e A_2 - \log_e A_1}{A_2 - A_1} \times \frac{\log_e W_2 - \log_e W_1}{W_2 - W_1}$$
$$\times \frac{A_2 - A_1}{\log_e A_2 - \log_e A_1}$$
$$= \frac{\log_e W_2 - \log_e W_1}{t_2 - t_1}$$

Relative growth rate does not necessarily have to be based on weight changes, although this is clearly the absolute index of growth, but can be based on other methods of scoring plant development. A useful measure which can be obtained without destroying the plant is leafiness or leaf area, and it has been found that the leaf area shows trends in time which often closely parallel those of

weight in the seedling stages of growth. Relative leaf area growth rate (R_a) is an interesting quantity because it is in fact the same as the net leaf area assimilation rate (NAR_a) thus:

$$R_a = \frac{\log_e A_2 - \log_e A_1}{t_2 - t_1} \quad \text{(by analogy with (1) above)}$$

$$NAR_a = \frac{A_2 - A_1}{t_2 - t_1} \times \frac{\log_e A_2 - \log_e A_1}{A_2 - A_1} \quad \begin{array}{l}\text{(by analogy}\\ \text{with (2) above)}\end{array}$$

$$\therefore NAR_a = R_a.$$

As explained, a growth index which is used to assess seedlings at different stages of growth should be reasonably constant in time before it can be used for comparison.

The growth of oil palm seedlings under Malayan conditions has been personally investigated, and the results are shown in Fig. 1.15 for four groups, each consisting of twenty-five oil palm seedlings. Initially the value R_a was found to rise to a maximum and then decline to a steady level. Ideally therefore, if R_a is used to compare seedlings and estimate growth and yield potential, then the comparison should be made at the late nursery stage.

Vegetative propagation

The production of vegetative material for planting is generally more time-consuming and costly than using seeds. However, in some instances (e.g. sugar-cane, and bananas) seeds are not normally produced by the plant and vegetative propagation is the only practical means of establishing large

Fig. 1.15 Relative leaf-area growth rate (R_a) for the four groups, each consisting of twenty-five oil palm seedlings

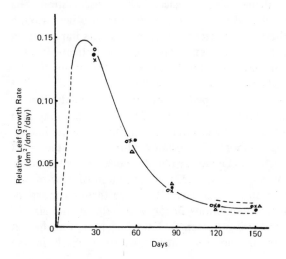

acreages. In other instances (as in rubber production, tea and sometimes cocoa and other trees) vegetative propagation is used as a means of securing high-yielding, genetically-uniform field plantings. High-yielding plants which are used to produce planting material are often far from homozygous so that considerable segregation occurs in seed material, as for example in the oil palm discussed above. Thus many planters prefer to use vegetative material for planting even though its production may be laborious. The following is a brief outline of the principal methods of vegetative propagation:

Graftage and budding. Grafting has a special place in the propagation of rubber, citrus and in some cases cocoa and coffee. The growth hormone auxin plays a part in graft establishment which however is not clearly understood, for horticulturists have long recognized that buds must be grafted into the stock the right way up in order to achieve a good take. Cambial union can occur in any position but the merging of vascular stands only occurs with difficulty in the inverted graft; it is necessary to assure that vascular cells can meet with the correct polarity.

Graftage onto a suitable stock is often preferred to propagation by cutting because cuttings do not develop the deep root system with a central tap root which is obtained when the stock is grown from seed so that losses due to wind may be high (see Rubber, Chapter 9). Furthermore, tree architecture may be affected by the origin of the cutting (see Cocoa, Chapter 7).

In addition to the above considerations, the 'compatability' of the stock and scion is important in determining the success of the association. The stock and the graft both affect each other through their own inherent properties. Factors important in determining compatibility are the uptake of nutrients and water by the stock and the production of elaborated material by the scion on which the stock must depend. Differences in metabolism or the ability to produce certain substances may also affect the stock/scion relationship. In some instances, elaborated substances formed by the graft partners may not be able to move across the union (Williams *et al.* 1953). A restriction of flow of certain substances may also occur through morphological conditions in the graft (e.g. in the size of conducting elements). Dwarfing in apples has in one instance been attributed to a check in the

movement of phosphorus (Dickson and Samuels, 1956). In other cases, incompatibility may result from the production of growth-inhibiting substances which pass from one graft portion to another. From the practical point of view the success of particular associations can usually only be determined by trial and error. Stock/scion relations are considered in more detail in Chapter 13.

In grafting, the 'scion' can be in a number of forms. Quite often this consists of a small branch shoot in which wood has already formed. This is placed into a wedge-shaped cut in the stock so that the cambial surfaces are in contact, and bound in place. After some time the cambial and vascular tissues grow together. Alternatively, the scion material is simply a bud peeled off with the surrounding cortical tissues and consequently this type of graftage is referred to as bud grafting.

In the production of wood for buds, the nursery plants are pruned back to promote the development of numerous shoots. When a bud is removed for grafting, care must be taken not to wound the bark strip by bending or rough handling.

Propagation by cutting. The technique of propagation by cutting is used in the commercial planting of many tropical tree crops. As with budding and grafting, the primary objective is the establishment of genetically-uniform plantings.

Plants from which cuttings are to be taken are usually raised in a nursery and cuttings from recently produced flushes have often been found to be most successful. Generally the cut should be clean and neat for successful rooting to occur. (It is well known in horticulture that green-wood cuttings are more successful than older brown-wood cuttings.)

Cuttings are often treated with rooting hormone (usually napthyleneacetic acid or indolylbutryic acid) because this promotes the initiation of roots at the cut surface. Following this they are planted in beds or in individual polythene bags or baskets of a light well-draining potting soil, or in a suitable rooting medium of composted palm fibre, rotted saw-dust etc. as determined by experience. Some cuttings (e.g. tea, Chapter 8) will only take in a sandy medium.

In establishing cuttings moisture supply plays a central role in the early stages before the roots are formed, and steps must be taken to ensure that water loss does not exceed the ability of the cuttings to absorb water. For this reason it is customary to reduce the leaf area of the cutting. Watering must be frequently carried out although conditions of soil water-logging must be avoided for this affects the initiation and development of the roots. Cuttings have generally been raised in shaded glasshouses or polythene-covered beds in which a high relative humidity is maintained. After cuttings have rooted, the shade provided is progressively removed. Single-leaf cuttings have also been used successfully for propagation of various crops (e.g. cocoa and tea, Chapters 7 and 8). This method is useful when propagation material is in short supply.

A technique closely related to the production of cuttings is to induce the growth of roots on a branch stem prior to its removal from the tree. This is known as marcottage. In this way the new roots of the cutting are produced *before* the root supply from the main trunk is severed. Roots formation is promoted in the marcot by the application of soil and moisture or by the binding of inert materials such as coconut husk to the branch. When a stem section is covered and is kept moist, root initiation and development usually take place in that portion of the stem. Another method of producing roots before removal of the cutting from a tree is known as 'layering'. By this method a branch is bent down to the soil and held in this position until root formation takes place.

In sugar-cane and tapioca planting, stem cuttings produced from an earlier crop are used. Each cutting must consist of at least one node from which the new shoot and roots will develop. As with other cuttings, the rudimentary buds are already present at the node. Such cuttings are planted completely beneath the soil surface and therefore are not subject to the conditions of water stress encountered with other crops.

Propagation by branch shoots. Two important tropical crops which are propagated by branch shoots are bananas and pineapples (Chapters 2 and 3). These slips or branch shoots develop on the plants of the previous planting and shoots for planting should be collected from plants with desirable yield characteristics. Pineapple slips tolerate desication and can be stored in the dried condition for some time.

Material used for banana planting varies considerably from place to place (see Chapter 2); in each case however the planting material morphologically consists of a side shoot or shoots.

WEED CONTROL AND HERBICIDES

If weed growth can be checked in short-term annual crops from the initial slow-growth phase to the period when the crop itself enters the fast-growth phase, weed competition can be much reduced. Once the crop has achieved a good ground cover it will prevent rapid weed growth by shading. (In this connexion, planting distance is important so that open spaces which will need continuous weeding are minimized.) In tree crops which predominate in tropical regions, there is a long period until the crop cover is sufficient to reduce weed competition effectively, and this calls for periodic weed control over an extended period. In the past this has usually been carried out by hand labour, particularly in tropical countries where normally a circle around each tree or seedling is cleared. The areas in between the planting points also require periodic weed control, usually of a lower standard, and this is usually achieved by cover crop cultivation or intercropping in which the planting of the cover crop or intercrop reduces noxious weeds, or by periodic shallow cultivation. In recent years, however, herbicides have largely replaced cultivation in inter-row weed control, and have also been largely adopted in circle weed control around the trees; thus the level of mechanical working of the soil has been greatly minimized or completely avoided by herbicides use. Even in field crops such as rice, minimal cultivation and zero cultivation techniques are being developed (for example using paraquat) which offer advantages in reducing the time required for field preparation (Fua Jee Mok, ICI Ltd., personal communication). In long-term tree crops, the success of herbicides has largely been due to the development of effective and long-acting herbicidal treatments for both circle and inter-row weeding. In circle weeding in particular, the use of mixtures of contact herbicides such as paraquat to kill standing weeds, and soil treatment herbicides of low solubility such as the ureas and triazines (see following list) to kill subsequently germinating weed seeds, has given very effective control for periods of up to six months. In tree crops these herbicides are usually applied by means of fan jets such as the 'polijet' of ICI, which deliver a confined swathe of herbicide with minimal drift. The low-solubility herbicides remain in the upper soil layers for considerable periods of the time. A general feature of weeds is the possession of small-sized seeds compared to crop species in which seed or other propagules have usually been selected for large size, and there is a general direct relationship between depth of emergence (the distance between seed location and point of emergence) and seed size in weeds (e.g. Kenji Noda, 1971); hence the selective activity of soil-incorporated herbicides in circle weed control.

There is yet another way in which herbicides may be used and that is in granular form. Granulated herbicides have been developed particularly for rice grown under flooded culture where the herbicide slowly dissolves and spreads in irrigation water (e.g. Jesinger *et al*. 1971), but have also been used in forestry where convenience in transport is an important feature (Nobuo and Takayuki, 1971). For the same reason there should be considerable scope for the use of granular herbicides in plantation tree crops.

The following list describes the characteristics of the herbicides under consideration in this book (see also Holly and Steele, 1968):

MCPA [4-chloro-2-methylphenoxyacetic acid]: very selective for broad-leaved weeds in cereals, including rice; translocated in the plant, it acts on cell membranes causing exosmosis, affects the nucleus, enzyme systems and causes leaf epinasty and disorientation of growing shoots; effective at very low rates ($\frac{1}{2}$ pt to 2 pt/acre commercial preparation); used as a selective herbicide in cereals (including rice) by overall foliage treatment and as a directed herbicide in inter-row weed control and tree circle weeding in a variety of crops; broken down in the soil in a few weeks; low mammalian toxicity.

2:4-D [2:4-dichlorophenoxyacetic acid]: as for MCPA above, used as sodium or potassium salt or the amine.

2:4:5-T [2:4:5-trichlorophenoxyacetic acid]: as for the above, but more effective on woody species; used as a directed herbicide either sprayed on the foliage or painted in an oil carrier on the trunk or injected; often used for killing trees in clearing, or thinning shade trees (e.g. in coffee and tea); much more persistent in the soil than the two previous herbicides. Its use is banned in certain countries because of long persistence but it can be recommended for tree killing when this is followed by burning.

2:4-DP [2:4-dichlorophenoxyproprionic acid]: developed by the US Rubber Co. in the USA; this herbicide is used as a spray of the soil surface; hydrolyses slowly to 2:4-dichlorophenoxyethanol which kills germinating weed seeds in the soil, used at rates of 2-4 lb/acre; fairly short persistence in the soil; low mammalian toxicity.

PYRICLOR [2:3:5-trichloro-4-pyridinol]: developed by DOW in USA; unselective but used mainly for grass weeds; readily absorbed by foliage and roots and translocated; causes plants to become chlorotic; used as an overall herbicide in pre-clearing and at lower doses as selective herbicide for inter-row application in pineapples and sugar-cane; dosage range from 1 lb to 30 lb/acre; long persistence in soil; significant mammalian toxicity.

PICHLORAM [4-amino-3:5:6-trichloropicolinic acid]: developed by DOW in USA; fairly selective against broad-leaf weed species; translocated in the plant with an action similar to MCPA above; effective on some weeds at very low doses of a fraction of an ounce per acre; usual dosage range 0.5 oz–2 lb/acre; persists in soil for long periods and also in straw; drift problems acute; low mammalian toxicity.

TOK granular [2:4-dichlorophenyl p-nitrophenyl ether]: used in granular form in flooded rice applied a few days after transplanting or when seed-sown rice is a few inches high, but before weed seed germination; applied at rate of about 30 lb/acre granular with 7 per cent active ingredient; very effective against grasses; acts in presence of light at photosynthesis centres; low mammalian and fish toxicity.

CNP granular [2:4:6:trichlorophenyl p-nitrophenyl ether]: similar to TOK granular above.

HE-314 [metatolyl-4-nitrophenyl ether]: developed by Tokyo Organic Chem. Ind.; similar to TOK and CNP above.

RP 17623 [5-t-butyl-3(2:4 dichloro-5-isopropoxyphenyl) 1:3:4-oxadiazoline-2-one]: developed by Rhone-Poulenc, France; similar to TOK and CNP above in action though not in structure, active at low rates of 2 lb/acre of granular containing 2 per cent active ingredient, for transplanted rice; mammalian and fish toxicity not clearly established.

CHLORAMBEN [3-amino 2:5-dichlorobenzoic acid]: developed by Amchem, USA; widely used for pre-emergence grass and broad-leaf weed control in soybeans and other legumes and maize; recently used for soil surface treatment in transplanted rice, active at 1-1.5 lb/acre; low mammalian toxicity.

BENTHIOCARB granular [S-(4-chlorobenzyl) N,N-diethylthiol carbamate]: selective against grasses in pre- and post-emergence treatments in rice. Used in granular form in flooded rice culture; active at low rates of $\frac{1}{2}$ lb/acre active ingredient.

PROPANIL [N-3:4-dichlorophenylproprionamide]: developed by Rhom and Haas, Monsanto and Bayer; this is a contact herbicide with limited translocation, acts primarily through the foliage with little soil action, used primarily as a directed herbicide in inter-rows; effective on young weeds; at 3 lb/acre has been used as a selective herbicide in rice, but the soil must be drained before use; low persistence; low mammalian toxicity.

AMINOTRIAZOLE [3-amino-1:2:4-triazole]: developed by Amchem, USA; very unselective as contact herbicide against both grasses and dicot. species; translocated; produces chlorotic and albino shoots through action on the photosynthetic mechanism, ammonium thiocyanate enhances activity in some cases; used at 4–8 lb/acre for pre-clearing; low persistence in soil, moderate mammalian toxicity.

DALAPON [2:2-dichloroproprionic acid]: developed by DOW; translocated and particularly toxic to grasses; acts both on foliage and through the soil, used principally as a directed herbicide; plants should be in an active stage of growth for effective kill; used at rates of 4-20 lb/acre; persists for a few weeks in soil; low mammalian toxicity. Has also been used in granular form and in circle weeding.

SIMAZINE [2-chloro-4:6-bisethylamino-1:3:5 triazine]: developed by Geigy, Switzerland; used as herbicide in maize and other crops; affects photosynthesis; action *via* the soil; rates of 1–2 lb/acre; low water solubility and high persistence in surface soil layers; low mammalian toxicity. Has been used much in tree circle weeding.

ATRAZINE [2-chloro-4-ethylamino-6-isopropylamino-1:3:5 triazine]: developed by Geigy, Switzerland; as with SIMAZINE above acts mainly through the soil but also on foliage; selective use in maize; rates of 1–2 lb/acre; high persistence but less than SIMAZINE, therefore more useful under drier conditions; low mammalian toxicity.

AMETRYNE [2-ethylamino-4-isopropylamino-6 methylthio-1:3:5: triazine]: developed by Geigy, Switzerland; similar to above, acts on both foliage and soil; used in pineapples and sugar-cane inter-rows at 4–8 lb/acre, somewhat selective at 1–2 lb/acre in cereals; moderate persistence in soil; low mammalian toxicity.

TRIAZINES: general name for the above three herbicides.

PCP [pentachlorophenol]: a contact herbicide with limited residual action in soil; used at 1-4 lb/acre; short persistence; low mammalian toxicity.

MONURON (CMU) [N'-(4-chlorophenyl)-N,N dimethyl urea]: developed by DuPont, USA; non-selective for soil use as a directed herbicide; seedlings killed by root uptake as they germinate; much used in circle and strip weed control in tree crops; action on photosynthetic mechanism; rates from 0.25 lb–3 lb/acre; restricted downward movement in soil and long persistence; low mammalian toxicity.

DIURON (DMU) [N'-(3:4 dichlorophenyl)-N,N-dimethylurea)]: developed by DuPont, USA; similar to MONURON above but with greater depth of action in the soil.

LINURON [N'-(3:4-dichlorophenyl) N-methoxy-N-methylurea)]: developed by DuPont, USA and Hoëchst, Germany; similar to the above two herbicides, being absorbed by the roots but acts also through the foliage with wetting agents; 1–2 lb/acre, fairly long persistence; low mammalian toxicity.

FLUOMETURON [N'-(3-trifluoromethylphenyl) N,N-dimethylurea)]: developed by CIBA, Switzerland; used in pre-and post-emergence treatments of cotton weeds and in citrus; low mammalian toxicity.

CHLOROXURON [N'-4-(4-chlorophenoxy phenyl) N,N-dimethylurea]: developed by CIBA, Switzerland; a soil-applied herbicide that kills germinating weed seeds; low-solubility and long persistence; dosage up to 5 lb/acre, in common with other urea herbicides, it has low mammalian toxicity.

BROMACIL [5-bromo-6-methyl-3-S-butyluracil]: developed by DuPont, USA; affects photosynthetic mechanism; action primarily through soil but also foliage with addition of wetting agent; used as a directed herbicide; dosage 2–15 lb/acre, long persistence in soil; low mammalian toxicity.

TERBACIL [5-chloro-6-methyl-3-t-butyluracil]: developed by DuPont, similar to BROMACIL above.

PARAQUAT (Grammoxone) [1:1 dimethyl-4:4' bipyridyliumdichloride]: developed by ICI, Britain; a very effective contact herbicide for foliage especially on grasses; very limited translocation; action on the photosynthetic mechanism depends on presence of light; rapidly sorbed and de-activated on clayey soil, therefore this herbicide has been much used in minimal cultivation techniques and for circle and strip weed control in combination with urea and other herbicides; used at 0.5–2 pt/acre commercial preparation; moderate mammalian toxicity.

DIQUAT [9:10-dihydro-8A:10A-diazoniaphen-anthrenedibromide]: developed by ICI in the UK; similar to PARAQUAT above, but more effective on dicot. species.

TRIFLURALIN [2:6-dinitro-NN-dipropyl.-4-trifluoromethylaniline]: developed by Eli Lilly in the USA; a soil-applied herbicide used prior to germination of weed seeds; persists for a moderate period of time if worked into the upper soil layers; dosage 0.5–2 lb/acre; low mammalian toxicity.

DICHLORBENIL [2:6-dichlorobenzonitrile]: developed by Shell, UK, and Philips, Holland; used to control germinating weed seeds in the upper soil layers, principally in perennial tree crops; moderately persistent if worked into the soil by cultivation, irrigation or rainfall; dosage 1–6 lb/acre; low mammalian toxicity.

SODIUM ARSENITE: A very effective contact herbicide for killing of a wide range of weed species by directed application to the weeds in inter-rows and circle weeding; strongly sorbed at anion-binding sites in certain soils; high mammalian toxicity however is a major disadvantage and its use has been prevented by legislation in many countries.

DSMA [disodium methylarsonate]: developed by Ansul in USA; has contact activity on plants and used as for sodium arsenite above, but with lower oral toxicity.

MSMA [monosodium methylarsonate]: as for DSMA above.

TCA [trichloroacetic acid]: a general herbicide which acts by root uptake; used primarily in pre-clearing at rates from 5–30 lb/acre; very effective against grasses. Oil palms are particularly sensitive to TCA which produces little-leaf symptoms even at very low rates.

SODIUM CHLORATE: a general contact herbicide with desiccating action on plant foliage; can be added to translocated herbicides such as 2:4-D to reduce their residual action on sensitive crops; Inflammable under dry conditions; low mammalian toxicity.

The effective use of herbicides and other methods of weed control depends much on a good knowledge of the biology of the weeds concerned; their methods of reproduction (whether by seed or vegetative means), their phases of development, longeivity and dormancy of seeds, methods of dispersal and so on. These aspects have often been neglected in tropical regions; as have questions of hygiene in preventing the introduction and spread of weeds in crops. One important weed, for example, *Mikenia cordata* in plantation crops, was actually introduced by planters in Malaya as a cover crop in coconuts. Weeds propagated vegetatively by means of rhizomes are often best dealt with by cultivation or by an actively translocated herbicide (e.g. Dalapon) which will be translocated to the underground parts. An example is *Imperata cylindrica*, a truly pan-tropical weed; except in the Markham valley of New Guinea where it is used as a pasture grass for cattle. Vegetatively propagated weeds are usually more restricted in their rate of spread than those which are spread by easily dispersed seed. In the latter, the generation time is of central importance in preventing spread; this affects frequency of weeding rounds.

The use of herbicides also requires careful attention to the effect, both immediate and delayed, on the crop itself. This is particularly important in the case of the introduction of a new herbicide. Herbicides may also bring about a number of basic changes in the crop and the environment which deserve mention. A significant change that may occur when herbicides are used to replace cultivations in weed control in perennial crops is the distribution of the root system. Cultivation will limit surface roots to a lower horizon, but little is known of the effects of soil-applied herbicides in this respect. This has been considered for the temperate regions by Clements (1968). Herbicides may also affect soil micro-organisms and organic matter breakdown, but generally such effects have been found to be transitory (see Fletcher *et al.* 1968). Little is known of herbicidal effects on other soil organisms. There is no doubt, however, that herbicides have and are having an important influence on the agronomy of tropical crops. The aspects of agronomy which are most immediately concerned with increasing herbicide use are the changes associated with minimal cultivation as mentioned above, and the potential changes in nursery techniques, and even the very existence of a nursery phase with the discovery of herbicides and herbicide mixtures capable of giving weed control for extended periods of six months or more.

2 BANANAS [Musa spp.]

INTRODUCTION, SPECIES AND VARIETIES

THE cultivated bananas which are grown for their fruits belong to the genus *Musa* in which two sub-groups are recognized, the *Eumusa* series, which includes all of the well-known banana cultivars, and the *Australimusa* which are grown chiefly in the Pacific regions. The *Eumusa* clones are all considered to have arisen from two species *M. accuminata* and *M. balbisiana* (Kurz, 1865; Cheesman, 1948). They possess 22, 33 or 44 chromosomes (n=11) and the most widely-cultivated clones of commerce are the triploids which have more vigorous growth characteristics and higher yield than diploids. Tetraploid clones are rather rare but diploids are often cultivated in tropical countries for local consumption and some are valued for their good-flavoured fruit.

The *Australimusa* series have 20 chromosomes (n=10) and the cultivated forms have apparently arisen from a number of species. The Tahiti cultivars are the only ones that are well known (see Simmonds, 1966).

As stated above, all of the cultivated clones of commerce and clones grown for local consumption outside the Pacific area belong to the *Eumusa* series of bananas and have been classified according to their genetic composition or the presumed contribution of the two species *M. accuminata* and *M. balbisiana* to their genotypes. Simmonds (1966) designates cultivars with the symbols A and B indicating the contribution of *accuminata* and *balbisiana* genes determined from morphological features; thus:

AA = *accuminata* type diploids
AAA = *accuminata* type triploids
AAAA = *accuminata* type tetraploids
AB = *accuminata* × *balbisiana* diploid hybrids
AAB & ABB = triploids derived from both species
ABBB = tetraploid derived from both species.

Within each class there are numerous bud mutants (sports) and all commercial clones are mutants which have seedless parthenocarpic fruit. Charac-

teristics of some of the more common bananas belonging to these various groups are shown in Table 2.1, drawn from Simmond's data and table. In addition there are many clones of local importance. AA clones are particularly abundant in New Guinea. Many of the upland bananas in East Africa are *Accuminata* triploids as are the red-coloured bananas which are known by various names.

BOTANY

The vegetative plant

The aerial parts of the banana plant consist of the crown of foliage leaves and their bases which combine to produce an aerial stem-like structure called the pseudostem. The true stem is small and composed of compacted internodes in the vegetative state, but becomes elongated within the pseudostems at the time of flowering to protrude and eventually bear the fruit bunch. At the base of the pseudostem is the corm from which the pseudostem, roots, the flowering shoot (or true stem) and the branch shoots arise (Fig. 2.1d). Several types of shoots are recognized as shown in Fig. 2.1b, c and d. For planting material the 'swords' or young 'peepers' or parts of the corm with buds are often used. 'Maidens' which are immature shoots and water suckers which do not bear fruiting shoots under normal conditions may also be used for planting. The corm is largely composed of compacted internodes of the true stem, root bases and buds.

The pseudostem, which is in fact the functional stem of the plant, is composed of interfolded leaf bases arranged in a phyllotaxy which varies from 1/3 to 4/9. Leaf blades arise from the pseudostem in the manner illustrated in Fig. 2.1a. The top of the sheaths gradually contract into the petiole. Internally the tissues of the pseudostem have large air spaces and although the surface is heavily thickened it still bears stomata. The laminae however form the main functional assimilation apparatus and bear stomata on both surfaces although they are three to five

TABLE 2.1 Characteristics of common banana clones and their assumed genetic origin

GROUP	NAMES	COMMENTS
AA	*Sucrier* Pisang Mas (Malaya) Ladies finger (Hawaii) Honey (W. Indies), etc.	The most important diploid banana cultivated for its thin-skinned highly flavoured fruit. Resistant to Panama disease but yields poorly with smaller fruit than triploids. Susceptible to leaf spot.
AAA	*Gros Michel* Pisang Embun (Malaya) Bluefields (Hawaii), etc.	A big vigorous high-yielding clone bearing cylindrical bunches of bottle-necked yellow fruit. The most abundantly produced banana clone and formerly the basis of trade from Central and South America and the West Indies. Susceptible to Panama disease.
	Dwarf Cavendish Canary banana Pisang Serendah (Malaya) Johnson (Canary Islands) Chinese (Hawaii), etc.	The most widespread clone. Better adapted to adverse climatic condition and resistant to Panama disease, but very susceptible to leaf spot. Fruit is good but the bunch is not as uniformly cylindrical as Gros Michel. It is the basis of trade in the Canaries, Australia, South Africa and Israel (see 'Climate', p. 29 below).
AAA	*Giant Cavendish*	Less widespread, Australia and Martinique.
	Robusta Pisang buai (Malaya) *Lactan* (W. Indies) Pisang masak hijau (Malaya) Pisang embun lumut (Malaya)	Similar clones somewhat shorter than Gros Michel and possessing some resistance to Panama disease and wind damage, but not adaptable to poor soils. The fruits ripen greenish and spoil more easily than either Cavendish or Gros Michel.
AAAA	*Bodles Altafort*	This is an artificially 'bred' banana which has excellent qualities of yield and disease resistance but is not widely tried. From Gros Michel and Pisang lilin (an AA clone) from Malaya.
	I. C. 2	Another 'bred' banana which is more widely distributed in the West Indies, Honduras and the Pacific. It is from Gros Michel and a wild *M. accuminata*. Resistant to leaf spot and somewhat resistant to Panama disease.
AB	*Ney Poovan* Ladies finger (Hawaii, W. Indies), etc.	The only certainly-known member of this group. Less vigorous characteristics of diploids but with pleasant-flavoured fruit. Highly resistant to Panama disease and leaf spot.
AAB	*The plantains* French plantain Horn plantain Pisang Raja (Malaysia)	There are numerous clones; vigorous plants resistant to Panama disease but almost all susceptible to stem borers. The fruits are starchy and used mainly for cooking. All have very large fruits, some markedly curved, e.g. horn plantain.
	Mysore P. Keling (Malaya), etc.	A very vigorous plant which is resistant to leaf spot, Panama disease and borers and possessing drought resistance. The fruit is attractive and it is an important banana in India. This is a very promising clone for widespread commercial use.
	Silk (P. rastali—Malaya) Sugar (Queensland) Silk-fig (W. Indies) Apple (W. Indies)	Widely distributed, moderately vigorous clone, moderately yielding. Resistant to leaf spot but not Panama disease. Pleasant-flavoured fruit but skin prone to splitting.

	Pome P. Kelat jambi (Malaya) Ladies finger (Queensland) Brazilian (Hawaii), etc.	Vigorous but not very high yielding. Resistant to leaf spot and Panama disease. Fruit not as attractive as silk. Of some importance in South India, Hawaii and Eastern Australia.
ABB	*Bluggoe* P. Abu Keling (Malaya) etc.	Vigorous clone with resistance to leaf spot and Panama disease. Few hands of large, angular, nearly straight fruits, green-skinned.
	P. awak (Malaya) Klue namura (Thailand), etc.	A most vigorous and hardy type. Immune to leaf spot and slightly resistant to Panama disease. If pollinated it is seeded. Common in Thailand. Various mutants exist, some pink-fleshed.
ABBB	Klue teparod (Thailand) P. Batu P. Abu Siam (Malaya)	The only known natural tetraploid. Robust and disease resistant with massive and blunt grey fruits with spongy, fibrous flesh. Used for cooked sweet-meat in Thailand and Burma.

times more numerous at the lower surface.

Internally the laminae contain some 50 per cent of air spaces. The laminae are produced successively from the centre of the pseudostem throughout the vegetative phase in the manner shown in Fig. 2.2. Thus the distribution of the leaves varies from vertical in the youngest emerged leaf to below horizontal in the oldest. Average elongation rates varied between 3.5 and 7.3 cm per day in one clone studied by Trelease (1923) who also noted a maximum rate of 20.7 cm/day with more growth occurring at night (which seems to be a general feature of most plants). The rate of leaf extension is also greatly affected by temperature, with little growth taking place below about 10°C (Turner, 1970).

The young leaves bear transverse corrugations which have a strengthening function (Fig. 2.2). Mature leaves show diurnal movements which are motivated by pulvinal ridges adjacent to the petiole. Turgor change causes the angle between the two laminal halves to change diurnally; bending downwards so that the angle between them varies from about 180° in early morning and evening to about 40° at noon. The degree of closure is less under shaded conditions and it seems likely that this is an adaptation to reduce heating and moisture loss from the leaf surface.

Splitting of the laminae transversely is a common occurrence in older leaves and under windy conditions. The edges of the split tissues become suberized and the laminae come to resemble the compound leaves of palms.

Leaves are produced at rates which vary considerably with climate, an average rate being prob-

ably of the order of one per week under warm conditions to one in 20 days under sub-tropical winter conditions (Summerville, 1944; Turner, 1971). Initially a series of scale leaves are produced on a young shoot followed by the functional foliage leaves. Estimates of the numbers of foliage leaves produced before fruiting vary from 23–45 (Champion, 1961) to 35–50 (Summerville, 1944). There are probably about 10–20 functional leaves present at any one time, the older ones dying off and decaying at the base and being replaced by young leaves. Leaves increase in size to a period just before flowering and then decline abruptly (Fig. 2.3). The development of leaf area seems to be central in determining the onset of fruiting (Summerville, 1944), and since the former is affected by climate, the time of fruiting also varies considerably with climate from about 18 months under sub-tropical conditions to less than 12 months under warm humid conditions. Sanchez *et al.* (1970) report a fruiting time of 21 months at elevated sites in Puerto Rico. Fig. 2.4 indicates how temperature affects the rate of increase of leaf area under Australian conditions (see 'Growth Quantities', p. 28).

The root system

The root system is adventitious, arising from the corm. Roots often arise in groups of four and are 5–10 mm in thickness. A corm may bear 200–300 roots or up to 500 (Summerville, 1944; Robin and Champion, 1962). Lateral spread extends to 5 metres (17 ft) and roots may reach a depth of 2½ ft. The majority however are in the top 6″ (Fawcett, 1913).

Fig. 2.1   Vegetative morphology of the banana

a. Pseudostems near maturity

b. A young 'sword sucker'

c. Young shoot 'peeper' near a matured
pseudostem

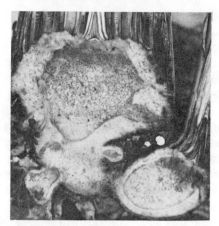

d. Section through corm and lower
pseudostem showing origin of a side shoot

e. Vegetative morphology of the banana. Transverse section
through pseudostem showing interfolded leaf bases which
form supporting structure. The true stem is at the centre

a. Newly-emerged 'spear' leaf

b. Commencement of unfolding after emergence

c. Fully unfolded leaf

Fig. 2.2   Stages in the unfolding of a banana leaf

Fig. 2.3   Leaf size and height of insertion of the climax leaves of a banana plant

Fig. 2.4   Temperature and relative growth rate of leaves of Dwarf Cavendish banana in Queensland

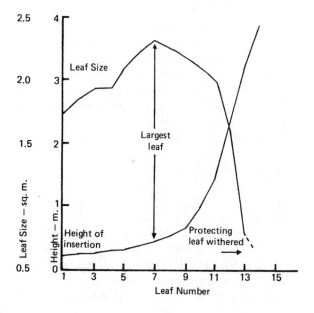

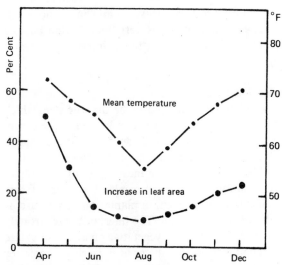

Source.   After Simmonds, 1966.

Source.   After Summerville, 1944.

Fig. 2.5a Young emerged inflorescence on the determinate pseudo stem

Fig. 2.5b Pseudostem cut open to reveal inflorescence shoot (the true stem)—note length of internodes

Fig. 2.5c Young banana fruits, enclosing sheath removed

Numerous small lateral roots bear root hairs and are probably the primary centres for uptake of nutrients and water. These occur more frequently away from the centre and thus uptake of fertilizers is likely to be more effective if it is distributed in a zone a foot or two away from the pseudostem or central ratoon stems. Davis (1961) has demonstrated that root pressure is very high in the banana plant and it is likely that this is important in water uptake.

Floral morphology

The inflorescence is borne on the true stem which until the flowers are formed is hidden in the centre of the pseudostem. The formation of the inflorescence makes the main shoot determinate so that its life will end with the maturation of the bunch. Perennation occurs as a result of growth of side shoots (suckers) which, in plantation agriculture, is called the ratoon crop. The internodes are strongly compacted during the vegetative phase but elongate to lengths of up to 1 meter at flowering, thereafter becoming compacted again at inflorescence maturity (see Figs. 2.5b and 2.7). Morphologically both male and female flowers are produced, the females

near the base of the inflorescence (Fig. 2.5c); but edible bananas do in fact develop parthenocarpically with most of the pulp forming from the tissues of the inner side of the skin. Parthenocarpy appears to result from a number of causes including female sterility, and in the case of many of the edible clones from the triploid nature of the chromosome make up. Some commercial bananas produce a few seeds and others are notably seedy particularly when out pollinated with wild types. Auxin appears to be central in causing parthenocarpic development, as well as possibly other growth factors (Simmonds, 1966). Gibberellins cause splitting.*

The stem which bears the inflorescence can be be either positively or negatively geotropic or diatropic (Fig. 2.6). This is a factor which affects market qualities since the banana fruit itself is negatively geotropic and the fruit curvature determines the bunch shape differently according to its orientation. Fei (*Australimusa*) bananas have negatively geotropic fruit stems. Those of Gros Michel and others are positively geotropic and produce the best-shaped bunches. Irregular-shaped bunches are

*R.G. Lockard, personal comm.

Fig. 2.6a&b   Diatropic and geotropic inflorescence stalks

Fig. 2.6c   Irregular-shaped bunch produced from diatropic stalks

produced from diatropic fruit stalks (Fig. 2.6c). The negatively geotropic reaction of the banana is, in most varieties, located in the body of the fruit, producing the characteristic curved shape.

Growth quantities, yield components

Lassoudiere (1972) reports that until the emergence of the inflorescence the stalk elongates little, but following this, elongation rate in *M. accuminata* reached 15 cm/day. Fruit growth increased from the fourth day before flowering (opening of the sheath). There is a negative correlation between fruit size and fruit number, and fruit growth rates of the larger-fruited types are correspondingly faster. In addition to this, triploids and tetraploids generally possess much faster fruit growth rates and larger fruits than diploids (Table 2.2). (The physiological reasons for this have not been closely examined.) With Robusta bananas, Lassoudiere and Maubert (1971) showed that from the time of uncovering for about 30 days, fruit growth was rapid with length and diameter growth increments of 4 mm and 0.3 mm/day. After this a slower growth rate pertained for a further 60 days and the diameter increased at a rate of 0.19 mm/day.

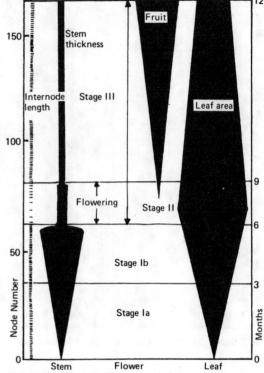

Fig. 2.7 Diagrammatic illustration of the growth phases of the banana

Source. After Simmonds, 1966.

TABLE 2.2   Growth rates of banana fruits

	Diploid	Triploid	Tetraploid
Size of ovary (cc)	9.0	19.3	19.1
Growth rates cc/week	2.9	7.2	6.9
Relative growth per unit size of ovary	0.3	0.37	0.36

Source. After Simmonds, 1966.

The number of fruits developing and the bunch weight are both correlated with the size of the assimilation apparatus and with growing conditions. Simmonds gives the following formulae for weight of the bunch in Dwarf Cavendish:

$$W = 1.5 L - 4$$
$$W = 1.2 L - 2$$

where W is bunch weight (kg) and L is number of leaves. Fernandez Caldas and Garcia (1972) report that there is also a significant positive correlation between the diameter of the pseudostem and the weight and the number of hands on the bunch.

Fig. 2.7 (from Simmonds, 1966) illustrates diagrammatically the various growth phases of the banana plant. Summerville showed that nutrient supply (fertilization) was most effective in the early development stages of the canopy (Stage I) and that deficiencies could not be made good at late stages. The climate during Stage II has a marked effect on fruit number and apparently affects the development of the last hands of the bunch at which the transition between female and male flowers occurs.

In commercial estates the bunches are cut some time before they are ripe and the interesting fact is that the fruit continues to increase in weight for about forty days after cutting, presumably at the expense of reserves and moisture in the stalk (Fig. 2.8).

Other factors which affect bunch size are concerned with interplant competition and therefore spacing and the extent of ratooning (see later).

Turner (1972 a and b) has studied the growth of the Giant Cavendish clone in Australia and reports that crop growth rate reached 6780 kg/ha in a period of 6 months near maturity (very likely suboptimal conditions) which indicate that assimilation rates in bananas could be relatively high for a perennial species, under optimal conditions.

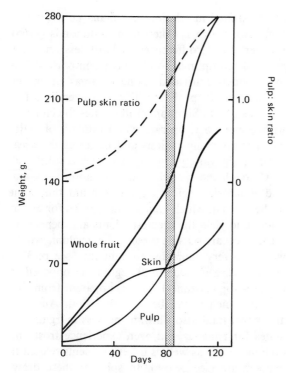

Fig. 2.8 Growth of the Gros Michel fruit
Source. From Wardlaw *et al*. 1939; after Simmonds.

In regard to root growth, Lassoudiere (1971) studying bananas in glass root chambers, reports that growth is very sensitive to moisture conditions; a 10 mm deficit and a 25 mm surfeit in soil water both retarded growth.

CLIMATE AND SOILS

Temperature and rainfall

All the evidence of experience indicates that the optimal climate for the banana occurs in the humid tropical lowland zones of equatorial regions. A temperature of 80° seems optimal for both growth and yield and an even distribution of rainfall without periods of moisture stress is also favourable for bananas. Simmonds (on the basis of general experience) considers that 2″ of rainfall per month is a limit below which bananas are seriously affected by drought and that 4″ is adequate on all but the most porous soils. The figure of 2″ seems to be a generally derived one by tropical agriculturists in defining a 'dry-season' but possibly a better way of considering plant water relations for such species as the banana, in which there appears to be no developmental requirement for moisture

stress for either flowering or ripening, is from rainfall and evaporation and the moisture storage capacity of the soil. As a general guide, we may consider 6″ of water to be that which is evaporated under tropical conditions (see Nieuwolt, 1965), and for tropical soils, 4″ of storage water would be a maximum figure for bananas since many of the monsoonal and humid region soils possess rather low moisture storage capacities. Moisture balance diagrams such as those of Nieuwolt (Fig. 2.9) would probably be suitable for determining periods of moisture stress and sufficiency in bananas, and indeed Arscott *et al*. (1965 a and b) have shown positive response to irrigation water as high as 2.6″ per week. Bredell (1970) considers that moisture use in bananas is of the order 0.9 of open water evaporation (0.9E).

Temperatures of 80°F (and above) seem ideal for bananas. A temperature of 70°F reduces growth

Fig. 2.9 Example of moisture balance diagram assuming 4″ of soil storage water

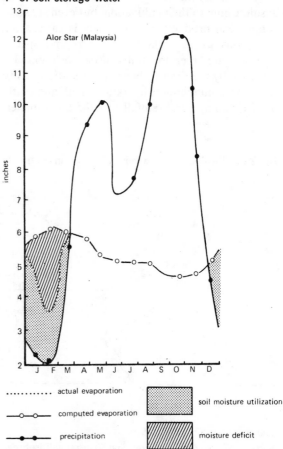

Source. After Nieuwolt, 1965.

and a practical minimum for growth is often considered to be about 60°F. Green and Kuhne (1969) and Turner (1970) both report that extension growth ceased at 10°C (50°F). Sub-optimal temperatures increase the time to maturity from about 9 to 10 months in equatorial regions to 18 months in sub-tropical latitudes such as Queensland. Even small differences in the temperature can cause considerable delays in maturity. The following table (from Daudin, 1955) shows the effect of altitude in Martinique on maturity together with the environmental decrease in mean temperature.

Altitude (feet)	0–450	600–1200	1300–2100
Months to shooting	6–7	9–10	11–13
Environmental drop (°F)	–	2	5

However, Sanches *et al.* (1970) report that bunches are somewhat heavier at high elevation sites. The relationships of banana growth to more useful temperature parameters such as degree-days (see sugarcane) have not, to my knowledge, been published.

Frost causes the death of bananas and marks an absolute limit to their cultivation, but even so, some commercial production of bananas is carried out in regions where slight frost occurs and special measures are taken to combat it (see below).

Generally the limit to banana cultivation can be shown by a minimum temperature of 60° and (without irrigation) a 50″ rainfall and these parameters

are illustrated in relation to banana cultivation in Fig. 2.10, which is taken from Simmonds (1966). Further, clonal differences in hardiness and tolerance of sub-optimal conditions are important. The most widely-cultivated clone in areas of marginal climate is probably Dwarf Cavendish. Fig. 2.11 (also from Simmonds) illustrates the climates, with respect to temperature and rainfall, of major banana-producing regions and indicates the areas in which the variety Dwarf Cavendish is cultivated.

Although the best climates are those of the humid equatorial regions, it is significant that some of the areas in which bananas are grown for export trade (and yield the greatest profits and benefits to the farmer) are outside these limits (cyclonic winds, which are very damaging to bananas—see p. 32—are also rare in equatorial regions as opposed to the sub-tropical zones). This is apparent from Fig. 2.12 in which it can be seen that both Australia and Israel (and also Taiwan) possess marginal climates for bananas and even experience frost and cyclonic winds as well as other problems related to temperature (see below). In spite of these drawbacks however, these countries enjoy a good banana trade with consumer countries. In the case of Israel and Taiwan success is undoubtedly related to the proximity to consumer countries, the former with Europe and the latter with Japan. But even so, there are some obvious lessons to be learnt by

Fig. 2.10 Distribution of banana cultivation in relation to climate

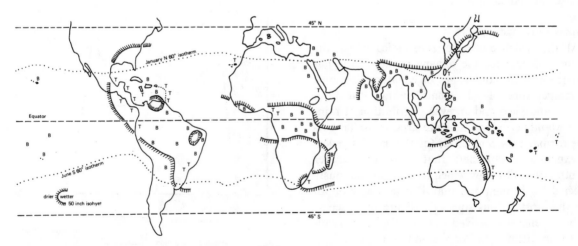

Source. After Simmonds, 1966.

Notes. Winter 60°F isotherms and 50 in rainfall isohyets are shown; B—banana grown for local consumption, T—banana grown for export trade.

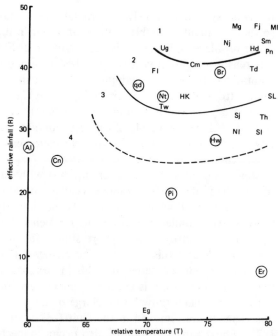

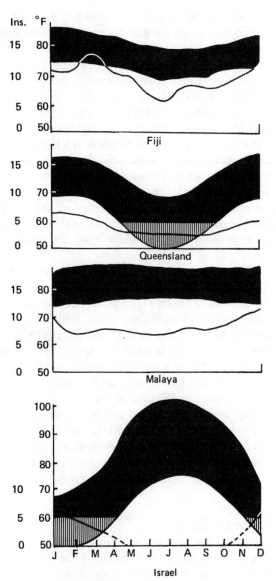

Fig. 2.11 Climate: rainfall against temperature of major banana-producing areas

Source. After Simmonds, 1966.

Abbreviations. Al—Algeria, Br—Brazil, Cm—Cameroons, Cn—Canary Islands, Er—Eritraea, Fj—Fiji, Fl—Florida, Hw—Hawaii, Hd—Honduras, HK—Hongkong, Mg—Madagascar, Ml—Malaya, Nt—Natal, NI—North India, NJ—North Jamaica, SL—Sierra Leone, Sm—Samoa, Tw—Taiwan, Td—Trinidad, Th—Thailand, Ug—Uganda. Dwarf Cavendish cultivations are ringed.

Fig. 2.12 Monthly rainfalls and temperatures (blacked, shaded below 60°F) of various banana-growing regions

Source. From Simmonds, 1966.

some developing countries. Take, for example, the case of Malaysia. Fig. 2.12 shows the important climate parameters of Malaya (in this case western Malaya) together with the climates of Israel and Queensland. It would be hard to find a climate more suitable for banana growing than that of Malaysia (including Sabah and Sarawak). Furthermore, this country is close enough to Japan (which is a major banana importer) to facilitate shipping, and all the commercial clones of banana are already present in the country and are grown for local consumption. Probably the only reason that Malaysia has not taken over the banana trade with Japan is that no organized effort has been made to do so.

Low temperature and wind

The damaging effects of low temperatures are as follows. Firstly, low temperature causes a delay in the emergence of leaves and reduces leaf size so that the formation of the assimilation apparatus is

delayed. Bunch production depends on the attainment of a certain leaf area (Summerville, 1944). Secondly, low temperatures produce a number of adverse effects on the fruit. If the leaf surface is reduced significantly, sun scorching of the fruit may occur leading to an unmarketable product. Furthermore, low temperatures may inhibit growth of the bunch stalk (the true stem) so that the bunch does not emerge fully, giving rise to a condition known as 'choke'. In areas affected by this condition the usual hardy Dwarf Cavendish is replaced by Gaint

Cavendish which shows less tendency to choke. Various fruit abnormalities are also caused by low temperatures (Fahn *et al.* 1961), and ripening and quality are also affected. Thirdly, frost is lethal to bananas and where it occurs special measures are taken to reduce its effects, including the burning of fires, the creation of artificial fog to reduce night radiation and temperature drop, and the sprinkling of the crop with underground water at a higher than surface temperature. Sometimes, since frost develops from the ground upwards, taller varieties are grown, but these are more susceptible to wind damage which usually also affects crops in sub-tropical regions. In northern New South Wales, bananas are grown on hillsides where cold air drainage occurs.

Wind is very damaging to bananas. Moderate winds such as those which occur with the monsoon and trade wind seasons (15–20 mph average) cause leaf damage and crown distortion, while stronger winds of 40 mph cause considerable damage. Winds of 60 mph are considered to be capable of causing almost complete loss and furthermore, lesser winds may be followed by the outbreak of diseases or subsequent poor development of fruit caused by root damage incurred. In regard to wind as well as temperature and rainfall, equatorial countries such as Malaysia are much better situated, and another damaging weather condition, namely hail, is also virtually absent in equatorial regions.

Soils and drainage

Bananas are grown on a great variety of soils ranging from alluvial flood plains to inland latosols of the tropics; from organic peat soils to volcanic loams and grumusols or 'black cotton soils' and the sandy soils of Israel and the Canary Islands. The most important feature in banana soils is probably drainage, so that heavy clay soils with poor drainage should be avoided. Free draining hill soils do not require artificial drainage but flat alluvial plains should be drained with field drains of depth about $2\frac{1}{2}$ ft.

The suitability of a soil for banana cultivation is clearly related to rainfall in as far as its depth and moisture-holding capacity are concerned, unless supplementary irrigation is carried out. Most tropical and monsoonal inland soils are adequately drained but as indicated earlier, do not possess great water holding capacity. In addition, the bana-

na root system is largely superficial (see p. 23), so that the plant is unable to draw on water in the lower profile even if the soil is deep structured. For this reason irrigation is often practised in modern plantations to obtain maximal yields even in the humid tropical zones. For example, in the Cameroons a dry period of 3 months was formerly considered to have negligible effects on bananas but now irrigation trials have shown yield increases of 15 per cent (Melin and Marseault, 1972).

Some tropical soils are better supplied with nutrients than others (for example volcanic basaltic soils and some lowland coastal soils, particularly those derived under a marine environment), but by and large nitrogen is in short supply in most tropical soils and this is one of the elements that frequently produces increased yields in bananas in the tropics. Potassium is another element very important to banana growth since large quantities of this element are removed in the fruit (see Joseph, 1971). Phosphorous responses also occur. Aspects of the nutrition of bananas are considered in more detail later in the chapter.

It is sometimes stated that a high soil reaction is best for bananas but some very successful crops are grown on rather acid soils. One factor which seems to be of importance is that Panama disease is more prevalent in acid soils so that susceptible varieties such as Gros Michel would be better confined to soils of less acid reaction, or to those that are appropriately limed.

PLANTING
Land preparation, spacing and yield

In tropical countries bananas are often planted on land newly claimed from jungle. In such cases felling and burning of the timber is first carried out. Usually only rough clearing is practised and planting is carried out by hand, since the cost of complete clearing is very high. In such 'cleared' land the course of time brings about decay of root stumps etc. so that in subsequent crops access to the land is much easier and wheeled vehicles can be used more readily for harvesting and other operations.

In old estates and in land which has been cleared of roots, mechanical cultivation can be used in planting. But even so, in recent years, trends towards minimal cultivation have been obvious and it may probably be correct to say that mechanical cultivation of the land is best avoided unless the soil is heavy and needs cultivation for adequate drainage.

The spacings adopted for the banana crop vary considerably and are to some extent affected by the way in which the fruit is marketed. If the fruit is sold by weight higher gross weights are more often obtained by closer spacings but if by grade (the number of hands per bunch) then wider spacings are used which favour large bunches. Spacing is also affected by clone, large varieties requiring more space than those that are small. For example, Wills (1951) recommends 9×12 feet square for Dwarf Cavendish and 10×14 for Giant Cavendish. Soil fertility also affects optimal spacing and according to Simmonds, both fertile and good soils should support denser spacings (contrary to the experience with oil palm, see Chapter 12). The pruning regime for the ratoon crops (see below) is also affected by spacing since plant density tends to increase following the initial crop.

Sharma and Roy (1972) report highest yields of Dwarf Cavendish at 2×2 metre spacing (c. $6\frac{1}{2}' \times 6\frac{1}{2}'$). Matos (1970) in a trial with Giant Cavendish with one plant and one follower per hole, found that spacings denser than 3×3 m (c. $9\frac{1}{2}'$) retarded

growth and yield and bunch quality. With a smallish-sized plantain in Puerto Rico, Caro Costas (1968) and Champion (1970) found that the highest yields were obtained at spacings of 1.5×1.8 m (c. $5' \times 6'$), a population density of c. 3500/ha. However, even spacing was necessary; when plants were crowded in rows at the same density but with less than 1.2 m between individual plants, yields were reduced. Thus avenue planting is not suitable for bananas.

In general the experience seems to be that commercial spacings are probably more often not dense enough. Table 2.3 gives an indication of the effect of spacing on both the initial crop and subsequent ratoons. The conclusion here is that up to very dense spacings much greater than the conventional, $(11' \times 11')$, gross yield is significantly increased even in the ratoon crops. But grade and individual bunch weight and time to shooting are all poorer, particularly in ratoon crops. The latter is undoubtedly due to light limitation and its effect on shooting as discussed earlier in relation to the assimilation apparatus. Actual yields vary considerably with (in addition to spacing) other factors particularly clone, manuring and irrigation. In India

TABLE 2.3   Relative yields, bunch characteristics, and time to sprout shoots in 'Lactan' in Jamaica

Spacing ft.		11	$9\frac{1}{2}$	$8\frac{1}{4}$	$7\frac{1}{2}$	$6\frac{1}{2}$
Plts./acre		360	480	620	780	1020
Yield	P	100	126	156	188	228
per	R_1	100	116	136	141	150
acre	R_2	100	110	132	134	147
Hands	P	100	96	96	93	94
per	R_1	100	96	94	89	86
bunch	R_2	100	95	92	86	84
Bunch	P	–	–	–	–	–
weight	R_1	100	89	84	75	71
	R_2	100	89	83	74	72
Time	P	100	101	103	101	101
to	R_1	100	112	117	126	131
shooting	R_2	100	105	110	115	115

Source.   Adapted from Simmonds, original data from Gregory 1952, 1954 and 1955.

Notes.   100=yield at lowest spacing. P, R_1 and R_2 refer to the initial crop and the two subsequent ratoons.

yields ranging from 14–27 tons/acre are reported under irrigation and at high densities from Dwarf Cavendish (Nayar and Bakthavathsalu, 1955). In the Canaries, the range of yield from Dwarf Cavendish (Holmes, 1930) seems to be from 16–19 tons. In West Africa and Central America yields from 'Gros Michel' range between 3 and 20 tons/acre depending on the standard of husbandry. Melin and Marseault (1972) report yields of 44 tons/ha under irrigation in the Cameroon Republic. Under humid tropical conditions high yields can be expected and the crop can be harvested in a shorter period of time than elsewhere.

Planting material and planting

All types of shoots can be used for planting but there are differences in the time to fruiting. (This may be utilized in commercial production in timing the crop to a favourable marketing time or to avoid adverse seasons). The absolute time for shooting and maturity varies, as indicated before, with climate but in general it seems that large shoots, including the corm ('head' or 'cabbage') and large swords mature slightly earlier. But there seem to be, seasonal interactions as well. In Jamaica (with a moderately warm climate), heads and sword suckers mature sooner than maidens and peepers and water suckers are not used at all; but in Israel, water suckers matured sooner than swords. This could be the result of a temperature interaction.

The multiplication of planting material is a major problem in the establishment of a banana industry and importation of suckers is often necessary. Minimal plant populations of 1,800,000 have been suggested as necessary for establishing trade (Simmonds, 1966) and this can take several years to achieve. Nurseries for propagation, rather than for fruit production, are necessary to build up plant populations and dense spacings are used in these.

Shoots can be stored for limited periods. Nasharty (1969) reports that storage for 30 days in Egypt resulted in earlier production, but yield was reduced.

Most planting is carried out by digging holes but in Australia furrow planting is used. Deeper planting (18″) and earthing reduces wind damage and deeper planting is also to be recommended in areas of marginal water supply.

Manuring and irrigation

Much of the early work on the manuring of bananas has been reported by Croucher and Mitchell (1940) and there is a recent review by Freiberg (1965). Obviously fertilization should be related to natural soil fertility and newly-cleared jungle areas are more likely to be sufficient in plant nutrients. Continuous cropping, however, as in commercial estates, calls for the use of fertilizers to replace those removed by the crop and lost by leaching and erosion, and to obtain maximum economic yields. Joseph (1971) reports that a moderate banana crop of 16 tons removes 38 kg N, 8 kg P, 285 kg K (about 2/3 of this in the pseudostem that may be returned to the soil) 30 kg Ca and 49 kg Mg.

Nitrogen is an important nutrient in bananas probably because N losses from the soil are often considerable and because adequate N nutrition is essential for the formation of the photosynthetic apparatus of the plant. Croucher and Mitchell (*loc. cit.*) recommended heavy applications of ammonium sulphate (540 lb/acre/year for Gros Michel at 360 plants/acre). Furthermore, fractionation of the N dose to minimize leaching is also recommended, but bearing in mind that the early establishment phase is the most critical for nitrogen supply. Nitrogen affects the time to maturity to the extent of reducing it by 20 per cent, so that this could be of importance in determining marketing time. Urea has been recommended for the rapid correction of N deficiency both in soil application and as a foliage spray up to 5 per cent strength (Ho, 1970). At higher densities heavier rates of N have been recommended. In Australia 225 kg N/ha is the standard recommendation (Leigh, 1969; Turner and Bull, 1970). Marques (1971) recommends similar rates in Mocambique and Champion recommends 560 kg/N in Puerto Rico. The highest level of 900 kg has been given for Dwarf Cavendish in India (Sharma and Roy, 1972).

Potassium (which is removed in the crop in significant quantities) is also an important fertilizer with continuous cropping and is recommended at rates between 90 and 360 lb/acre of potassium sulphate, depending on the soil K status (see below) by Croucher and Mitchell. More recent recommendations at high plant densities range from 480 kg (Sharma and Roy, 1972) to above 700 kg (Champion, 1970) as K_2O.

Phosphorus has been recommended at rates between 255 and 360 lb acre as super-phosphate by Croucher and Mitchell. The use of rock-phosphate on acid tropical soils that fix phosphate does not appear to be widely adopted for bananas. Australian and Mocambique recommendations are generally between 50 and 150 kg P_2O_5/ha but Turner and Bull (1970) report the use of up to 1700 kg P_2O_5 on phosphate-inactivating basalt soils.

Magnesium deficiency has often been reported in bananas and dolomitic limestone is recommended to correct this. Broadcasting of limestone produces slow response but placing it in the planting hole is rapid (Chalker and Turner, 1969). Moreira (1969) also recommends magnesium limestone dressings to raise soil pH, particularly in the case of low pH arising from use of heavy N fertilizer from ammonium sulphate.

Marked responses have been reported to Ca, Mg, Mn, B, Zn and Cu supplied in addition to the usual N, P and K in Puerto Rico (Hernandez and Lugo Lopez, 1969).

Application of fertilizer is usually by broadcasting around the plant sometimes with light working into the soil to minimize washing off, and split application on 3 or more occasions is widely recommended. When considering the initial crop (as opposed to ratoons) and under conditions where cropping is seasonal the N should be supplied early at the time of canopy development (cf. Fig. 2.7). K is required throughout but particularly during bunch development—see below. Under tropical non seasonal conditions fertilizers should be applied at regular intervals, but avoiding peak rainy and dry spells.

Soil analysis has been used in determining K and P requirements with 20 ppm, P_2O_5 being regarded as critical for P and 300 ppm as critical for K. Probably the technique of foliage analysis, combined with fertilizer trials to establish critical levels could be a more reliable technique for determining fertilizer applications. Levels of N, P_2O_5 and K_2O which have been regarded as critical in the foliage are respectively, 2.6, 0.3–0.45 and 3.3–3.8 as a percentage of dry matter. A higher optimal level of 3 per cent has been recommended for N on the basis of nutrient studies by Lacoeuilhe and Martin-Prevel (1971) and Ho (1969) reports yield responses with leaf K up to 5 per cent in the three months preceeding harvest.

The need for irrigation in bananas is obviously determined by the rainfall as indicated above. About 6″ of water per month would seem to be adequate under humid, tropical conditions on soils of moderate (4″) water-holding capacity. Under more arid conditions 8″–10″ may produce economic yield increases and reduction in time to maturity. Arscott et al. (1965) obtained yield increases with 10″ of water per month.

The actual quantities of water supplied by irrigation vary considerably. In the Canary Islands 3″–6″ is commonly supplied and yields are high (Champion and Mounet, 1962). The highest irrigation frequencies and quantities on record appear to be from India where Oppenheime (1956) reports application of 22″ per month at 3 to 6-day intervals.

Both furrow and overhead irrigation are used but the overhead method seems to be more adaptable and is now much used in plantations.

Pruning and timing of the crop

Probably the most important management function in banana growing is related to the timing of the crop for shipping and marketing and to avoid environmental stress. Under humid tropical conditions this is relatively simple but under sub-tropical and less favourable conditions the timing of the crop to avoid stress is most important. Furthermore, timing is affected by fluctuating market prices and under seasonal conditions this can become quite a problem. It is not practicable to consider every situation in detail here, but simply to give some indication how the time to maturity is affected by cultural practices. The effect of N fertilizer in hastening maturity has already been mentioned and the effect of planting material has also been considered. Under sub-tropical conditions planting is usually done in the spring. Plants grow throughout the summer, become dormant in winter and mature over the following summer and autumn seasons. The following ratoon crop is more difficult to control and the most critical operation is the setting of 'followers' or the new stems of the ratoon crop. The old trunk is cut back and unwanted suckers are pruned with a heavy chisel or a similar tool. More often two suckers are selected for the ratoon crop. These should be chosen to achieve a spread of time between their maturity (so as not to over-tax the root system). The selection of suckers for maturity time is in fact very difficult and in some cases growers replant yearly in order to time the crop for the market more accurately. At the other extreme, where continuous production is desired (as in Uganda), little or no pruning is carried out and the crop is allowed to ratoon indefinitely. In general, in the equatorial tropics, plantations are kept for between 5 and 20 years before replanting.

Pruning of leaves is sometimes carried out for various reasons. The pruning of all leaves which is called 'trashing' in the banana business is sometimes carried out to reduce the spread of leaf diseases, particularly 'leaf spot'. Leaves which rub against the bunch following shooting are also removed, otherwise they cause abrasion of the fruits whose market value would then be reduced. The use of pruned leaves as a mulch is a practice which is to be much encouraged where the spread of disease is not involved. The cutting of old pseudostems at ground level and chopping up and spreading in the interplant spaces is also recommended both for mulching purposes and for hygiene (particularly to reduce borers—see p. 37). This practice also returns K to the soil.

General husbandry and weed control

Apart from mulching with pruned leaves and pseudostem, mulching the soil with black polythene has been found to reduce production costs in the Canary Islands as a result of the suppression of weeds and a better use of irrigation water (Garcia, 1968).

Earthing up the stems is carried out sometimes as a protection against wind and when the roots and the parts below the ground are exposed by erosion. The use of wooden poles to prop up the stems is also practised in windy locations such as Australia.

As mentioned before, frost protection is necessary in marginal climatic zones and probably the most generally economical method is the burning of heaps of smoldering sawdust, where this is available. Other methods of frost protection have been considered before.

Removal of the old flower parts (style and perianth) at the apex of the fruit is carried out with some varieties of banana in which these parts are not completely deciduous. In Dwarf Cavendish for example, it is necessary to brush off the flower parts a few days after anthesis. The male bud at the end of the inflorescence is also sometimes pruned off and this is said to increase bunch weight although there appears to be little evidence to support this idea.

The bagging of the inflorescence with plastic or hessian bags is carried out in cold climates to reduce fruit spoilage and this also prevents sun burn-ing of the fruit. Blue polythene tubes are commonly used and sometimes newspaper is used to cover the upper fruits to reduce sun-scorch.

Weed control by mechanical cultivation is best avoided in bananas because of the superficial root system. Hand slashing of weeds or light cultivation with hand tools is commonly practised in tropical countries where labour is available. However this is becoming increasingly expensive and there is an increasing tendency to use chemical methods of weed control. Furthermore, hand slashing tends to encourage a grass-dominant weed cover which is undesirable because of the property of grass-weeds of bringing about nitrogen deficiency in the crop.

In Australia arsenical herbicides are used (MSMA and DSMA) and effect good weed control. 2:4-D and 'dalapon' have been thought to be unsuitable because of crop sensitivity but in recent years these have been used extensively. Chambers (1970) recommends dalapon at 2.75–5.5 kg active ingredient/ha for grasses such as *Imperata cylindrica*. Kasasian and Seeyave (1968) recommend 2:4-D amine (2.2 kg active ingredient/ha) once the crop is well clear of the ground and foliage contact does not occur. 2:4-D amine is best premixed with sodium chlorate (15 kg/ha) which seems to reduce any harmful residual effects of this hormonal weed killer. Care, however, must be taken to prevent the spread of the herbicide to the foliage of the crop and for this reason mist jets should not be used. Jets of the type which produce a line of small droplets (such as the 'polijet') are more suitable. Paraquat is also used for inter-row weeding. Simazine, atrazine and diuron are suitable for long-term weed control (Moreau, 1971; Kasasian and Seeyave. 1968, etc.). Bromacil and terbacil are toxic to bananas and should not be used (Lassoudiere and Pinon, 1971).

In mature estates, the shading effect of the crop itself greatly reduces weed competition.

The growing of leguminous cover plants has been recommended for bananas but is not widely practised. Various legumes including *Calopogonium mucunoides*, *Pueraria spp.*, and *Centrosema pubescens* have been used. These fix nitrogen under tropical conditions but also compete for moisture with the main crop under marginal rainfall conditions.

MAJOR DISEASES AND PESTS

Wardlaw (1935 and 1961) has reviewed the diseases of bananas and Simmonds (1966) also gives a good account of these diseases. The following account gives only a brief outline of the major diseases and pests and their significance. The most serious disease of bananas in the past has been Panama disease or banana wilt. In equatorial regions 'leaf spot' is sometimes a serious disease in plantation bananas. Resistance of various clones to these two diseases have been mentioned in Table 2.1. Other important diseases are 'bacterial wilt' and 'bunchy-top'.

Panama disease (banana wilt). The causal organism of Panama disease is *F. oxysporium f. cubense*, pathogenic strains of which also cause diseases in other crops (e.g. wilt of oil palms). This disease caused catastrophic losses in tropical America in the past, particularly with Gros Michel, which is susceptible to it. The organism attacks the roots and underground portions of the plant giving rise to wilting of the foliage and death of the plant. The 'Lactan' banana and the Cavendish possess resistance to the disease and the former is now largely grown in place of Gros Michel in tropical America. The work of Rishbeth (1957) shows that resistant strains limit the pathogen to the rootlets.

Apart from genetic factors, the occurrence of the disease is affected by developmental and environmental factors. Young plants are more susceptible than old, and heavy soils with poor drainage combined with high rainfall favour attack, apparently because of reduced host vigour under these conditions. In addition, wind damage, which causes the proliferation of young roots favours attack and a low soil pH, as mentioned earlier, also favours the parasite. In addition to these factors nutrition also affects the severity of Panama disease in that fertile soils appear to favour the survival of the host. Furthermore, in common with most soil pathogens, the continued cultivation of susceptible varieties leads to a build up of the pathogen in the soil.

Control of the disease by phytosanitary techniques (i.e. removal of diseased plants and soils sterilization) are of limited use but soil treatments to increase pH and improve drainage are useful. The most important way of reducing the disease however is by the planting of resistant clones.

Leaf spot (sigatoka disease). This disease is caused by the fungus *Mycosphaerella musicola*, which is largely specific to bananas and attacks the leaves both by means of conidiospores and ascospores. It produces yellow and then necrotic leaf lesions. Since the lesions take several weeks to develop the older leaves appear most infected. The more humid regions of the tropics are most susceptible to this disease since the spread of the fungus and infection is favoured by high humidity and temperature. Poorly-drained acid soils also favour the pathogen.

Control is affected by the use of resistant strains and by the spraying of copper fungicide and finely atomized oils. In certain plantations in the Philippines where continuous use of copper fungicides has led to a toxicity problem the fungicide 'Dithane' has been successfully-used to control leaf spot.

Bunchy Top Virus. This disease has appeared periodically in Australia and other locations. The initial symptoms are a thin, green streaking on the secondary veins of the lamina followed by dwarfing and curling of subsequent leaves which may also show chlorosis. Satisfactory control has been achieved by well organized phytosanitary procedures.

Bacterial wilt (moko disease). Caused by *Pseudomonas solanacearum*, the symptoms of wilt are similar to those of Panama disease. The organism is widespread in the tropics and occurs also on other hosts. It is favoured by moist conditions and poor drainage. Spread of the disease and control are achieved by soil improvement, the sterilizing of pruning instruments which can transport the organism to other plants and by fallowing if the area is badly infected. Formaldehyde solution is used to sterilize pruning instruments.

Bananas suffer greatly from nematodes. Populations build up under continuous cropping in a logarithmic manner (Maas and Imambaks, 1971). Flooding can be used to reduce populations under conditions of annual replanting in alluvial areas, and fallowing, but chemical control is most widely practised. DBCP and other nematocides are used. Vilardebo *et al.* (1972) report that 'Mocap' and 'Nemacur' at 3–5 g/m^2 gave yield increases from 25 to 150 per cent. The dipping of planting material is also recommended.

Of insect pests, the most widespread is the banana borer *Cosmopolites sordidus* (to which the plantains are particularly susceptible). It can be controlled by 'aldrin' or 'dieldrin' at 2 lb/acre at 6 monthly intervals sprayed onto pseudostems, or by trapping (see Simmonds, 1966). PANS has published a recent manual of banana pests (*Pest control in bananas*— Ed. S.D. Feakin, 1972).

3 PINEAPPLES [Ananas comosa]

INTRODUCTION, VARIETIES

PINEAPPLES originated in the Americas and became widely distributed in the tropics only in the mid-sixteenth century when the cultivated species *Ananas comosa* became recognized for the excellent qualities of its fruit. Interest in pineapples as a commercial crop began in the early nineteenth century and by the mid-and late nineteenth century sizeable industries existed in Australia, South Africa, Florida, Malaya and elsewhere. The big producing areas today for trade purposes are tropical Australia, Hawaii, Central and South America, South Africa, Malaysia, the Philippines and others.

The cultivated forms of *Ananas comosa*, of which there are several of importance (see below), are self-sterile and are propagated vegetatively to produce the seedless fruits of commerce. Cross pollination between different clones and wild species to

TABLE 3.1 Characteristics of the major pineapple varieties

Cayenne (smoothe cayenne)	The most important variety, particularly for canning. Large fruits (2.5–3.5 kg). Yellow-fleshed with good flavour. Leaves are spineless. Grown in Hawaii, Africa and other areas.
Red Spanish	A hardy variety with smaller fruits (1.–1.5 kg). Flesh pale and rather acid, particularly outside the equatorial zones. Grown chiefly in Cuba and Puerto Rico.
Sugar loaf	Generally recognized as the sweetest of the white-fleshed pineapples, with fruits of about 1 kg. Grown widely in tropical America.
Queen (Natal Queen, Palembang, MacGregor)	Sweet fruit and good flavour. Size $\frac{1}{2}$ to 1 kg. It produces numerous suckers and there are many strains. More adapted to sub-tropical conditions.
Abakka (Abachi, Abacha, golden Abachi)	Sweet well-flavoured and yellow-fleshed fruits of size 1.5–2.5 kg. Grown in Indonesia, Florida and Venezuela. Produces characteristic basal slips on the fruits.
Elenthera (Pernambuco, Abacaxi, English)	Similar to Abakka but without basal slips on fruit. Double or multiple crowns are common.
Singapore Spanish	A good quality fruit of weight about 1–2.5 kg. Yellow-fleshed and grown mainly for canning in Malaysia.
Cabezona (cabezone)	Large fruits up to 7 kg. It is recognized as a triploid (Collins, 1933) and is a vigorous clone.
Congo Red	Yellow-fleshed, medium-sized fruit of good flavour.

produce seeded fruits occurs readily; therefore the growing of mixed fields is avoided.

Commercial varieties are mostly diploid (2n = 50) but some are triploid (e.g. 'Cabezone'—Ramirez, 1966).

Characteristics of the common varieties of commerce are indicated in Table 3.1 (from Ochse *et al.* 1961; Mortensen and Bullard, 1964; Cobley, 1956 and other sources).

New varieties are produced from time to time, for example, P.R. 1.67 which is a variety derived from Red Spanish and Cayenne (Ramirez, 1970). Numerous varieties have been produced in Hawaii but few to equal the quality of Cayenne. Furthermore, natural segregation takes place frequently and many races are recognized in, for example, the Cayenne variety. One of these, the crown-of-slips mutant, is described in some detail by Collins (1968) as being an undesirable mutant Cayenne which has been concentrated in fields in some areas by virtue of the large numbers of slips it produces. The ring of slips beneath the fruit is an undesirable feature and furthermore, leads to lowered fruit weight. The weights of fruits are often used to characterize varieties (see above), but in fact these figures refer to average conditions only since fruit size is much affected by spacing and growth conditions. One Sarawak variety, for example, if grown at wide spacing will produce fruits of weight 10 kg (*var.* Sarawak.)

BOTANY

The vegetative plant

The aerial part of the pineapple consists of a compact rosette of leaves arising from a central stem which perenniates by means of side shoots as illustrated in Fig. 3.1. (after Collins, 1968). Fig. 3.2 shows a section of a mature plant bearing a terminal fruit. In the vegetative phase, which lasts from about 10 months to 2 years (see later), a plant may reach a height of about 1 metre with a spread of 130–150 cm and bear from 60–90 long, assicular, smooth or spined foliage leaves.

The central stem which bears the apical meristem, and to which the foliage leaves are attached, is illustrated in Fig. 3.2. This maintains a fairly uniform diameter throughout the vegetative phase but increases in length to 20–25 cm. The leaves bear axillary buds and some of these may develop into side shoots (Figs. 3.1 and 3.2).

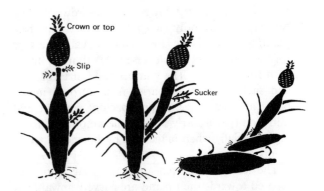

Fig. 3.1 Diagram to show the manner of perennation of the pineapple, following production of the terminal inflorescence
Source. After Collins, 1968.

Fig. 3.2 Cross-section of a pineapple plant showing crown, immature fruit, slip and suckers

The plant is usually propagated by means of the side shoots or the terminal crown shoot (top) of the fruit. Two classes of side shoots are recognized; the slips which develop from the base of the fruit or the peduncle, and suckers which develop from buds in axils of the lower or mid-crown leaves. Suckers are generally recognized as being the best planting material (see later). Anatomically the slips formed on the peduncle resemble abortive fruits with leafy crowns. The numbers and sizes of slips and suckers produced varies from zero to many, being affected by genotype, spacing, sunlight and nutrition (particularly N) and the amount of fruiting, especially when influenced by hormonal treatments (see p. 48).

The leaves of the pineapple are long, thin and tapered towards the tip and are curved upwards in cross section which appears to be a strengthening adaptation to keep the leaves rigid (Krause, 1948–9). In addition, the fluted nature of the leaf creates a channel which conducts water to the axils in which axillary roots occur. This is considered to be an adaptation to arid conditions since dew collects in the axils and constitutes an important source of water under low rainfall conditions (Collins, 1968). The pineapple has a restricted root system which reflects its epiphytic origin.

The pineapple leaf possesses a number of xerophytic adaptations. The upper epidermis consists of compact interlocking cells which are thickened and stratified and coated with a waxy cuticle. Stomata occur only on the lower surface within longitudinal furrows. Krause (1949) reports an average of 70 to 85 stomata per mm^2, rather lower than numbers reported by Collins, 1968 (see 'Growth quantities', p. 41).

The lower and also, to a lesser extent, the upper surface bears mushroom-shaped hairs (trichomes) which effectively increase the thickness of the boundary layer at the leaf surface and increase air resistance to the diffusion of gases and loss of water. The trichomes have also been described as 'reflecting harmful rays and preventing excessive heating', and suggested as organs for water absorption, although this latter function seems improbable (see Krause, 1949).

Internally, the leaf possesses a layer of water-storage tissue below the upper epidermis consisting of columnar, colourless cells which occupy about half the thickness when fully turgid. These cells collapse and decrease in thickness under dry conditions.

Functionally the pineapple leaf is also adapted to survive drought. Joshi et al. (1965) showed that stomata opened mostly at night and in the early morning, and suggested that the pineapple is adapt-

ed to store CO_2 which is later used when the stomata have closed with the advance of the day. Water loss during the day was estimated at about 0.3–0.5 mg/cm^2/hr. (much less than, for example, corn which can transpire 26 mg/cm^2/hr. (Spector, 1956)). Ekern's studies on turgidity conform to these observations since leaves were found to reach maximum turgidity in the afternoon.

The phyllotaxy of the leaves is 5/13 (number of whorls and number of leaves per spiral or whorl).

The root system

The pineapple has a very restricted root system which reflects the epiphytic origin of the species. Many of the wild species of pineapple are epi-

Fig. 3.3a Longitudinal section of the growing point during (1) the vegetative phase and (2) during flowering and fruit formation
Source. After Kerns *et al.* 1936.

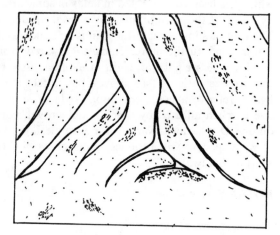

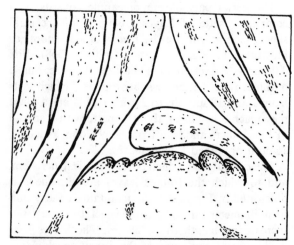

Fig. 3.3b Longitudinal section of pineapple stem (leaves removed) (after Krause, 1948-9).

phytic and although the cultivated pineapple is terrestrial, its root system is nevertheless very confined. Also, it seems likely that the xerophytic nature of the leaves is more related to the restricted root system than reflecting ability to survive severe moisture stress.

The roots are shallow and few. In a lysimeter study, pineapple roots established to a depth of only 30 cm as opposed to 3 ft or more for other species (Ekern, 1965). Nforzato *et al.* (1968) state that in a latosol in Sao Paulo, 95 per cent of the root system occurred in the top 20 cm of soil.

Root growth occurs under moist conditions, but when conditions are dry, it is very much reduced and damage can occur to the superficially-placed roots (Black, 1962). However, other workers consider that the pineapple is well adapted to drought (e.g. Chandler, 1958), but Collins (1968) points out that in low rainfall regions, high humidity and dew formation are necessary for good growth since the roots are able to absorb the axillary water (see above).

According to Cobley (1956), pineapple roots are associated with a mycorhizal fungus.

Floral morphology

During the vegetative phase, the stem produces compacted internodes and leaves. It is about 5.5–6.5 cm in diameter in the vegetative state. Under natural conditions flowering is irregular and is marked by an increase in the diameter of the meristem, one year or more after planting (depending on environmental conditions) which then produces a series of expanded floral organs and longer internodes. After this, the diameter again decreases until purely vegetative leaves are produced, which together with the short starchy stem forms the 'top'

TABLE 3.2 Time required for completing different stages of the development of the pineapple plant: Hawaiian conditions

Stages	Days
Planting to inflorescence initiation	427
Formation of inflorescence	37
To first open flower	43
Period of flowering	26
From last open flower to ripe fruit	109
Total period of fruit development	215
From planting to ripe fruit	642

Source. After Kerns *et al.* 1936, cited by Collins, 1968.

of the fruit or inflorescence. After formation of the crown (or top) the plant becomes dormant. Table 3.2 (after Kerns *et al.* 1936) gives an indication of the length of the development phases of the pineapple in Hawaii and Fig. 3.3 illustrates changes in the growing point.

During inflorescence formation the phyllotoxy of the apex changes to 8/21 and rows of fruitlets are produced which develop parthenocarpically in plantation culture to give the large multiple-fruit of the pineapple (Fig. 3.2a).

Flowering is promoted and fruiting becomes more even when pineapple fields are treated with the artificial hormone napthaleneacetic acid (NAA), or with ethylene or acetylene. Such treatments should not be carried out too soon or small fruits and reduction in sucker formation would occur (e.g. Nyenhuis, 1967). A period of up to 6 months before the setting of the natural fruit appears to be satisfactory for treatment. Hormones are also used to bring fruit into maturity to satisfy particular market demands (e.g. Ali and Talukdar, 1965). Furthermore, treatment is needed particularly at dense spacings and in ratoon crops (Dodson, 1968). Fig. 3.4 illustrates the effects of NAA treatment at different stages of development. The hormone is either poured into the crown of the plant or sprayed on the crop. Further details are given under 'Planting', p. 46 below.

The role of NAA in flowering is not certain but it seems most likely that it affects the sink activity of the apex and the developing inflorescence in some way.

There is some evidence that some pineapple varieties may show short-day characteristics (Py *et al.* 1968, Py, 1970) although they are normally regarded as being day-neutral.

Growth quantities

As with many of tropical crop species there have been very few analytical studies on growth and assimilation in pineapple, but it can be assumed from the low stomatal conductivity and stomatal number of the pineapple that it conforms to those perennating species which have relatively low assimilation rates. Probably the best way of increasing yield potential in the pineapple is to breed for varieties with more responsive stomata which could function under conditions in which irrigation or adequate rainfall prevents severe moisture stress. It seems likely that the low stomatal conductivity

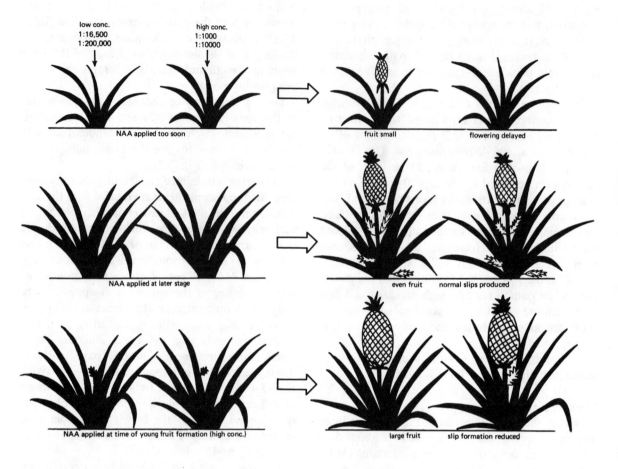

Fig. 3.4   Effects of NAA on flowering and fruit development in the pineapple

Source.   After Williams and Joseph, 1970.

in the day and the protected nature of the stomata are adaptations to the small root system, so that breeding for increase in the size of the root system could also be a feature making for higher productivity. This is recognized as a desirable character by Collins (1968) who discusses breeding features in some detail. The rate of mutation and production of sub-clones seem to be sufficiently high to be able to select and concentrate many genetic characteristics in pineapples.

The basic canopy structure of the pineapple should be highly efficient in relation to sunlight penetration, and the species should be able to support a high productive LAI with the improvement of assimilation potential.

Like the banana, growth and vigour in the pineapple appears to be related to ploidy. Triploids (3n=75) seem to be generally vigorous plants and

this is also true of the only cultivated triploid (Cabezona) which is recognized as a very vigorous clone and in which the vigour is also reflected in fruit size. Another vigorous-growing unnamed triploid clone was described by Baker and Collins (1939). Triploids produce a low percentage of viable pollen (e.g. Ramirez, 1966).

Tetraploids are even more vigorous generally, though not necessarily in relation to field performance. Tetraploid 10120 has been described by Collins (1968). The leaves are about twice as long as the diploid and the fruits are larger than either of the diploid parents. Tetraploids produced from 10120 and Cayenne grow more rapidly and develop into larger mature plants (Collins, 1968). The stem is generally taller (Kerns and Collins, 1947) due to greater internode length. Cell size has been reported to be larger in both triploids and tetraploids and

TABLE 3.3 Comparative size and number of stomata of diploid, triploid and tetraploid pineapples

Plant Type	Mean DIA (μ)	No. per sq. mm.
2n Cayenne	24.62±0.16	180.4±3.07
3n (8846)	28.67±0.27	132.5±3.05
4n (10405 & 10401)	31.02±0.13	105.3±3.55

the size of the stomata is also greater as shown in Table 3.3 (taken from Collins, 1968), while the number of stomata is reduced.

Both the size and condition of the planting material are important in pineapple growth. A comparison of the performance of small and large slips (Linford et al. 1934) showed that small slips produced 10 per cent smaller fruits when planted alone (in pure stands), and 22 per cent smaller fruits when planted in a mixed stand of large and small slips (due presumably to shading out). Small slips also took longer to produce fruit. It seems likely that the importance of size of the starting material would be accentuated in species with inherently low assimilation rates. Similarly, the origin of the planting material, its climatic history and conditions of growth are important in determining the growth of the crop, possibly because of differences in reserve

material etc. (see discussion by Collins, 1968, pp. 218–23 and later under 'Planting'). Primarily, however, it appears to be the assimilation apparatus and the effect of slip size on this which affect fruit production, and at any given temperature fruit size is proportional to the leaf area developed (Van Overbeek, 1946). Fruit size is much affected by spacing, denser spacing producing a greater number of fruits of lower individual weight (see later).

Moisture reserves in the leaf are important in growth of the fruit. According to Sideris and Krause (1955), fruit development can continue even without water for a period of three months.

Growth responses in relation to environmental factors are considered in greater detail below.

CLIMATE AND SOIL
Climate

The geographical areas in which pineapples are successfully grown and in which industries are established for trade purposes are shown in Fig. 3.5. By and large, distribution of the crop is similar to that of the other two pan-tropical species, bananas and sugar-cane, but there is evidence that pineapples can be grown under drier conditions than either of these two crops.

As regards temperature, the optimum for pineapples appears to be close to that of the humid

Fig. 3.5 Map showing the areas throughout the world where pineapple is grown and where pineapple canning has become an important industry

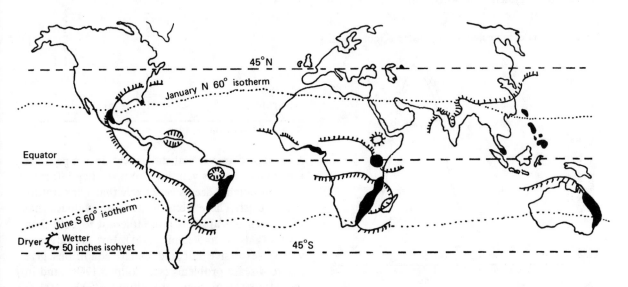

Source. After Collins, 1968.

tropics in that a crop can be obtained in 18 months in areas such as Malaysia when good planting material is used (Collins, 1968) or the Cameroons (Gaillard, 1969), while a crop may take up to two or more years in sub-tropical areas such as South Africa, Australia and Taiwan. (The effect of planting material on maturity time is considered in greater detail under 'Planting' below.) From Table 3.2 which refers to Hawaiian conditions, it can be seen that a crop matured in about 21 months in this area. Furthermore, sub-optimal temperatures produce fruits of inferior quality, being smaller, more acid and often with a lower sugar content (see Collins, 1968). Collins compares Cayenne plants grown at both low and high altitudes in Hawaii, in which the former produced smaller plants with leaves which were smaller and less green, and fruits of relatively low quality. He attributes this to lower temperatures alone, and in regard to the small size of leaves produced at higher altitudes, this would be in agreement with general plant reactions to cold. However, it seems likely that low sugars could more appropriately be associated with low sunlight occurring at high elevations. Low temperature also induces excessive slip formation at the expense of fruit size.

Ochse *et al.* (1961) state 21°C to 27°C as being the optimum range for pineapples and suggest that altitudes of about 2,500 ft would be best in equatorial regions. They consider above 27°C to be too high because excessive transpiration occurs for the limited root system. However, in view of the drought resistance characteristics of the pineapple (see above), this seems unlikely and in any case, transpiration would be a function of humidity also. These authors consider 10°C to 16°C as being minimal for growth. Gaspar and Diniz (1965) state that mean temperatures should be above 20°C for optimal growth. Fig. 3.6 (after Nightingale, 1942) shows the response of root extension growth to temperature which indicates that the optimal for this growth reaction is around 30°C.

In Hawaii, in the past, the use of black paper mulch resulted in a higher-than-normal soil temperature in the cool season which favours root growth.

Some idea of the temperatures encountered in principal pineapple-producing areas is given in Table 3.4. It indicates that considerably sub-optimal temperatures are sometimes tolerated, primarily because of the availability of consumer markets for this fruit.

Collins (1968) reports the growing of pineapples at 1555 m (c. 5,000 ft) at a latitude of 14.5°N in Gautemala, and up to 700 m (c. 2,000 ft) in South Africa where winter frost is encountered (*cf.* Table 3.4).

TABLE 3.4   Temperature range in selected pine-apple-growing areas (°C)

Area	Average maximum	Average minimum
South Africa		
Site 1	47	5
2	43	2
3	41	3
4	42	−1.1
Hawaii	32	10
Malaya	29.3	25.9
Australia	31.7	11.6

Source.   After Collins, 1968.

In regard to varietal differences in temperature sensitivity, there is very little comparative information, although it seems very likely that temperature strains exist. There seems to be some evidence that the Queen pineapple is more tolerant to cold than others (Gaspar and Diniz, 1965). For the Cayenne variety, excessive hot and humid conditions give rise to disease problems (see Collins, 1968), and in regard to this variety, the altitudes 500–2,500 ft suggested by Ochse *et al.* in the equatorial tropics

Fig. 3.6   Effect of temperature on pineapple root growth

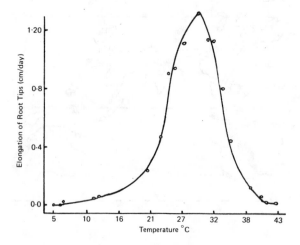

Source.   After Nightingale, 1942.

would probably be nearer to optimum than at sea level.

Pineapples will tolerate short and light frosts but not prolonged periods below freezing. They are fairly resistant to cyclonic winds. Queen variety again seems more tolerant to frosts (Gaspar and Diniz, 1965), and potassium nutrition markedly reduces frost susceptibility (Nyenhuis, 1967a).

As described above, the pineapple possesses definite drought resistance characters, but also has a very restricted root system. The xerophytic characters are probably linked with the restricted root system, but even so, the pineapple is successfully grown in some relatively low rainfall areas without irrigation, compared to those required for bananas for example.

According to Collins (1968) the rainfalls of major pineapple areas range from about 60 cm (25″) to 254 cm (100″), virtually all without irrigation. Green (1963) considers pineapples to be well adapted to drought, but Collins (1968) states that in dry regions, the plant depends on dew which is condensed on the leaves and percolates to the leaf axils and roots as described before, and that such areas are often in humid maritime zones where much dew formation is possible. Nyenhuis (1967a) also comments on the collecting of dew water by the pineapple leaf system. Black (1962), however, describes severe root damage (accentuated by nematode attack) as a result of planting in a dry period in Australia and Wee and Ng (1968) showed how drought at planting affected yield and maturity in Malaysia (on very free-draining organic soils). In a dry-season planting in Sao Paulo, Nforzato et al. (1968) state that no roots formed for 4 months but when the dry season lifted marked root growth occurred.

A number of research workers have stated that pineapples need a regular supply of water (e.g. Groszman, 1948; Chandler, 1958, etc.) but there seems to be a general agreement that pineapples require less water than many other tropical species and 40″–60″ of rainfall per annum is probably the optimal on most soils. Excessively moist conditions lead to physiological and disease problems (Green, 1963). The early phases of growth are probably more critical since once the moisture reserves of the leaves are established, the plant can apparently grow for some period of time without additional water. Fruit development, for example, can continue for up to 3 months without soil water (Sideris

and Krause, 1955). Excessively moist conditions near flowering lead to reduced fruit growth and excessive leaf production, and furthermore, Van Lelyfield (cited by Green, 1963) states that excessive moisture at flowering leads to reduced fruit quality and production of a large core in the fruit.

The use of mulch to conserve water is often undertaken in pineapple growing (e.g. Py, 1968a). Chandler (1958) reports that in some regions, crops can only be produced when the soil is mulched. In Hawaii, black tar-impregnated paper was used for mulching in the past but this has been replaced by polythene in recent years. Mulch also reduces leaching in regions of high rainfall.

Sunlight is obviously essential for a good crop but evidence suggests that the leaf surface might be saturated by relatively low sunlight values (Collins gives average sunlight values as a percentage of maximum possible in what is considered to be a good pineapple region (Wahiawa, Oahu, Hawaii) as between 17.5 and 26.9 per cent, and pineapples are often grown as spaced plants as an intercrop under coconuts (e.g. Kotalawala, 1968). However, it seems likely that at the denser than conventional spacings being adapted in recent years (see later) and particularly in ratoon crops, sunlight could become a severe limiting factor to high yield.

Damage of the fruit by the sun occurs at maturity (e.g. Dodson, 1968) and in Taiwan it is the practice to cover fruits with straw as they approached maturity.

Soils and drainage

In common with all pan-tropical crop species, pineapples are grown on a wide range of soil types, most commonly however, on latosols of varying parent material, and in Hawaii on dark-red volcanic soils. In Malaysia pineapples are grown on almost pure organic peat soils.

The most essential soil feature for pineapples is good drainage and heavy soils are unsuitable unless extensively drained (e.g. Tree, 1966; Gaspar and Diniz, 1965).

Calcerious soils (Gaspar and Diniz, 1965) and heavily-limed soils of high pH are unsuitable for pineapples and on such soils a type of chlorosis may occur.

In some Hawaiian soils high manganese levels lead to iron deficiency (by complex formation) particularly at pH levels above 4.5, which necessitates foliar application of iron.

Some inland tropical soils inactivate phosphate so effectively that frequent P application is necessary (e.g. Su, 1965a), even though the pineapple requires relatively little P for growth and production.

Pineapples do well on sandy soils with appropriate nutrition and it may generally be said that if the soil is acidic, well drained and permits root penetration in the upper horizons, it will be suitable for pineapple growing. The optimum clay content and moisture holding properties of the soil will be a function of climate.

Questions of nutrition are considered in greater detail later on. Most of the acidic tropical soils are highly leached and require heavy fertilization for optimal yields.

PLANTING

Land preparation, spacing and yield

Because of the low stature of the plant and the need for frequent field access, land preparation for pineapples is normally more exacting than for tropical tree crops. In some countries (e.g. Hawaii) a high degree of mechanization is employed. Soil is usually worked to a depth of 25–30 cm, chopping and ploughing in the old pineapple stand. Heavy soils may be ploughed to a depth of 45–50 cm and drains constructed to achieve adequate drainage. As explained earlier, mulch is used in the Hawaiian islands and is particularly advantageous in areas of marginal rainfall. Cover crops of legume species are sometimes grown prior to planting and between the rows (Tisseau, 1969). Legume covers (*Crotolaria usaramoensis, Flemmingia* and *Stylosanthes*) are also beneficial in reducing nematode populations (Guerout, 1969).

Spacing markedly affects yield and fruit size and close spacing is used to increase yields and produce small fruits for canning; for the fresh-fruit market, a somewhat larger fruit is often preferred. Blommenstein and Osthuysen (1970), for example, recommend a spacing of $100 \times 60 \times 30$ cm (39,000/ha) for market production and $100 \times 45 \times 30$ cm (43,500/ha) for canning. The relationship between fruit weight and spacing is similar to that encountered in the banana. Furthermore, wide spacing promotes the production of multiple crowns and faciated fruits (Fig. 3.7). Dense spacing has been used in the tiploid variety Cabezona to greatly increase yield at the expense of fruit size, which is ordinarily very large.

Fig. 3.7 Multiple crowns and fasciations in pineapple

Traditional spacing varies from country to country but usually spacing within rows is less than spacing between rows. In South Africa, in the past, a single-row system of 38 cm in rows 122 cm apart for Queen variety (*c.* 9,000/acre), and 30 cm in rows 61 cm apart for Cayenne (*c.* 14,000/acre) was used. More often, in recent years, a double-row system is adopted with 100–120 cm between the double rows which are 45 or 60 cm apart and have 30 cm spacing within the rows, giving densities of *c.* 16,000–18,000/acre. The double-row system is most convenient for harvesting and field entry, and seems generally the best system for pineapples. Experiments with triple-row systems have not proved advantageous (e.g. Wee, 1969). In Hawaii, the home of the modern pineapple industry, spacings of 16,000 to 18,000/acre are used.

In very recent years, closer spacings than this have been recommended. Wee (1969) reports that 72,000/ha (30,000/acre) gave 50 per cent increased yield for the plant crop of Singapore Spanish, but not for the ratoons. Many other research workers have recommended denser spacings (e.g. Mitchell and Nicholson, 1965; Nyenhuis, 1967a; Dodson, 1968). The advantage of dense spacings is generally seen only in the initial or plant crop and not in the ratoons. Reynhardt and Heerden (1968) have recommended the pruning of initial stand to three suckers per plant for the best ratoon yields.

Actual yields vary according to husbandry, fertilization etc., and with climate, particularly in regard to the time to harvest. Yields of the plant crop of up to 40 tons/acre are possible with good husbandry etc. Often the ratoon crops yield less than the plant crop.

Planting material and planting

Planting material markedly affects the time to harvest. In Hawaii large shoots will mature into fruit in about 17 months, slips (from the peduncle or 'collar') in about 20 months and crowns in 22–24 months. All these materials are used, but they are planted in separate fields and the staggering of the crop is considered advantageous. At wider spacings, yield is also affected by the planting material. Generally shoots or 'suckers' from the axils of foliage leaves are considered better planting material (e.g. Dodson, 1968; Gaillard, 1970; Py, 1970 etc.) and performance appears to be related to size, the initial leaf area and the reserve materials present. Leigh *et al.* (1966) in Australia, however, consider large slips to be superior to crowns or 'tops', and after these material can be graded in the order of small slips, suckers and butts (the stems of the plants already flowered which are trimmed of roots, leaves and the peduncle). Nyenhuis (1967) reports that butts have a tendency to rot in wet weather.

Planting material (excluding butts) is removed and laid on the leaves of field plants to dry before storage and planting, or planted directly. In Malaysia and in other humid localities, planting material is sometimes treated wilth Bordeaux mixture to reduce rot.

Planting material has usually been imported to established a pineapple industry. Rapid propagation can be achieved by the use of stem sections (Fig. 3.8). These should be treated with a mercurial fungicide prior to planting and planted to a depth

of 1–2 cm, in sand or soil. About 25 shoots can be obtained from a single plant by this method, maturing in 2–3 years. Seow and Wee (1970) used buds cut off with the foliage leaves to multiply material of Singapore Spanish. They obtained more than 40 buds per crown of which about 75 per cent developed into plants.

Planting of slips etc. in the field is done with a flat-pointed implement and shallow-upright planting is often best when planting is carried out in an appropriate wet season (e.g. Reynhardt and Blommestein, 1966).

Manuring

On virgin soils good crops can be obtained without fertilization but for continued cropping the pineapple requires heavy fertilization. For example, Guidian (1970) reports yields of 68 tons with 230 kg N, 138 kg P_2 and 430 kg K_2O per hectare, but only 12 tons from unfertilized soils.

In Hawaii usually two or three applications of N are made at 0, 3 and 6 months making a total of about 200 kg/acre (400 kg/ha). Potassium is supplied at 100–200 kg/acre K_2O and phosphate at 68–114 kg/acre as ammonium phosphate, since the pineapple does not tolerate high Ca^{++} levels from super-phosphate or other P sources (Collins, 1968). Su (1965a) reports that 2–4 g. of P_2O_5 (as ammonium phosphate) per plant (40–80 kg P_2O_5/acre) is required on phosphate inactivating soil in Taiwan.

In Australia Mitchell and Cannon (1953) recommended a 10:6:10 N:P-K mixture at rates of 300 kg/ha at 3 and 6 months and supplemented with about 200 lb ammonium sulphate between these times, and in South Africa up to 2000 kg ammonium sulphate is needed on poor soils (Reynhardt, 1970).

Fertilizers are either applied to the soil, the leaf axils or in the case of urea, as a foliar spray. Axilliary ammonium sulphate application is widely practised (e.g. Nyenhuis, 1967a) but has been reported to cause damage and to lead to invasion by rot organisms (Grice and Proudman, 1968). Urea spray is becoming popular for supplying nitrogen. Mitchell and Nicholson (1965) recommend a basal dressing of K_2SO_4 and approximately 2-monthly sprays of urea at a rate of 200 kg/ha/year, and up to 400 kg on the ratoon crop.

Nitrogen should neither be applied near the time of flowering since it induces vegetative growth and may reduce yield (e.g. Dunsmore, 1957) nor after

Fig. 3.8 Stem sections used for rapid asexual propagation

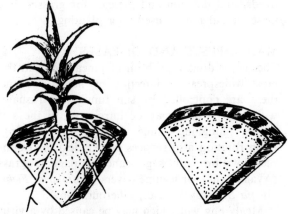

Source. After Collins, 1968.
Notes. Right: Stem section ready for planting.
Left: Section with plantlet about four weeks after planting.

6 or 9 months depending on the time of maturity. Nitrogen, at the correct time, increases fruit size (e.g. Gaillard, 1970, Dodson, 1968) but reduces fruit sugar and acid through promotion of vegetative growth (cf. sugar-cane.) Potassium generally increases sugar and acid levels of the fruit (e.g. Tay *et al.* 1960; Nyenhuis, 1967b).

Phosphorus has been reported to reduce yield and fruit quality (e.g. Tay *et al.* 1969; Reynhardt, 1970) and the ratio of P to K in the plant should not exceed 1:12 (Kanapathy, 1958).

Pineapples suffer from Cu deficiency on some soils, as well as Zn and Fe as mentioned before. These deficiencies are corrected by foliar sprays. The determination of nutrient requirements by foliar analysis is widely practised. Kanapathy (1958) records leaf levels of N $-1.2\pm$ 0.2 per cent, P $-$ $0.22\pm$ 0.03 per cent and K $-3.2\pm$ 0.4 per cent on a dry-weight basis taking leaf tips as samples. A critical level for Mg is considered by Su (1965) to be 0.22 per cent on a leaf dry-weight basis.

Irrigation is seldom practised in pineapple growing, presumably because of the pineapples' camel-like ability to store water and its specialized stomatal behaviour which conserves water.

Timing of the crop and use of flower-promoting substances

The timing of the crop and the time to maturity is intimately linked with the use of artificial flower-promoting agents to bring on earlier and synchronous flowering (cf. Fig. 3.4), and to time the fruits for the best market period. The pineapple, under natural conditions, does not bear fruit at even and regular intervals. With hormonal treatment, however, virtually every plant can be brought into fruit at the same time. However, as indicated in Fig. 3.4, treatment should not be given too far ahead of the normal time of fruiting. For example, Nyenhuis (1967), in South Africa states that NAA treatment should not be carried out more than 6 months ahead, normal fruiting occurring between 18 and 28 months depending on the planting material used.

The discovery of artificial flower-promoting agents to induce the pineapples to fruit goes back to the days when they were grown in glass-houses in the Azores. After fumigating a glass-house with wood smoke to expel insects, it was found that pineapple plants matured prematurely and this was subsequently attributed to the unsaturated hydrocarbon content, particularly ethylene, of the smoke.

Today the synthetic plant hormone, napthalene-acetic acid (NAA), is widely used in commercial production, usually by applying a solution to the heart of the plant (eg. 20–50 ml of NAA of strength 5–60 ppm, e.g. Nyenhuis (1967); Seeyave (1966); Das and Baruah (1967) etc.). Sometimes acetylene (in the form of 2-chloroethane phosphoric acid which releases ethylene in solution e.g. Py and Guyot (1970) or acetylene from carbide e.g. Gaillard (1969) is used instead of NAA.

Weed control

The use of hormonal weed killers such as 2:4-D is avoided in pineapples because of the upsetting effect on flowering and sensitivity of the species. The technique of mulching combined with pre-planting cultivation used in Hawaii greatly reduces weed growth and in the past has been used in combination with periodic hand eradication of weeds. In recent years certain herbicides have come into use. Pantachlorphenate (PCP) is used widely, particularly with oil as a carrier (e.g. Reynhardt *et al.* 1967; Godfrey-Sam-Aggrey, 1969). It gives a good and fairly long weed-free period used as a pre-emergent. The substituted ureas (monuron and diuron) are also widely used as long-acting herbicides following cultivation, and simazine has also been used to delay weed invasion after cultivation. The urea herbicides and simazine can be applied periodically but wetting of the leaves should be avoided, with application concentrated upon the inter-rows.

For pre-planting treatment Py (1968b) recommends 2:4:5-TP and ametryne for broad-leaved weeds, and dalapon and pyriclor for grasses, but these should not be used after planting.

MAJOR PESTS AND DISEASES

The major disease problem of pineapples and the most widespread is infection by nematodes and these are controlled by soil fumigation (usually with D-D) which may be introduced to the soil during ploughing—some weeks before planting. Crop rotation also reduces nematodes; for example, rotation with sugar-cane or pangola grass (Ayala, 1968), or legume covers (*Crotolaria, Flemmingia Stylosanthes* etc.; Guerout, 1969).

Mealy bug wilt which may be caused by a virus (Carter, 1952) is also a serious pineapple disease and control of mealy bugs by periodic spraying is recommended. Interestingly, border zone spraying has been practised to keep fields clean of mealy bugs following an initial spray (see Collins, 1968). Some varieties of pineapple, notably Cayenne, are more susceptible to this disease.

Heart and root rots, caused by *Phytopthora cinnamomi* and *P. parasitica*, particularly in wet weather and at cooler sites and on poorly-drained soil, also affect pineapples. Improved drainage and the use of raised beds are recommended, together with planting in a warmer season.

Butt rot, as explained before, is also a problem in wet conditions.

Various fruit rots are troublesome to the pineapple industry, notably when fruits are left too long before processing and become over-ripe.

4 SUGAR-CANE [Saccharum spp.]

INTRODUCTION, SPECIES AND VARIETIES

SUGAR-CANE is recorded as having been cultivated as early as 400 B.C. and at the present time there are numerous cultivated varieties comprising both hybrids bred in recent years and varieties of *Saccharum officinarum* (see below) selected in ancient times. The cultivated canes come from either the species *officinarum* or its hybridization with *S. barberi*, *S. sinense*, *S. spontaneum* and *S. robustum*, or combinations of these. *S. officinarum* is a thick-stemmed species with many varieties which have been referred to as 'noble-canes'. These formed the basis of the early plantation industries in most parts of the world. They gave good yields and are still recognized as excellent canes but have proved to be poorly adapted to inferior and less fertile soils and to possess little resistance to certain diseases and pests.

Sugar-cane is largely vegetatively propagated and does not ordinarily produce fertile seeds. However seeds can be produced from suitable intervarietal and interspecific crosses. Modern disease-resistant varieties are largely based on hybridization between *S. officinarum* (n=40) and the species *spontaneum* and *robustum* (with varying chromosome numbers) from which they derive their properties of disease resistance and hardiness. Hybridization occurs easily between these species with the *officinarum* parent usually supplying 2n chromosomes to give resulting polyploids. The same result occurs on back crossing of the hybrid with another *officinarum*. In a typical breeding programme a cross between *S. officinarum* and one or another of the other species is first produced. This is then back-crossed one or more times with *S. officinarum* varieties to obtain the high sucrose yields characteristic of this species, while at the same time retaining the desired hardiness of the other species.

Sugar-cane is typically a short-day plant in that flowering is hastened by short photoperiods and this poses special problems in both breeding and in the production of varieties for particular latitudinal regions. Sometimes artificial manipulation of the day length by means of lights and black-cloth shades is necessary to produce synchronous flowering in varieties to be crossed. The control of temperature, particularly night temperature which should not be too low, is also important in flowering (e.g. Daniels and Krishnamurthi, 1965; Antoni, 1965). Furthermore, varieties which flower poorly are desirable for field canes since flowering reduces sugar yields.

The following table (from Ochse *et al.* 1961 and other sources) lists commercial varieties commonly grown in tropical countries:

TABLE 4.1 Tropical sugar-cane varieties grown in various regions

Countries	Varieties Grown
Argentina	Tucuman varieties
Australia	Q 50, 57, 58, 66, 75, 83 and others
Brazil (southern part)	Co 421, Co 290
Brazil (central part)	CB varieties, Co 421
Cuba	POJ 2878, ML 318
Hawaii	H 37–1933, H 32–8650, H 32–1063 and others
India	Co 421, Co 419, and other Coimbatore varieties
Jamaica	B 41227, B 42231, B 4362 (Barbados varieties)
Java	POJ 3016, POJ 2967, POJ 2878
Mauritius	M 134–32
Natal	N:Co 310 and others
Puerto Rico	POJ 2878, M 336, PR 902.
Louisiana	CP varieties
British West Indies	Barbados varieties
Malaya	N:Co 310 and others.

BOTANY

The vegetative plant

Sugar-cane belongs to the grass family and the vegetative stem forms the harvested part of the plant which consists of a series of expanded inter-nodes containing the sugar sap. The early-formed internodes near the base of the stem and beneath the soil surface are compacted while those of the aerial stem increase in length with height and then contract again because of immaturity as the grow-ing point is approached. Fig. 4.1 (after Houtman, 1916) and Fig. 4.2 indicate the nature of the change in the length of internodes of a typical cane stem near maturity. The size of the internodes is also strongly influenced by climate; they become shorter during periods of moisture stress or cold (this has a bearing on sugar yield as discussed on p. 60).

Between the internodes, the nodes bear buds which may develop into side shoots. The crop is planted from stem cuttings bearing buds (see later), one of which develops to form the first main shoot. In the establishment phase of the crop several prom-inent side shoots will form from the buds at the compacted part of the first main stem and from similar buds on the side shoots, giving rise to both secondary and tertiary stems. Fig. 4.2a (after Mar-tin, 1938) illustrates the basal portion of a clump of shoots (referred to as a stool in the sugar busi-ness), and it can be seen that the compacted nature of the internodes at the base of the stem results in a high concentration of buds for this tillering pro-cess. Once internode extension becomes marked, side shoot development does not ordinarily occur, the shoots exercising an apical dominance effect on the growth of side shoots.

The tillering of the initial shoots is important in building up the stem population following planting of the cuttings. The tillering phase is characterized by exponential growth and the post-tillering period by linear and declining growth. During post-tillering (called the 'grand period' of growth) height incre-ments are closely correlated with weight increments since growth consists primarily of the extension of the established stems which are fairly uniform in thickness (see Fig. 4.3).

The stem may be circular or oval in cross section and considerable variation occurs in the average shape, colour and texture of the internodes. Such characters together with characteristics of the nodes and the buds are used in classifying varieties.

Internally the stem consists of parenchyma con-taining the sugary sap and typical monocotyledon-ous bundles. The outer layers are variously streng-thened and this is an important property in deter-mining resistance to stem borers.

The leaves, produced at the apex of the stem, become displaced laterally with growth in height. They consist of two parts, blade and sheath (Fig.

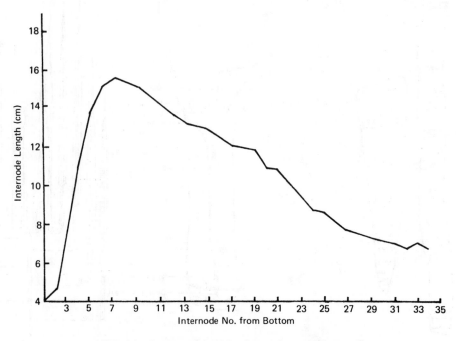

Fig. 4.1 Pattern of length change in cane internodes
Source. After Houtman, 1916.

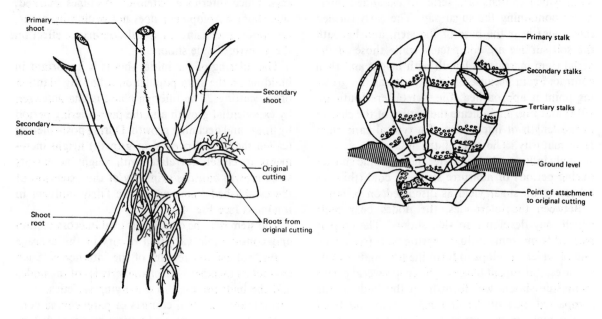

Fig. 4.2a   Basal portion of cane stool
Source.   After Martin, 1938.

Fig. 4.2b   Characteristics of cane internodes
Source.   After Geerts, 1917.

Fig. 4.2c&d   Characteristics of the leaf
Source.   After Artschwager, 1940 and Clements *et al.* 1941.

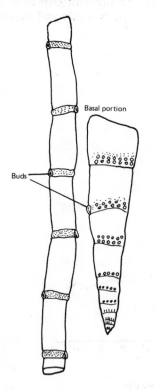

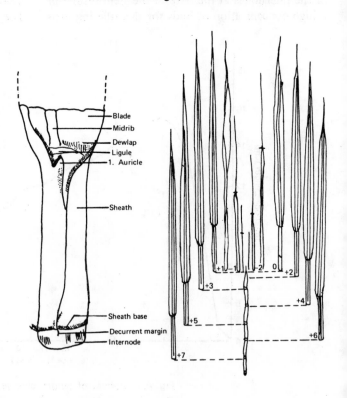

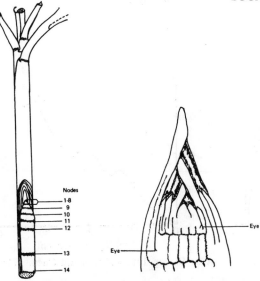

Fig. 4.3 Apical region of stem showing changing diameter of the internodes
Source. After Martin, 1938, from Dillewijn, 1952.

4.2c and d), the latter encircling the upper internodes. The leaves of the older nodes have usually fallen away revealing the stem in maturing sugarcane. There are considerable differences in leaf shape and orientation (Fig. 4.4) which could have a bearing on yield as in other species, and as discussed in the first part of this book.

Leaf size varies from the apical region to the base, as indicated in Fig. 4.5 (after Barber, 1918). At any one time, following establishment, about ten mature expanded leaves are on a stem with a combined leaf area of about $\frac{1}{2}$ sq. meter on the average, although there is considerable varietal range. Leaves are alternately arranged as in most grass species and production rate varies from about one per five days to one per twenty days depending on variety and, particularly, on climate. Leaf area reaches a maximum after a variable period ranging between five to ten months (also depending mainly

Fig. 4.4 Two cane varieties with different degrees of leafiness and leaf orientation from Malaya

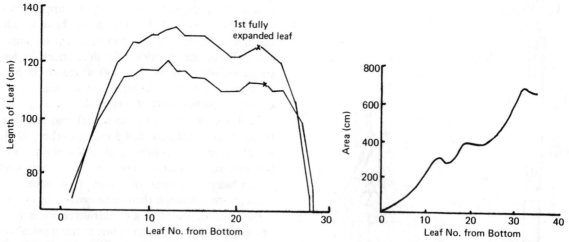

Fig. 4.5 Pattern of change of leaf length and area in various cane varieties

Source. After Barber, 1918 and Loh and Chen, 1948.

on climate) and thereafter remains constant or even declines. Net growth after this period consists mainly of stem material and accumulation of sugar.

Stomata are most numerous on the blade of the leaves and the lower surface bears about twice as many as the upper. There are apparently considerable differences in the number of stomata between varieties, and this most probably results in differences in both assimilation rate between varieties and in drought resistance. Size and protection of stomata also vary (Thuljaram, 1951). Rolling of the leaves occurs in many varieties and is considered to be a drought-resistance character.

The root system

The roots of sugar-cane are typically adventitious. Initially roots develop from the cutting (or set) but after development of the shoots they arise from the lower internodes (Fig. 4.2a) and ramify to a considerable depth in strong-rooting varieties and in permeable soils. However, there is a concentration of roots in the upper layers of the soil. Three types of roots have been recognized by Evans (1935 a and b) as illustrated in Fig. 4.6. The buttress roots have a supporting function. Under humid conditions and particularly in some varieties, aerial roots may form from the lower nodes and this is found to be detrimental to sugar accumulation. The set roots, which form the initial root system of the plant, die off as a result of competition from the shoot roots since they can be experimentally induced to develop further and maintain the plant to maturity (for discussion see Dillewijn, 1952).

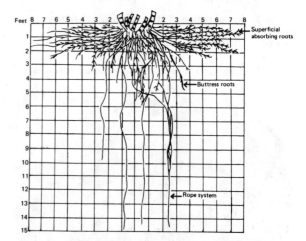

Fig. 4.6 A root system of sugar-cane
Source. After Evans, 1935a.

As mentioned earlier, differences exist in the extent of root development between varieties, in particular between the original noble canes and its hybrids. Fig. 4.7 illustrates the extent of the root systems of two varieties, which also shows the effect of soil texture on root development. Weller (1937) and Evans (1934) have demonstrated the existence of significant root pressures in sugar-cane which seems to be a common feature of large monocot species. Secondary and tertiary roots are formed from the primaries of the superficial roots, the latter being the main centres for uptake.

Growth quantities, tillering

The growth pattern of the sugar-cane crop is typical of crops in which accumulation of dry matter continues for a significant period after the achievement of a 'ceiling' leaf area index which is below compensation point. Fig. 4.8a (from Kobus, 1900) shows the changes in accumulation of dry material (which is typically sigmoid in form) and distribution into the various plant parts. Fig. 4.8 also shows the increase in height and changing height growth rate which is similar in form to the dry weight curve.

The logarithmic part of the sigmoid curve (Fig. 4.8a) corresponds to the tillering phase of growth of the crop, and linear and declining growth to the 'grand-period' of growth. During the early tillering) phase, growth is best described by the relative growth rate formulae; and during the grand-period height increment is strongly correlated with weight and this is the usual measure of growth.

The time to maturity in sugar-cane varies between one and two years, mainly depending on climate (Fig. 4.8c) but also on variety.

Leaf area index at the stage of maximum development of the leaves ranges from about eight to twelve or more and is typically high as characteristic of graminaceous crops. There would appear to be differences in peak LAI values in different clones particularly in relation to leaf size and orientation (Fig. 4.4), with the more narrow-leaved clones with erect foliage capable of developing higher productive LAI value and requiring close spacing (see discussion by Dillewijn, 1952). The photosynthesis of leaves of sugar-cane is capable of responding linearly to high light values approaching half that of natural sunlight, unlike many other crops in which leaves show threshold photosynthesis values at much lower light levels. After germination the tillering phase lasts between four and eight months, the time to achievement of maximum tiller number being very dependent on climate and sensitive to light, the whole process of tillering be-

Fig. 4.7 The main root systems of two clones of sugar-cane

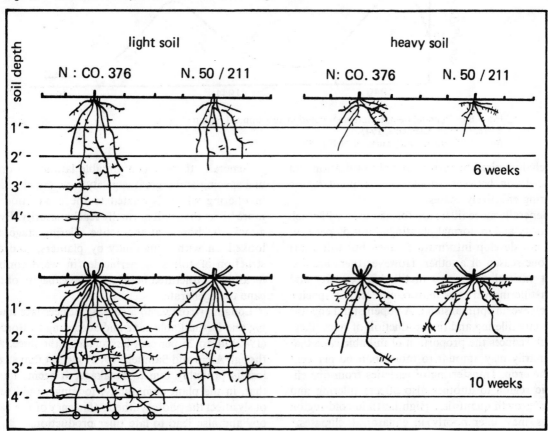

Source. From *Annual Report, 1965–6, South African Sugar Association Experiment Station.*

Note. The spacing between plants is 4′ and the circles at the bottoms of the left-hand figures indicate that roots have grown below 4′ depth.

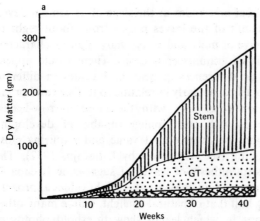

Fig. 4.8a Dry matter accumulation of plant parts. GT= green tissue.
Source. After Kobus, 1900.

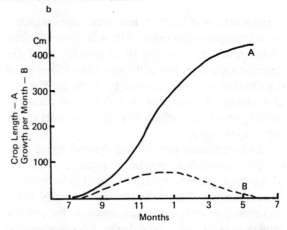

Fig. 4.8b Height changes and monthly height increment
Source. After Dillewijn, 1952.

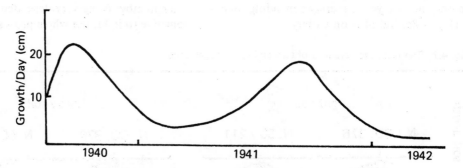

Fig. 4.8c Bimodal height increment curve in sub-tropical region
Source. From Clements, 1943.
Note. Compare with curve B of Fig. 8b.

ing characterized by increasing light utilization and competition between shoots to the point where tillering effectively ceases.

Generally more tillers are formed than will reach maturity and in normal planting about 50 per cent will not develop into mature canes but will abort for one reason or another. However there are distinct varietal differences in tillering capacity and poor-tillering varieties may. require closer spacing to achieve an optimal stand. An open stand encourages late tillering and the production of stem tillers which reduces the proportion of millable cane that ordinarily may amount to more than 60 per cent of the crop. Transferring of varieties from one climatic region to another also affects tillering and other growth quantities. High latitude cold-region canes may tiller poorly in equatorial climates.* Short photoperiods tend to reduce tillering.

*This will probably limit yield of sugar in the humid tropical lowlands until locally adapted varieties are produced.

Generally there is considerable latitude in optimal spacing and a great capacity for compensation in tillering which is related to the production of more tillers than will normally reach maturity. Even attacks by borers at the active tillering stage are looked on with equanimity by planters, and destruction of tillers by herbicides in weed control measures is tolerated because of the capacity of the cane to regenerate.

Lodging (from various causes) induces late tillering and less millable cane and according to Borden (1942) up to 25 per cent less sugar in the juice. Although Alers and Samuels (1962) suggest that sugar loss is probably due to breakage, it seems likely that, in the plant community, lodging would seriously affect the photosynthetic efficiency of the canopy and also lead to late tiller production.

The cane crop is very much affected by climate and furthermore, response to climatic and environmental factors such as light, temperature and fertilizers is much greater in the early phases of growth (tillering and the early linear phase), and this has significance in plantation practice. Response to these factors may be several times as great in the early phases of growth than near maturity and it would seem likely that this is due to the approach of ceiling yield with maturity. In sugar-cane much effort has been put into the characterization of climate parameters and their relationship to growth and yield as discussed below.

CLIMATE AND SOILS
Rainfall, temperature and light

The geographic zones in which sugar-cane is planted are almost identical to those for bananas as indicated in the following map (Fig. 4.9) adapted from Humbert (1968), and as far as growth is concerned the moisture and temperature requirements appear to be the same as those of bananas, the most suitable climate being that of the humid equatorial regions with adequate sunshine (see below). Estimates of rainfall from between 40″–50″ annually (Ochse *et al.* 1961) to 60″ (Wadsworth, 1949) are considered necessary for optimal yields. One significant difference, however, is that sugar-cane matures best with a slight dry season which permits a high sucrose content to develop in the stalks. It appears that competitive sinks between cane extension growth and sugar accumulation are involved in cane maturity and that these two processes are differentially affected by moisture stress. The reducing effect of moisture stress on extension growth appears in general to be more severe and to occur sooner

than on assimilation. Cessation in extension growth occurs at wilting point (e.g. Fig. 4.10 after Shaw, 1937) while assimilation can continue below wilting point (e.g. Burton *et al.* 1957). Investigations with other varieties indicate that extension growth is restricted at soil tensions below 4 atmospheres (Cornelison and Humbert, 1960), and Mascaro (1962) states that irrigation is necessary at tensions of 0.25 atmospheres for extension growth. Swezay and Wadsworth (1946) and Raheja (1951) however, both report that sugar yields are not always affected even by severe moisture stress. It is apparent also that there are considerable genetic differences in resistance to moisture stress. Tanimoto and Nickell (1965), for example, describe marked clonal differences in some Hawaiian canes and also report a strong correlation between salt tolerance and drought resistance (a fact which they claim can be used in screening for drought resistance).

In general it could be said that the optimal sugar-cane climate is one in which there is little moisture stress throughout the germination, tillering and grand period of growth, but a short dry season prior to harvesting. Fig. 4.11 (after Das, 1936) shows the relationship between sugar content and cane elongation in Hawaii. Apart from the requirement for slight moisture stress at maturity, an excessively humid climate is also unsuitable in that it encourages the growth of a high tonnage of cane with a low sucrose content, and much late tillering which is particularly due to the growth of buds on the stalk and this is detrimental to sugar yields. Altogether a less than humid tropical climate with

Fig. 4.9 Sugar-cane growing areas of the world

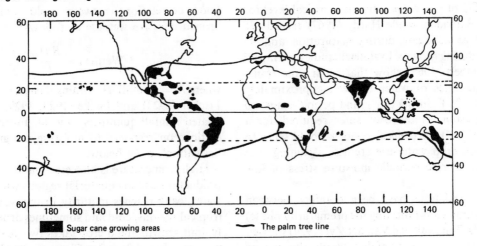

| ■ Sugar cane growing areas | — The palm tree line |

Source. After Humbert, 1968.

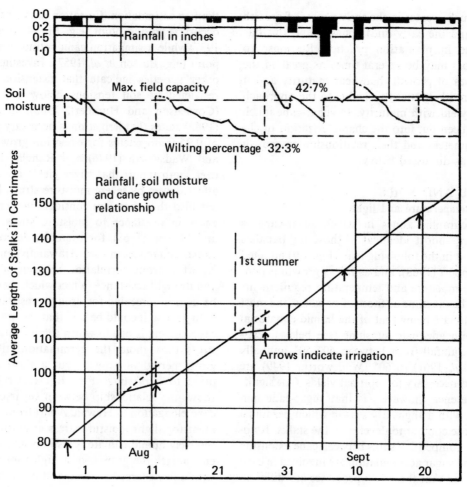

Fig. 4.10   An example of the relationship between soil moisture and cane elongation
Source.   After Shaw, 1937.

much sunshine is probably best for sugar-cane. Oguntoyinbo (1966) considers that rainfall amounting to 0.73 of pan evaporation is a critical level for cane in Barbados, while Hardy (1967) regards 75 per cent as adequate during maturation and 95 per cent during growth. Lysimeter studies in Hawaii (see Humbert, 1968) show that sugar-cane transpired quantities of water between approximately 0.90 and 1.25 E during the 'grand-period' stage, so that the estimates above of water required for a crop (50″–60″) for tropical regions would seem to agree with crop water-use studies, assuming that cane can tolerate periodic moisture stress of low order.

High correlations have been obtained between rainfall parameters (affecting the prematurity phase) and yields. Tengwall and Van der Zyl (1924) showed correlations between rainfall towards the end of the dry season in Java (when moisture stress would be greatest) and yield, and Walter (1910) showed good correlations between yield and what he termed effective wetness as follows:

$$\text{Wetness} = \frac{R.t^1}{t}$$

where R = rainfall, t^1 = rainy days, t = total days. Fischer (1923) and Leake (1928, 1929, 1934) also related rainfall quantities to yield. Further considerations on response to moisture are given under 'Irrigation', p. 64 below.

High temperature is also essential for sugar-cane production and in equatorial regions with adequate water, cane reaches maturity in one year. In subtropical regions, cold and sometimes drought greatly limit growth and a crop takes two years to reach maturity. Fig. 4.8c (after Clements, 1943) indicates

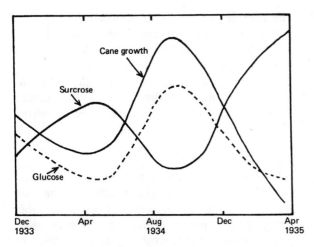

Fig. 4.11 An example of the relationship between sugar-content and cane elongation

Source. After Das, 1936.

the effect of warm and cool seasons on crop height increment (approximately crop growth rate—see p. 55 above).

All phases of growth are affected by temperature and optimal temperatures for growth appear to be between 80 and 100°F, with varietal differences corresponding to the cane region in which the clone was developed. About 70°F is considered to be marginal for germination and tillering and 50°F has been stated to be harmful (see Dillewijn, 1952) although Sartorius (1929) observed sprouting to occur at 43°F. Development of the buds however is greatly reduced below the optimal temperature range. The growth process most sensitive to temperature is extension of the stem and close correlations with temperature are observed (*vis.* Fig. 8c). Temperatures between 60–70°F are probably the lowest at which measurable extension occurs. Extension growth has been related to 'day degrees'. Halais (1935) cited by Das (1936) considered 15°C as an absolute minimum for cane growth and in relating temperature to growth calculated the number of days and the degrees of temperature above 15°C, each day at which the temperature exceeded 15°C by one degree being one 'day-degree'. Above 20°C two day-degrees were assigned for each degree above this temperature. Fig. 4.12 (after Halais,

Fig. 4.12 Cane elongation rate and 'day-degrees'

Fig. 4.13 Construction of cambered bed on very heavy soils

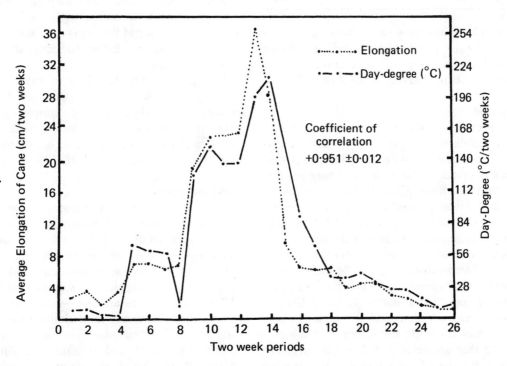

Source. After Halais, 1935.
Note. See text.

1935) indicates the close relationship that can be obtained with this integrated measure of temperature and extension growth. Day-degrees can also be corrected for moisture stress, discounting days in which the soil moisture level was below wilting point.

A certain amount of cold is probably advantageous to sugar accumulation at maturity in a manner similar to moisture stress. Sun and Chow (1949) have demonstrated an inverse relationship between temperature and sugar content in Taiwan. It seems likely that slight cold (not frost) can substitute for drought in the ripening of sugar-cane.

High sunshine is essential to high sugar yields and sugar-cane has been termed a 'sun-loving' species. An example of the effect of reduced sunshine on yield and sugar contents is given by the following table (after Borden, 1941) in which two sites are compared which differ predominately in the sunshine received by about 50 per cent.

	High Sunlight Area	Low Sunlight Area
tons cane	69	24
% sugar	11.1	8.7
tons sugar	7.8	2.1

Clements (1940) has shown a similar effect of sunlight in Hawaii, and Fogliata (1965) has shown sugar yields to be correlated with sunshine in the 6 months preceding harvest in the Argentine.

Photoperiod

An aspect of light which is of central importance in sugar production is photoperiod, since sugar-cane is predominately a short-day plant and flowering greatly reduces sugar yields and cane growth. Many varieties have particular photoperiods in which they flower (Mangelsdorf, 1950) and recent work (e.g. Daniels and Krishnamurthi, 1965) indicates that progressive reduction of the photoperiod by a small amount each week is most effective in flowering. Temperature, as indicated earlier, must not be too low for flowering to occur as well as for the production of fertile pollen (Lopez, 1963).

Breeding and selection work are aimed at producing clones which do not flower in the region for which they are developed. It is not surprising therefore that the adaptation of varieties to other geographical areas has not always been successful.

The problem of flowering, however, is two-fold since the breeder has to induce synchronous flowering in the varieties to be crossed, while the planter requires varieties which will not flower.

Flowering is also affected by age, the oldest tillers flowering first, and this has led to the practice of cutting back in some situations (see Humbert, 1968).

Frost and strong winds

Frost damage occurs in many sub-tropical localities where sugar-cane is grown, although the crop is more tolerant than some other tropical crop species (e.g. banana). Furthermore, varieties differ greatly in resistance to frost (e.g. Bourne, 1963). In South Africa, Ranger et al. (1969) report that freezing between −1.4 and −4.7°C for six days affected cane only in causing four to eight weeks loss of growth.

Slight frost causes leaf damage characterized by chlorotic bands on the leaves when they expand. More severe frost causes death of the young leaves and the stem becomes constricted at the corresponding internodes. Severe frost causes death of the apex with subsequent growth of side shoots which reduces cane millability and causes deterioration of sugar content (see Humbert, 1968).

Smoke screens to reduce night radiation have not been economical for sugar-cane but irrigation with ground water is known to reduce the severity of frost damage.

Strong winds are damaging to sugar-cane but less so than with bananas. Assessment of damage by a cyclone in Martinique for example was 90 per cent for bananas as opposed to 50–60 per cent for sugar-cane with pineapples being unaffected (*Marches Trop. et Medit.* 19, p. 2391, 1963).

Soils and drainage

In common with all pan-tropical crop species, sugar-cane is grown successfully on a wide variety of soil types. Drainage to about 60 cm is adequate for development of the root system (e.g. Sequera, 1965), and on heavy soils sub-soiling and the construction of 'cambered beds' is usually carried out. All types of drains are used—from normal field drains to subterranean mole drains (which facilitate the use of machinery in field operations). Bamboo has been used to make drains in South Africa (Boulle, 1961). Higher yields are obtained on better-structured soils and high acidity is often reported to reduce yields. However, this appears to be related to calcium availability rather than to pH itself (Humbert, 1968; Castro and Cortez, 1965; etc.). Liming also counteracts the acidifying effect of ammonium fertilizers which are the most widely used as a source of N. In cases of soils with high alluminium levels, application of lime reduces toxicity of this element (Rixon and Sherman, 1962).

Restricted K uptake is observed in heavy soils which require measures to improve drainage (e.g. Shaw, 1963), and soil compaction is sometimes a problem due to the use of heavy machinery in cane cultivation. In such a circumstance periodical sub-soiling becomes necessary to improve tilth and drainage.

Saline soils are sometimes used for sugar-cane and it appears that there are considerable varietal differences in tolerance to salinity, although at salinities of about 0.5 atmospheres most varieties are probably affected (e.g. Davies *et al.* 1966). Fogliata (1965a) emphasized the importance of Na/Ca ratio in saline soils. Soils with conductivity values below 1 millimohs/cm and with a Na^+ content of less than 12 per cent of the exchangeable cation complex were found to be satisfactory.

PLANTING
Land preparation, spacing and yields

In field crops such as sugar-cane, as opposed to tree crops, mechanical preparation of land and the mechanization of planting and harvesting operations is essential to economic production in modern times, although the extent to which this is most economically carried out is dependent on the cost and supply of labour. Land is usually prepared by conventional means of ploughing, *rotavating* etc. and the cane is planted in rows ranging from 1 to $2\frac{1}{2}$ metres apart (see Ochse *et al.* 1961) wider spacings being used when furrow irrigation is practised. The spacing of the cuttings (referred to as seed-pieces in the sugar-cane industry) within the rows varies from butting on to 1 to 2 metres apart. As indicated earlier, the optimal spacing is a function of leafiness and leaf characteristics of the variety and its tillering capacity which are determined by genotype, environmental conditions, fertilization etc. Suitable spacings are often determined by experience or may sometimes be a question of tradition or dictated by the needs of irrigation practices or machinery dimensions. There would seem to be scope in most cane areas for closer investigation into the question of spacing for optimal yield and its interaction with genotype, fertilizer, soil and climate.

In low-lying areas and on impermeable soils, drainage is necessary as indicated before. Field drains are most commonly used although mole drains are sometimes favoured mainly because they allow freer access of machinery. In very heavy soils deep cultivation and the construction of cambered beds are necessary.

Yields vary considerably from country to country and are also a function of maturity time and irrigation. In Hawaii, harvesting is carried out between 18–24 months from planting, giving yields of 60–80 tons of cane per acre. Barnes (1964) gives Hawaiian examples of a 16 months crop yielding 55 tons of cane and 6.325 tons of sugar and a 24 months crop yielding 80 tons of cane and 9.2 tons of sugar, which illustrates the age effect. Irrigation (see below) is also important in areas with periods of moisture stress and Leffingwell (1963) states that unirrigated fields in Hawaii average about 8 tons of sugar per acre as opposed to 12–14 tons with irrigation. The highest yields are of the order of 30 tons of sugar per ha for a 24 months crop (Ochse *et al.*). Generally lower yields of the order of 30 tons per acre of cane are obtained in some sugar-producing regions. Sugar recovery varies in the range 7–14 per cent and is very much a function of climate, particularly during the maturing period as discussed above. The practice of burning to remove trash prior to harvesting can lead to lowered sugar recovery if harvesting and milling are not carried out promptly (Foster, 1969).

Planting material and planting

Vegetative propagation using stem cuttings is used in sugar-cane planting since this is genetically homogeneous material. Furthermore, as indicated before, seed production is difficult and the vigour of sugar-cane seeds is generally poor.

If planting of new fields occurs at the time of harvest of the mature crop then the top internodes from the harvested cane bearing the sheath of the upper leaves (top seed pieces) form excellent planting material. If not, and generally this is impossible over whole estates, then special fields are usually prepared to grow planting material. Depending on the density of planting, the area of planting material required varies from one quarter to about one twentieth of the final field area.

Seed fields should be well fertilized and irrigated since this affects germination, tillering and yield of the crop. Seed fields mature at about 6–10 months and cuttings from old areas, except for top seed pieces, are less satisfactory as planting material. Young seed pieces and the nodes from the upper stalk germinate more readily. Generally seed pieces of between 25–35 cm length and with 2–3 buds are used and longer pieces and whole stalks are considered to be wasteful since the older buds develop slowly and may be inhibited by the presence of actively growing young buds. The presence of the sheath on cuttings impedes shoot growth but this can promote the development of more buds and increase tillering and even yield (McMartin, 1949).

Germination (or growth of the buds) is affected by nearness to the apex and by the orientation of the seed pieces, those stored horizontally germinating sooner than those which are stored vertically. The endogenous level of auxin is considered to be a controlling factor (for discussion see Dillewijn, 1952). However, such differences due to orientation are not so apparent under field-planting conditions. After planting, however, upward-pointing buds germinate faster than downward-pointing buds (see King et al. 1965).

With poor quality cuttings (from old stems etc.) various pre-treatments have been found to enhance germination. Soaking cuttings and treatment with running water for up to 48 hours sometimes enhances germination of old cuttings, and hot-water treatment (50°C) for 20 minutes usually greatly enhances germination. This however is difficult to control at a practical level. Treatment with running water has been suggested to act by removal of fermentation products and inhibitors from the cuttings. IAA (and NAA) treatment enhances root growth but delays bud development, while acetylene promotes general activity of the cuttings.

Lime and magnesium solutions enhance germination (Evans and Wieche, 1947) and appear to stimulate water uptake by the cuttings, and the swelling of the buds.

Some fungicides and insecticide also affect germination. Organic mercurial compounds may act (apart from the effects on pathogens) by inhibiting fermentation (McMartin, 1946, 1946a). BHC and chlordane have also been reported to be beneficial (Bourne, 1948). Various other substances including ethyl alcohol, ammonium phosphate, complete nutrient solution and ferrous sulphate have all on occasions proved beneficial to germination (see Dillewijn, 1952). Where labour costs are low, the topping of the seed cane a few weeks prior to cutting may induce lateral buds to commence development and give a better germination on planting.

Seed pieces are usually planted in furrows about 10″ deep which are usually formed mechanically, and then covered to a depth of about 5 cm or less in wet conditions. In Java the practice is to only partly cover the seed pieces in a moist planting season (which favour germination) and then to earth up when the plants have germinated. In cooler or drier climates, a somewhat deeper planting depth can be advantageous and the best depth to cover the seed pieces is largely a function of climate and soil drainage. In developed countries planting machines which are greatly labour-saving are used.

Sprouting occurs in one to several weeks and when shoots have reached a height of 6″, it is the usual practice in furrow planting, to bed down the cuttings with soil from the banks (generally by means of tractor or animal-drawn cultivators), and to repeat this process on one or more occasions as the shoots grow. This gives the canes a firm foundation resistant to lodging and also serves to control weeds during the establishment phase.

Since low temperature and moisture stress are detrimental to germination and subsequent establishment, the planting season in sub-tropical regions would desirably be in the spring, but in fact, where winter restricts cane growth or causes it to die back, planting material may only be available in autumn, thus necessitating pre-winter planting. In tropical regions, particularly where irrigation is not practised, a sufficiently moist season should be used for planting and establishment.

Propagation of planting material is essential to the establishment of a sugar industry and is also necessary for the introduction of new clones. Various techniques are used for the rapid propagation of clonal material:

1. Cuttings are planted at a wide spacing and watered and fertilized to encourage tillering. Following this the tillers are removed when they have developed roots and planted under similar conditions.

2. One-bud segments are cut from tillers which have grown to about the 4 internode stage and carefully planted as above.

3. Topping is carried out to induce lateral bud

development on the stem and these buds, together with their internodes, are planted and carefully nurtured.

Ratoon cultivation

Following the initial crop, the subterranean parts remain after harvesting and give rise to the ratoon crop. Although the ratoon crop is seldom as uniform and weed-free as the initial crop, it is common practice to grow several successive ratoons in sugar-cane production. Inter-row cultivation is usually carried out to straighten the rows and often a furrow is opened alongside the original row to expose the subterranean parts and encourage germination of deeply located buds. Cultivation of the inter-rows also loosens soil and encourages new growth.

Manuring

In the normal course of events a crop of cane of 50 tons will remove about 75–90 lb N, 50–60 lb P_2O_5 and 150 lb K_2O (Barnes, 1964). Part of this may be replaced in the filter-mud (filter cake) which is the fraction that settles out in the process of clarification, when this is returned to the fields, but by and large these quantities, together with leaching losses etc. must be made good in continued sugar-cane cropping.

Nitrogen in the form of ammonium sulphate is the most commonly used source of N in sugar-cane although urea and other forms of N are also used. However, urea is sometimes reported as being less effective (Tjioe Sien Bie, 1965). Ammonium sulphate has an acidifying effect on the soil and sometimes necessitates the application of lime (Castro and Cortez, 1965; Lopez et al. 1965 etc.).

Nitrogen will almost always produce a response in terms of increased tonnage of cane but sugar percentage decreases and it is necessary to determine the most optimal level in terms of sugar yield. In terms of sugar yields an N response is not always obtained. For example, in 18 trials carried out in Sao Paulo only 6 showed response to N fertilizer (Alvarez et al. 1963). Furthermore, excess N produces an excessively leafy crop with lower drought resistance properties, so that response to N is linked with irrigation practice (e.g. Verma, 1965). Green manure or ploughed-in leguminous crop residue is sometimes also used prior to sugar-cane planting, but since it must be incorporated in the soil several months before planting this is not very practicable

in sub-tropical latitudes. Nitrogen is best applied in split applications after sprouting at about 6 weeks and 12 weeks, at the times of earthing up, and quantities required are probably best determined by foliar analysis (see below).

K and P are necessary on deficient soils and are applied at the time of planting. Soluble forms of P are better on most soils. On phosphate-inactivating soils response is often marked, as for example on inland soils in Malaya (unpublished). K response up to levels of 180 kg/ha is reported on deficient soils (Alvarez and Friere, 1962). Muriate of potash is the most common form of K fertilizer used.

Sulphur, Mg and Ca are usually present in sufficient quantities in other fertilizers, although a sulphur response is reported from Rhodesia (Gosnell and Lang, 1969) who claim the N:S ratio should not exceed 17:1. Aluminium uptake was reduced by sulphur fertilization.

On soils requiring lime, as discussed before, rates of 1–3 tons per acre are often used. On volcanic ash soil of low pH, Rixon and Sherman (1962) report the application of 11 tons/ha of lime which reduced Al uptake and Al toxicity symptons.

Micro-nutrient deficiencies are not commonly reported in sugar-cane regions probably because of generally adequate nutrition, but they could occur on particular soils. Peat soil, for example, might be used for sugar-cane but would require various micro-nutrient fertilizers. Silica (slag) has been reported to increase yields in Hawaii (*Hawaiian Sug. Planters Assoc. Ann. Report*, 1968).

Montieth (1966) reports correlations between soil analysis figures and fertilizer requirements, but leaf analysis is more commonly used to determine fertilizer needs. Halais (1950) describes the technique in Mauritius, and Samuels (1969) has recently reviewed foliar diagnostic methods in sugar-cane. Optimal levels have to be determined by experiment in every climatic and soil region for good results. As a guide, levels for Mauritius are as follows: N, 1.66–1.85 per cent, P_2O_5, 0.45–0.55 per cent and K_2O 1.26–1.75 per cent, all on a leaf dry-weight basis. Values above and below this range are considered to be too high and too low respectively. Clements (1940) has developed a periodic sampling system (the upper leaf sheath) for determination or fertilizer and irrigation needs.

Irrigation

A general consideration of the water requirements of cane has been indicated above, but specific requirements must be considered in every region from studies of water consumption and water balance, and in relation to the stage of growth of the cane. A useful technique of determining when water is required has been developed by Clements (1940) in which leaf sheath material from the upper leaves is sampled for moisture determination (as well as nutrient level for fertilizer application).

Mongelard (1969) compares methods of estimation of leaf water including the density column technique. Soil moisture can be used as a basis of determining the time and quantity of application of water. Tensiometers have been used for this purpose Mascaro (1962) as well as conductivity blocks (Ewart, 1949).

Both furrow irrigation and overhead methods are used in the sugar industry. Pearse and Gosnell (1969) conclude that small V-shaped furrows gave a more efficient use of water than broad shallow ones. Many planters consider that there is an overall advantage in the overhead systems and on hilly land this indeed is the only system which is economic. Furthermore, water is more efficiently used with overhead systems since the exact amount required can be delivered and drainage losses are minimized.

General husbandry and weed control

The presence of large amounts of trash from the lower senescent leaves is a problem in sugar-cane cultivation and more frequently this is burnt, either on the stem prior to harvest or after harvesting. In dry climates (where irrigation is practised) trash does not rot quickly and burning is the only practicable means of removing it. In moist climates the trash may be used to mulch the fields where it will reduce weed competition. (e.g. Panje *et al.* 1965).

Grammoxone (paraquat) is sometimes used to dry off the tops prior to burning in the field.

Weed control in sugar-cane is effected both mechanically and with herbicides. Mechanical control is usually combined with the operation of earthing up (see above) and the careful field preparation necessary for planting gives a good initial kill of weeds. Where weeds are not adequately controlled by these operations, herbicides are to be recommended.

2:4-D and other translocated herbicides are commonly used at present and are selective in action against broad-leaves weeds. Sugar-cane is relatively resistant to 2:4-D. Long-acting (low solubility) herbicides, simazine, atrazine, monuron (CMU) and diuron (DMU) are all suitable for sugar-cane when sprayed on the soil surface after planting but before sprouting has occurred or between rows. As with all crops, the importance of early weed control must be emphasized.

MAJOR PESTS AND DISEASES

Sugar-cane is affected by numerous disease organisms of minor importance and local distribution which are not considered here. A few have caused major losses or are more widespread. Two virus diseases, 'mosaic' and 'chlorotic streak' (both affecting foliage leaves), are of considerable importance and have necessitated the production of resistant varieties of cane to combat them. A third virus disease affects ratoon crops ('ratoon-stunting disease') and is spread by means of mechanical implements. It is probably of considerable importance in sub-tropical regions of Australia, South Africa and America.

It is possible that 'Sereh' disease of Java is also caused by virus. The symptons of this disease are the bunching of leaves and the sprouting of roots on the stalk. 'Fiji disease' (also a virus) causes death of the stalk and affects susceptible varieties in Fiji and Australia.

The fungus *Helminthosposium sacchari* causes 'eye-spot'; reddish-brown elongated spots appear on the leaves. It affects cane in Latin America, Hawaii and Cuba. *Xanthomonas vascularum* which causes yellowish-brown stripes along the blade and reddish discoloration of vascular bundles is a bacterial leaf disease which was of considerable importance in the sugar cane industry until the development of resistant clones.

Various stalk diseases are sometimes serious in cane. 'Smut' (*Ustilago scitaminea*) occurs in South-East Asia, South Africa and America and causes thin attenuated internodes with greatly reduced sugar yield. 'Pineapple disease' causing a black internal discoloration of the stalks which gives off a strong pineapple odour is a widespread disease and is controlled by disinfecting the seed cane. 'Red rot' (*Colletotrichum falcatum*) produces reddish areas internally in the stalk tissue and can be serious as a consequence of stem borer attack which facilitates its spread.

The major pests of sugar-cane are the larvae of various beetles that attack developing roots and shoots on the sett and stems borers. The former are controlled by soil cultivation and/or by the use of BHC or other insecticides disced in or furrow placed near the setts. Certain varieties of sugar-cane such as NCo 310, Q58, CP29–116 and others possess good root regeneration and recovery from attack by soil insects. Varieties of sugar-cane with a hard rind are resistant to stem borers.

Leaf hoppers and aphids are responsible for the spread of mosaic, chlorotic streak and other viruses.

Crop rotation with an alternate crop at the end of a planting and rotooning cycle is recommended generally to reduce pest and disease build up.

5 RICE [Oryza sativa]

INTRODUCTION, VARIETIES

RICE is a crop which is cultivated over an extremely wide range of habitats extending from the equatorial tropics to high latitudes, the highest latitude being 49°N in Czechoslavakia (Kratochvil, 1956). It is of greatest importance, however, in tropical Asia where it is the main food crop and is adapted to monsoonal regions and cultivated under flooded conditions. There are two main sub-groups of rice (*Orysa sativa*) referred to respectively as the *Indica* and the *Japonica* varieties. There are numerous varieties within each group.

There are certain broad morphological and physiological differences between the Indicas and the Japonicas which are reflected in yield. These are discussed in some detail below. Furthermore, Indicas are predominantly photosensitive, being short-day plants (although exceptions exist) and are therefore strongly seasonal and adapted to particular regions and planting seasons. The Japonicas are predominately day-neutral and therefore not seasonal, although they are mostly planted in high latitude areas in the summer. Amongst the Indicas great variation in both critical photoperiods and sensitivity to photoperiod occur, from varieties which are virtually day-neutral to those which respond dramatically to short variations in photoperiod (see later).

Most rice varieties are adapted to flooded culture, while others are adapted to upland culture on drained soils. The varieties used in flooded culture are the most important economically and they are the only types to be considered here.

The basic chromosome number of rice is 24 but hybrids of 2n=48 exist (Ochse *et al.*). Both self-pollination and out-pollination occur, the latter being a central factor in the production of the wide range of genotypes that exist in rice.

There is much variation in grain quality but Indicas tend to be long-grained and non-sticky and Japonicas short-grained and of a sticky consistency. There are many complicated taste preferences among rice-eating peoples but there is an overall tendency to concentrate more on high yield in recent years.

BOTANY

Morphology and yield characteristics

The rice plant is a typical annual grass in growth form. It is grown almost exclusively from seed, either by direct sowing or by transplanting (see below) and depending on maturity time (variety), photoperiod and other environmental factors, ordinarily matures between about 95 and 170 days from sowing.

The plant develops from seed (Fig. 5.1a), producing alternate leaves successively from the apex and initially compacted nodes and internodes that bear side shoots or tillers which in turn undergo similar development during the vegetative phase until culm development and flowering occur. Secondary, tertiary and sometimes quartenary tillers are formed. Tillering and tiller development to produce the leaves and canopy are the main features of the vegetative phase of growth, the extent of tillering being a function of variety and maturity time, and environmental factors of nutrition, sunlight and temperature.

Initially the seed produces a radicle and a number of seminal roots. After this roots are produced adventitiously from the lower nodes of each tiller. The roots branch freely and are fibrous, often arising as whorls from the nodes at and above soil level. The depth of the rice root system in flooded culture is usually limited to the 20 to 25 cm depth of soil working or puddling that is carried out. Root formation and growth usually cease at the time of heading (e.g. Inada, 1967).

Stem (culm) elongation often occurs in the later stages of vegetative growth preceding floral initiation, and occurring in the primary and other early-formed tillers.

Floral initiation takes place either at about the time of culm elongation or later, and gives rise to

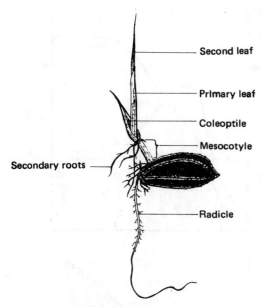

Fig. 5.1a Germinating rice seed
Source. From IRRI Rice Production Manual.

Fig. 5.1b Lower portion of the primary and secondary tillers

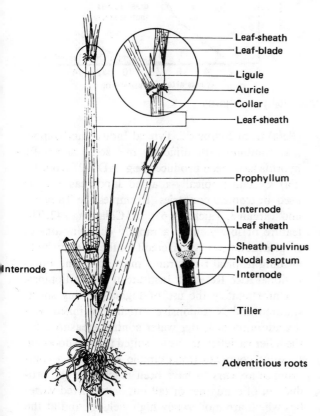

Source. From IRRI Rice Production Manual.
Note. Inserts show leaf parts.

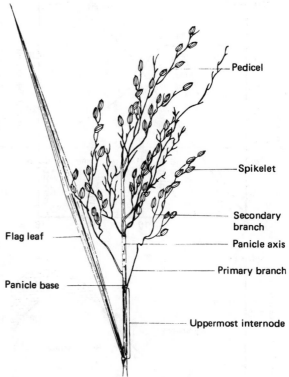

Fig. 5.1c Panicle of rice (partly shown)
Source. From IRRI Rice Production Manual.

the panicle and grains (Fig. 5.1c). The subtending leaf of a flowering culm is the flag leaf and in common with many grain species it is generally active in photosynthesis for grain development (see Tanaka, 1964b).

Probably the most important morphological feature from the view-point of yield is related to the production of an efficient non-lodging canopy of optimal leaf area index for assimilation. Indica varieties are typically tall and leafy, with high early tillering rates, low responsiveness to N in terms of grain yields and susceptibility to lodging. It seems likely that they have been selected for conditions of low fertility and deep flooding. Japonica varieties are typically short and sturdy (lodging resistant), less leafy and slow tillering and therefore N responsive, and with erect leaves which do not bend horizontally. Such varieties are the result of long years of selection in Japan and other countries for high yield responsiveness to manuring. For the reasons that have been discussed in the first part of this book, the Japonica type of leaf gives rise to a more efficient canopy which responds to high N nutrition (Tsunoda, 1962; Tanaka, 1964a and b,

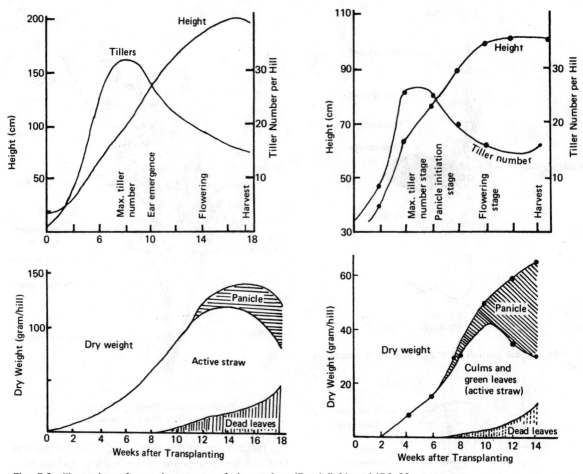

Fig. 5.2 Illustration of growth patterns of rice variety 'Peta' (left) and IR9-60

Source. After Tanaka *et al.* 1964 for 'Peta' and after Vergara, 1963 for IR9-60.

Notes. 'Peta' may be considered as typical of the Indica type of rice variety and IR9-60 of the Japonica or the improved Indica types possessing shorter stature and maturity time and high yield.

1966; Beachall and Jennings, 1964, etc.). Although there are many reasons for the low-yield characteristics of most tropical regions, one of the most important is the nature of the Indica genotype which has the tendency to produce an above optimal leaf area index with poor photosynthetic properties. However, in recent years, it has been shown in Taiwan, the USA, the Philippines (at IRRI) and in other countries that desirable foliage properties can be bred into Indica types and Indica/Japonica hybrids, together with low response to photoperiod which makes for wider adaptability. Short-statured Indicas which are moderately high-yielding are Taichung 1 and Dee-geo-woo-gen, both from Taiwan. The remarkable high-yielding variety IR8 was produced from the latter and an improved tall Indica

(Peta) from Indonesia. Typical Indicas and Japonicas combine with difficulty but some successful hybrids have been produced (e.g. ADT-27 from India). Certain tropically-adapted Japonicas are also used in tropical countries. Examples are Tainan 1 and Tainan 3, Taichung 65, and Chianang 242. The feature of low yield is a rapidly changing one in tropical countries. Much credit for the work on high-yielding tropical rice strains must go to the International Rice Research Institute in the Philippines.

Unfortunately the use of higher-yielding short-statured and N-responsive varieties is linked with the adoption of better water control measures, for the taller varieties are better suited to deep flooding. In some countries (for example Malaysia) a compromise appears to have been reached in the production of a number of tall but erect-leaved varieties which are moderately high yielding and at the same time resistant to deep flooding.

There would also appear to be a relationship, not

necessarily linked, between maturity time, yield and N-response, since late-maturing varieties become excessively leafy with high N nutrition.

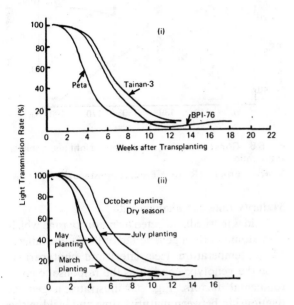

Fig. 5.3 Light transmission through the canopy of a number of varieties planted in the wet season in the Philippines and of variety 'Peta' (a typical Indica type) in both wet and dry seasons

Source. After Tanaka *et al.* 1964—*Mineral Nutrition of the Rice Plant.*

Growth quantities

The growth of rice can conveniently be divided into a number of phases as follows (Tanaka, 1964a):

1. *Active vegetative phase.* From planting until maximum number of tillers is achieved. This phase, predominantly exponential at first when considering the crop becomes linear later on.

2. *Vegetative lag phase.* A variable phase which may be long or short or even absent or negative, since floral initiation can occur during the preceding phase. In some varieties (see Fig. 5.2) there is a considerable lag between the achievement of a maximum number of tillers and the onset of flowering.

3. *Reproductive phase.* Initiation takes place and the panicles emerge from the subtending leaf sheaths. Prior to emergence, the panicle is apparent as a budge on the elongating culm within the surrounding leaf sheaths. During this phase the plant is most sensitive to environmental stress, see later.

4. *Ripening phase.* Panicle growth, grain development and ripening occur. Straw weight also increases.

Fig. 5.2 illustrates the growth patterns of two contrasting varieties, a typical Indica (in regard to canopy structure), and an improved type. Apart from the difference in the proportion of panicle (grains) to straw (which is an important yield factor and reflects the sink strength of the reproductive parts as well as canopy efficiency), it can be seen that the improved type matures sooner, produces fewer tillers and is shorter in stature. This figure does not indicate the nature of the canopy which, in high-yielding varieties, permits a better penetration of light into lower horizons of the canopy and is less prone to self-shading. Fig. 5.3 indicates the nature of light interception by the whole canopy of a number of contrasting varieties and an Indica variety (Peta) in both the wet and dry seasons. This latter figure shows an interesting feature of Indicas and the tropical environment, namely, that in some Indica varieties, the extra sunshine of dry-season cultivation affects canopy development favourably, and also yield. Plants are shorter in seasons with high sunlight, require closer spacing and are more responsive to N (Moomaw and Vergara, 1964). Consequently, the improvement of irrigation and the introduction of off-season growing seems to offer the most immediate solution to increasing yields in tropical countries.

The main objective in both the breeding for a short erect leaf structure and the husbandry practices of fertilization and obtaining the correct plant spacing is the achievement of an optimal canopy giving the highest net production, with good light penetration to lower layers of the canopy at the time of grain formation and inflorescence growth. This will allow the maximum yield in as far as the photosynthetic apparatus is concerned.

Another plant or genetic factor of central importance is the panicle/straw ratio. Although this ratio is also a function of light received and other climatic factors, it has a strong genetic component in that the panicle represents a sink for assimilates, and if the sink capacity is not present, then high yield, even with the most efficient canopy, would not be possible.

Grain yield is derived both from accumulated assimilates (in the vegetative plant) and assimilation occurring after the onset of the reproductive phase. There are basic differences between the Indica and Japonica and improved varieties in this respect. In Indicas, early rapid tillering is characteristic and carbohydrate accumulation does not occur till the late vegetative and initiation phases. This, however, is the main source of carbohydrates for the grains since net assimilation after flowering is reduced by the poor canopy structure. In Japonicas, with lower tillering rates, carbohydrate accumulates early but is present in relatively low amounts by the time of initiation. Much of the grain is formed from late assimilation taking place after initiation (Tanaka, 1964a).

For Japonica varieties the following equation for yield is probably accurate (Uchida, 1963):

YIELD OF RICE (100) = Accumulation before heading (20)+Assimilation after heading (120)— Respiration after heading (40)

Short maturity (see below) also results in a lower proportion of accumulated starch, as does high N nutrition which promotes early rapid tillering.

Uptake of N is more rapid in Indicas in the early stages, and slower but over a greater proportion of the growing time in Japonicas and improved varieties. Loss of N from Indicas and death of late tillers and leaves are also common occurrences (Navasero and Tanaka, 1966).

The yield of rice can be divided into a number of components, namely: the number of panicles per unit area which is related to tiller formation and photosynthetic efficiency and is markedly affected by N nutrition in the vegetative phase, the number of spikelets (grains) per panicle which is also affected by N availability in the late vegetative and early reproductive phases, the percentage of grains which ripen, and the individual grain weight (usually 1000 grain weight). These yield components and the effect of fertilization on them are considered later.

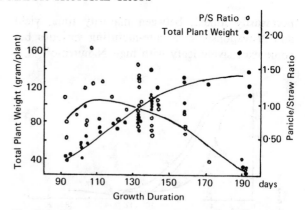

Fig. 5.5   Growth duration, total plant weight and panicle/straw ratio

Source.   From IRRI Rice Production Manual.

Maturity time and photoperiod

Yield is markedly affected by maturity time which is predominantly a genetic factor but is also affected by temperature (see later) and photoperiod, since the genotype conditions the response to environmental factors. Fig. 5.4 illustrates a general relationship between maturity time and yield in the tropical environment. It can be seen that there is an optimal maturity time in the region of 130–140 days from germination. Presumably the reason for reduced yields in shorter maturity periods is due to incomplete development of the canopy which could be affected by both nutrition and spacing, while reduced yields in later varieties is due to excessive vegetative growth to a stage above optimal LAI, and probably other factors such as stem borer attack and lodging. Reduction in the ratio of reproductive to vegetative parts is a feature of long maturity periods as is shown in Fig. 5.5.

In the photosensitive Indica varieties, seasonal changes in day-length greatly affect production and adaptability to other regions, and it is true to say that such varieties are strongly locally adapted. Photoperiod affects the length of the 'vegetative lag phase' mentioned before and consequently also affects maturity time and yield as indicated above. Fig. 5.6 indicates the nature of the effect of photoperiod on the various growth phases of photosensitive and day-neutral varieties.

As indicated before and in Fig. 5.6 photosensitive rice varieties are short-day plants, that is, short days reduce the maturity period. Equatorial varieties are particularly sensitive to even small variations in photoperiod as shown in Fig. 5.7 from Dore (1960). This figure compares the growth of a

Fig. 5.4   Relationship between growth duration and grain yield at 25×30 cm spacing and with 40 kg/ha nitrogen

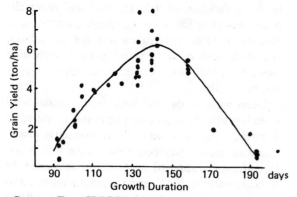

Source.   From IRRI Rice Production Manual.

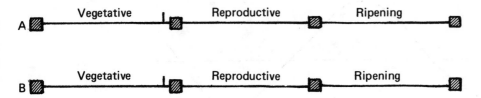

LOW PHOTOSENSITIVITY

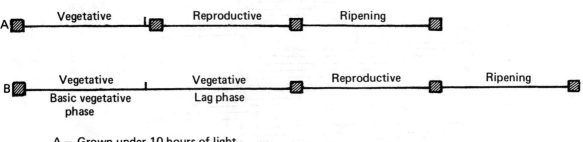

HIGH PHOTOSENSITIVITY

A — Grown under 10 hours of light
B — Grown under 13 hours of light

Fig. 5.6 Growth phases of sensitive and insensitive rice varieties in relation to photoperiod
Source. After Vergara and Jennings, 1963.

photosensitive variety at two localities in Malaya. The interpretation given by Dore is that in the north, a plant sown in December will receive sufficient short days to induce early flowering, but in the south it will receive longer day-lengths and consequently will not flower for an extended period of time. In the north, a sowing made in January and February will receive longer days resulting in delayed flowering. The shortest maturity periods occurred during decreasing photoperiods, progres-

Fig. 5.7 Number of days to ear emergence in rice variety 'Siam 29' grown in northern and southern sites in Malaya

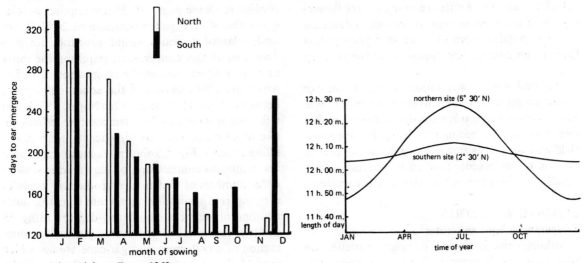

Source. Adapted from Dore, 1960.

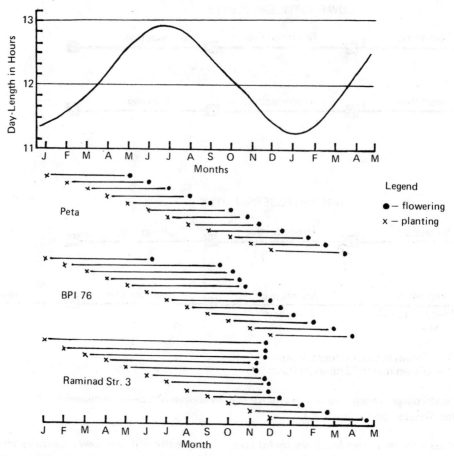

Fig. 5.8 Data on maturity period in three rice varieties of different responsiveness in relation to day-length
Source. From Vergara, 1963.

sively so until October. Fig. 5.8, illustrating data for three varieties in the Philippines, also emphasizes the importance of *decreasing* photoperiods (which are also more effective in sugar-cane flowering), where the most sensitive variety (Raminad Str 3) responded more to decreasing photoperiod than to the average photoperiod over the growing time.

The problem of adaptation of photosensitive strains to other localities, where they may flower either too soon or too late for optimal yield, makes inefficient use of breeding procedures for improved yield and disease resistance etc., and for this reason the aim of the breeder is often to produce a type which is not very sensitive to day-length variations.

CLIMATE AND SOILS
Temperature, light and water

Although rice is tropical in origin, varieties are adapted to a wide range of climatic situations, par-

ticularly in relation to temperature. However, rice does not survive freezing, but being an annual, is grown in the summer of high latitude regions. According to Ochse *et al.* (1961) rice requires at least 4 months of average temperatures of 20°C. Tropically-adapted varieties would probably not produce well at this temperature, requiring the fairly uniform temperatures in the range about 25–35°C which are characteristic of the equatorial regions between about 15°N and 15°S (Nagai, 1959), but reference is made to low temperature effects on rice in so far as they are significant in the grain-filling process. Fig. 5.9 illustrates mean, maximum and minimum temperatures of the growing seasons of a number of Asian rice-producing localities ranging from tropical to temperate climates. Summations of temperature in day-degrees using all positive centigrade temperatures (unlike the summation of day-degrees in sugar-cane studies which used 15°C as a base), show that northern latitudes

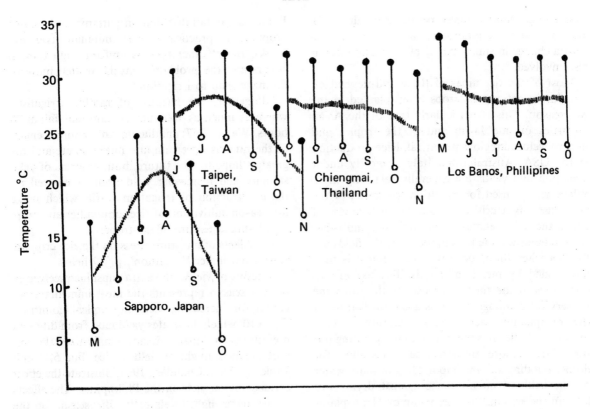

Fig. 5.9 Mean temperatures and temperature ranges during the main rice-growing months at four locations in Asia
Source.   From Moomaw and Vergara, 1964.

Fig. 5.10 Relationship between the average daily radiation during the Aug.–Sept. period and the yields of rice among different prefectures in Japan

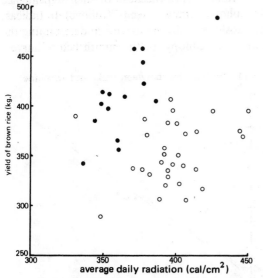

Source.   After Yoshio, 1966.
Note.   Closed circles indicate prefectures situated to the north of latitude 36°N and open circles show those to the south of it.

of Japan accumulate about 3000 day/degrees C over the growing season, while equatorial regions accumulate the order of 5000 or more (Moomaw and Vergara, 1964).

Low temperature does not favour tiller development and vegetative growth, as well as nutrient uptake (Ishiguka and Tanaka, 1961), and even in Japanese varieties, optimal temperatures for tillering are of the order of 30°C. However, lower temperatures are considered to favourably affect grain yield and mean temperatures above 22°C, particularly night temperatures, are considered to reduce yield (Matsushima and Tsunoda, 1958; Yoshio, 1966). Fig. 5.10 (from Yoshio, 1966) illustrates the relationship between radiation and yield at two latitudinal areas of Japan, and indicates that the low temperature effect on grain is independent of radiation, although high radiation levels associated with the long days of summer in Japan are probably also responsible for the high yields which are characteristic of that country. The effects of low temperatures on grain yield have been attributed to slower development and more complete grain filling over a longer photosynthesis period (Ramiah, 1954), and to reduced respiration (Chandler, 1963). Possibly a change in sink capacity of culm and grain is also involved.

Most of the rice areas of the world depend on rain for irrigation, the crops being planted in the wet season. This is particularly true of the Asian countries outside Japan and of the tropics and underdeveloped regions in general. Rice production in the USA, Australia and Italy is mostly under irrigation and yields are generally relatively high. Where rain is used for irrigation, the onset of the wet season is a critical factor. In such areas, of which the delta regions of the Mekong and other river systems are the most important, the fields are flooded when the flood water rises, and it is then impounded by means of bunds. Because of uncertainties in the time of arrival of the rains the nursery for seedlings, which is a central feature of the transplanting method of rice culture of Asia, serves as a buffer to some extent in the planting out time. Furthermore, the larger seedlings allow for deeper flooding and the impounding of more water on the fields as an insurance against drying out later in the season. (See discussion on 'Transplanting', p. 79 below.)

In absolute terms, rice transpires between 25″ and 50″ (c. 600–1200 mm) of water for each crop, which is similar to that transpired by dry-land graminaceous crops and depends mainly on the insolation. Under the high sunlight conditions of Australia the crop could transpire 50″ while under the wet-season conditions of the equatorial regions only about half of this quantity will be used. In terms of soil moisture, the point of field saturation is optimal for rice production. Deeper flooding is again simply an insurance against drying out and the loss of water through downward and lateral movement. (Some slight movement of water to prevent excessive reduction of the soil is better in many soils because of undesirable chemical changes that occur on reduction—see p. 76.) Generally between $3\frac{1}{2}$ and 6 feet of water is used to produce a rice crop, depending on the degree of water control, drainage and the prevailing evaporative conditions.

Within the tropics several types of rainfall are encountered which allow different cropping systems. By far the largest rice area is located in regions with a single 'monsoon' season alternating with a dry season. Only one crop is possible without extensive irrigation works. Other regions have bi-modal rainfall distribution patterns and double cropping is practicable with minimal irrigation works. In yet other regions uniform high rainfall occurs and the problems may be related more to drainage than to irrigation.

Altogether, the hazards of rainfall irrigation, where as much as 1000 mm of rain may fall in 24 hours (Watts, 1957) and the time of commencement of the rains is uncertain, are indeed acute and the greatest hope for the future, both in terms of yield stability and increasing yield itself, is centred on the construction of irrigation works which make dry-season cultivation of rice, or an alternate crop, a widespread feature of the tropics.

The effects of moisture stress on the rice crop are considered under 'Irrigation', p. 79 below.

In terms of yields, the extra sunshine received in the dry season represents the most immediate potential for increasing yield in tropical countries. Fig. 5.10, which illustrates yield data from different prefectures in Japan, already emphasizes the importance of sunlight in yield of rice. Fig. 5.11 and Table 5.1 (from Chandler, 1963) illustrate the effect of sunshine on yield in the Philippines. The effects of the higher light levels of the dry season on the form of the canopy and response to N have already been mentioned.

The period of about one month prior to harvest is the most critical in regard to yield in all rice varieties, but particularly in Japonicas and improved Indica types (because of their dependence on late photosynthesis—see p.70 above). In Indicas, early sunshine is also important in determining the nature of the canopy. In the month before harvest

Fig. 5.11   Relationship between yield and sunshine

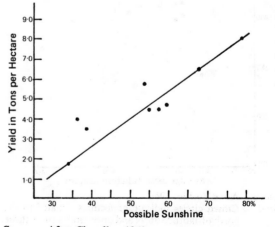

Source.   After Chandler, 1963.

TABLE 5.1 Comparison of rough-rice yields obtained from the same varieties planted in the rainy season and in the dry season. IRRI, 1962–63

Date of planting	Variety	Yield (kg/ha)	Average per cent possible sunshine during last 60 days crop was growing (approximately)
1, May 1963	Tainan 3	4,300	57
26, October 1962	Tainan 3	5,960	53
1, May 1962	Peta	4,320	55
26, October 1962	Peta	6,040	69
June 1963	Milfor-6(2)	3,915	57
January 1963	Milfor-6(2)	5,454	84
29, May 1963	Chianung 242	4,249	38
10, January 1963	Chianung 242	8,440	84
29, May 1963	Milfor-6(2)	3,808	38
10, January 1963	Milfor-6(2)	7,356	84
13, June 1963	Peta	1,970	34
20, December 1963	Peta	7,680	70

Source. Chandler, 1963.

the accumulation of more than 200 hours of sunshine (14,000 cal/cm^2) seems to be critical (Aspiras, 1964). Sato (1955, 1956) studied the effects of light and temperature on the yield of rice in Japan and from correlation studies concluded that the highest yields were associated with the occurrence of 400 hours of sunshine(s) in September and October prior to harvest, and a mean air temperature of 27°C (t) in the establishment phase (July and August), thus:

Yield ($\%$)$=9.0+0.2s \times 25.4e-4.32 (t-27.2)^2$

The effect of photoperiod on yield of rice has already been discussed.

Wind damage and cold

Since rice is a short-term species, frost or extreme cold does not ordinarily affect the crop as it does with perennial species such as bananas and sugar-cane. Regarding wind, gentle winds are considered beneficial in renewing CO_2 at the leaf surface but heavy winds and typhoons cause lodging and may reduce yields considerably, particularly at heading time. The damaging effect of lesser winds is mainly caused by a reduction in the number of grains that mature (Matsubayashi, 1963) and in the number of spikelets.

Soils and flooding

Rice is grown on a great variety of soils but there are few clear relationships between soil type and yield (Grant, 1964). A common feature of the continued cultivation of rice, in which the top 25 cm or so is 'puddled' during field preparation prior to planting, is the formation of an impermeable layer just below the worked horizon which impedes downward movement of water. This occurs even in permeable sub-soils after long periods of cultivation and has been attributed to mechanical compaction and the leaching down of cementing compounds of Fe and Mn under reducing conditions.

The extent of drainage determines the amount of reduction which the soil undergoes and this in turn affects growth and yield of rice. Generally some degree of drainage or some lateral movement of water is desirable in preventing excessive reduction of the soil and consequent undesirable chemical changes that may occur, and in this regard sandy soils are often reported to give superior yields (e.g. Grant, 1964). Heavy soils become strongly reduced while well-drained soils may retain an oxidizing root zone for long periods. Fig. 5.12 (after Ponnamperuma, 1964) illustrates the nature of change of Redox potential on flooding, and in a well-drained situation. However, excessively-drained paddy soils are also undesirable because they require too much water to be practicable and also use excessive fertilizer because of leaching. Excessive drainage may also lead to iron becoming unavailable because of the insoluble nature of the Fe^{+++} hydroxide, and in addition, nitification is also affected (see below).

Rice is grown on soils ranging in pH from about 3.5 to 8.5 and there appear to be distinct varietal adaptations to pH. Furthermore, changes of pH up to 2 units can occur on flooding. Generally high pH is associated with saline or alkaline conditions and low pH with sulphuric acid formation on drainage (from oxidation of sulphides) and H_2S formation in the reduced state. Moderately acid soils are themselves apparently not harmful to rice, but they are usually associated with low nutrient availability. With sulphide soils excessive reduction produces toxic levels of H_2S. The standard treatment to combat excessive reduction is temporary drainage but in sulphide soils this causes extreme acidity.

The nature of the change in Redox potential on flooding is illustrated in Fig. 5.12, but differences exist with depth (the upper few millimetres will generally not be reduced) and with the organic matter and ionic content of the soil. Sometimes an oxidized horizon also occurs at some depth due to the trapping of air in the soil on flooding.

After flooding, nitrate is the first soil nutrient to become reduced and denitrification is the main mechanism whereby nitrate is lost from the soil. Pearsall (1938) showed that the presence of nitrate prevents reduction below E_7 of 200–400 millimohs.

In the presence of nitrate iron will be largely in the Fe^{+++} from (Ponnamperuma, 1955). Manganese in the soil also prevents an excessive reduction while organic matter encourages reduction in its supply of micro-organisms that can utilize these ions, particularly nitrate (Ponnamperuma, 1964). The ferric ion is not very effective in preventing excessive reduction. In the absence or exhaustion of manganese or nitrate, intense reduction occurs which may lead to the formation of the toxic sulphide ions or toxic organic compounds if these are present in the soil. The nature in the change of SO_4'' and S'' in relation to Redox potential is shown in Fig. 5.13 (after Yamane and Sato, 1961). In normal soils the presence of manganese, nitrate and other ions keeps the concentration of H_2S below 10^{-8} moles per litre (Mortimer, 1941; Sato, 1957), which is fortunate because H_2S is extremely toxic to rice. In Japan, the condition known as 'akiochi' is attributed to H_2S poisoning. Wen and Ponnamperuma (1966) studied the effects of manganese dioxide, nitrate and chlorate (which have relatively high Redox potentials), on the Redox changes of flooded soils and found that all three retarded the accumulation of the ferrous ion and reduced organic products. Nitrate itself would not be economical as an additive to prevent excessive soil reduction because of its loss by dentrification, but manganese dioxide is considered to be a soil ameliorant of

Fig. 5.13   Relationship between E_7 and sulphate reduction

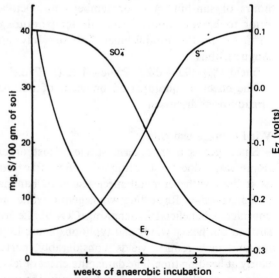

Fig. 5.12   Changes in Redox Potential (E_6) of a well-drained and submerged soil with time

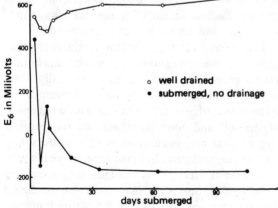

Source.   After Ponnamperuma, 1964.
Note. E_6 represents an oxidation-reduction potential measurement at pH6.

Source.   After Yamane and Sato, 1961.
Note. E_7 represents an oxidation-reduction potential measurement at pH7.

considerable potential for controlling physiological disorders associated with highly reduced soils. High yielding varieties with efficient leaf canopies also prevent excessive reduction through O_2 production by the roots.

PLANTING

Land preparation, spacing and yield

Land preparation for rice cultivation is complicated by the need for bunding to impound water, and levelling in an exact manner to permit even flooding. Where very deep flooding occurs, notably in East Pakistan and Burma, special floating rice varieties are grown which have root systems adapted to floating. The water itself supplies the nutrients for the floating rice crop (Grist, 1955).

The common method of cultivation throughout Asia, however, is by transplanting seedlings grown in special nurseries. This is rather less exacting in terms of land levelling and water control.

In some countries (notably Australia) the seed is first drilled in rows in the fields and then flooded. This requires more exacting water control and levelling. In America, germinated seeds are broadcast-sown by aircraft and this also requires exacting water control and levelling.

Since this book is chiefly concerned with the agriculture of tropical regions, the methods of land preparation for the transplanting of rice will be described in more detail. By the transplanting method the soil is first cultivated either by animal-drawn ploughs or by means of tractors. Where the size of fields is relatively small (as in many parts of Asia) small pedestrian cultivators are in common use. In rain-fed areas some water is impounded to soften the soil and bunds are made sound either by hand or by throwing up the soil with the plough. Following this the soil is thoroughly puddled, desirably to a depth of about 25 cm to mix in weeds and stubble and to aid in levelling. Usually the straw from the previous rice crop is burnt prior to ploughing. This is effective in disease and pest control and also prevents the occurrence of an excessively high C:N ratio of the organic matter of the soil, which could lead to N immobilization. The weeds and stubble are then allowed some weeks to rot with one or two subsequent workings which will destroy late germinating weeds. After flooding, the soil should be maintained in the flooded condition to prevent the formation of nitrate which on subsequent flooding would undergo anaerobic de-nitrification to nitrogen gas. (Estimates of N lost by this process are between 20 and 300 lb/per acre.)

The Rice Production Manual (IRRI, Philippines) gives detailed instructions on the preparation of rice fields and also considers the various merits and disadvantages of animal power and tractors.

The depth of tillage affects rice yields. High yields in Japan and other countries have sometimes been attributed to deeper cultivation (Casem, 1964), but the use of small pedestrian-cultivating machines and animal-drawn implements precludes very deep working of the soil.

Experiments carried out at IRRI studied the effects of tillage depth in the Philippines in unfertilized fields and the results are shown in Table 5.2.

TABLE 5.2   Effect of soil depth on yield per hill of Tainan 3 (30×30 cm spacing), 1964 wet season, non-fertilized

Soil depth (cm)	Average no. of tillers	Average height (cm)	Grain weight (g)	N content		
				Straw	grain	plant
4	5.2	65.5	8.2	0.035	0.101	0.136
8	6.1	67.7	12.4	0.062	0.142	0.204
12	7.5	70.2	18.2	0.068	0.206	0.274
16	9.9	72.4	20.8	0.097	0.243	0.340
20	10.9	74.2	23.2	0.137	0.272	0.409
24	11.6	76.8	26.4	0.149	0.341	0.490
28	12.7	78.8	27.9	0.198	0.398	0.596
32	13.1	78.8	26.8	0.228	0.373	0.601
36	13.7	79.5	29.8	0.233	0.459	0.692

Source.   From Rice Production Manual.

TABLE 5.3   Rice area, yield and production:
1963–64, paddy or milled

Country	Area (1,000 ha)	Yield (kg/ha)	Production (1,000 met ton)
Spain	64	6,230	399
Australia	23	5,910	142
United Arab Republic	345*	5,840*	2,039*
Japan	3,272	5,240	17,157
Italy	115	5,120	589
U.S.A.	717	4,440	3,187
Taiwan	749	3,500	2,623
South Korea	1,165	3,230	3,367
Malaya	380	2,290	869
South Vietnam	2,358	2,100	5,327
Hongkong	8	1,840	14
Ceylon	567	1,810	1,026
Indonesia	6,738	1,740	11,764
Pakistan	10,294	1,720	17,724
Thailand	6,387	1,590	10,168
Burma	4,789	1,560	7,457
India	35,474	1,540	54,734
Philippines	3,129	1,220	3,223
Cambodia	2,296	1,200	2,760
Laos	590	860	510*

Source.   FAO, Production Year Book No. 18, 1964.
*Based on 1962–63 data.

The increased yield seems to be mediated through an increase in the number of tillers and is probably related to the extent of the root system and availability of nutrients.

The optimal spacing for rice varies considerably with variety, nitrogen fertilization and season (particularly in relation to sunlight). Traditionally, seedlings are planted about 20–25 cm apart in a random manner. Improved methods involve planting in straight rows to facilitate weeding and other operations and the adoption of spacing suited to the variety, season and nutrient level of the soil. Strong tillering and leafy varieties should be planted at 30–35 cm or more using the wider spacing for wet seasons and/or highly fertile soils. Low tillering erect-leaved types (improved types) are best planted at 20–25 cm, again depending on the season and on soil fertility. Direct drilling of rice is done in rows 15 to 25 cm apart.

The quantity of seed required is usually between 80 and 120 lb/ha depending on the variety and assuming high (about 80 per cent) viability; rather more seeds are used in the broadcast method while transplanting requires relatively fewer seeds.

Yields of rice vary considerably from country to country for the reasons which have been discussed before, and because of variation in disease and pest control, and general standards of husbandry and fertilization. Table 5.3 (from FAO, Production Year Book No. 18, 1964) indicates yields and production from a number of tropical and temperate countries. Since then, the situation has changed with the introduction of higher-yielding varieties in many tropical countries.

Nursery, planting and seed production

Several techniques are used to raise nursery seedlings. The most common method is to construct raised beds 1–1½ metres in width and of appropriate length (300–500 sq.m for each hectare depending on the variety etc.—IRRI Manual). Adequate fertilizer (60–100g of N fertilizer per sq. metre) is worked in prior to sowing pre-germinated seeds. During development the soil is kept saturated and then the level is gradually increased up to about 5 cm as the seedlings grow. Seedling are ready in 3 to 4 weeks, the early maturing ones in the shorter period, and are pulled a few at a time, washed and stacked in bundles for planting.

Sometimes 'dry-beds' are used in which frequent watering is carried out, and sometimes the seeds are germinated on polythene strips (or other material) which keeps the root system on the surface and easier to raise. Such seedlings are ready for planting in 10 to 14 days.

At planting, several seedlings are planted at each point in the puddled soil of the field.

Although the nursery technique should be well timed for the best result and transplanting carried out promptly, in actual practice it is sometimes necessary to retain the seedlings for much longer periods in the nursery because of uncertainties in the time of the rains. In areas of deep flooding, very large seedlings, sometimes almost at the initiation stage, are used for planting.

When the seed has been drilled in the direct-seeding method, flooding is controlled very carefully and a permanent flood is brought in only 3–4 weeks after emergence. In the broadcast method the pre-germinated seed is usually sown on the well-prepared mud surface and flooding controlled accurately.

The advantages and disadvantages of the labour-intensive transplanting method have often been argued. The advantages in terms of uncertainty of rainfall and the less exacting requirements in seed-bed preparation have already been mentioned. In addition, a better standard of husbandry can be applied to young seedlings as well as protection from birds, pests and diseases. Less time is required in the field (which could allow multiple cropping under some circumstances) and less water is required for the crop. Furthermore, when compared with the broadcast method, inter-row cultivation is possible to effect weed control. The obvious disadvantage of transplanting is the high labour input required.

Sometimes it is claimed that higher yields are obtained by transplanting, although countries which use the broadcast or direct-drilling methods consistently produce the highest rice yields. Even so, these higher yields are probably more related to other factors such as the use of better varieties, more sunshine etc. as discussed before. The transplanting method often produces higher yields in the Asian environment. The following table (from the Department of Agriculture, Ceylon, 1956) summarizes rice experiments in that country up to 1956.

Broadcast   (hand-weeded) — yield 41.5
                                                bushels/acre
Row sown   ,,   ,,   — yield 49.0
                                                bushels/acre
Transplanted   ,,   ,,   — yield 54.4
                                                bushels/acre

Senewirante and Appadurai (1966) attribute higher yields by the transplanting method to the period of maximum availability of nutrients mineralized from organic matter, which occurs a few weeks after working the fields and coincides with the period of maximum of N requirement in transplanted seedlings more closely than with a direct-sown crop.

The importance of using the correct variety for any particular area is often emphasized. Seeds are sometimes available from government agencies but more often are produced in special seed-beds by the farmer. Such fields should be planted out of phase with surrounding areas to ensure self-pollination and should be well tended. About 150 sq. metres will produce seed for about 1 hectare for the transplanting method or for about $\frac{1}{2}$ hectare for the other methods of cultivation.

Manuring and irrigation

Manuring. Mention of nitrogen response in rice varieties has already been made, and in fact this is the main fertilizer element required in rice growing. Table 5.4 gives an example of the quantities of nutrients removed from a high-yielding rice crop. The relatively large quantities of K and silica are generally returned to the soil in the straw ash. Silica is important in rice in bestowing resistance to blast disease (Yoshi, 1941; Tasugi and Yoshida, 1958), and to reduce lodging. Furthermore, Okuda and Takahashi (1962) have shown that silica reduces manganese and aluminium toxicity in rice. In Japan, the critical level of available silica is considered to be about 10 mg per 100 g of soil. Kawaguchi (1966) considers that silica deficiency is fairly widespread in tropical Asia.

Phosphorus is more available in flooded soils and potassium is present in most rice-irrigation water in significant quantities. However, yield responses to these elements are not uncommon in the tropics and furthermore, the use of heavy nitrogen fertilization to raise yields also increases the requirements for the other elements (e.g. Allen, 1953).

Most of the nutritional work on rice has been concerned with N response, and the N/variety/climate interaction. Fig. 5.14 (from IRRI Manual) shows an example of N response in the two types of rice varieties described before. The high-yielding varieties which possess short erect foliage (and include most of the Japonicas and the improved tropical varieties) respond to high levels of N in terms of yield. Climate also affects the response to N as shown in Fig. 5.15 (from IRRI Manual). In this case the improved variety (IR 8) showed an almost

TABLE 5.4   Nutrients removed by a crop of rice producing 4,074 kg of dry grain per hectare

Nutrient element	Amount of nutrient in the rice plant at harvest (kg/ha)	
	Total	Panicle
Nitrogen	90	48.0
Phosphorus	20	13.0
Potassium	219	11.0
Calcium	34	12.0
Magnesium	25	9.0
Iron	12	1.6
Manganese	12	2.3
Silica	1780	371.0

Source. Rice Production Manual.

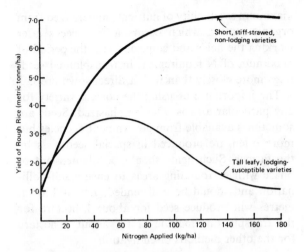

Fig. 5.14 Comparative performance of low nitrogen-response and high nitrogen-response varieties
Source. Rice Production Manual.

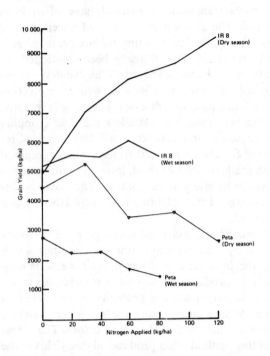

Fig. 5.15 Effect of plant type and season on the nitrogen response of Indica rice
Source. Rice Production Manual.

linear response up to 120 kg N per hectare in the high sunshine dry season but failed to respond in the wet season. This is related to the production of an efficient canopy of high LAI which is only possible when abundant light energy is available.

The two most common forms of N fertilizer used in rice are ammonium sulphate and urea, both of which are in the reduced form. Under flooded conditions they do not readily undergo oxidation to nitrate and become lost by denitrification. However denitrification does occur in the oxidized zones of the rice soil as indicated in Fig. 5.16 (from Mikkelsen *et al.* 1958). Deep placement of fertilizers

therefore gives generally better results (e.g. Table 5.5).

Leguminous crops are sometimes used as a source of N and should be ploughed in after flooding to prevent denitrification losses.

TABLE 5.5   Grain yield as affected by forms and methods of nitrogen application, IRRI, 1965 dry season

	Grain yield (kg/ha)		*(average of 4 replicates)*			
	Chianan-8		*Peta*		*Milfor-6(2)*	
Fertilizer	*Broadcast and incorporated*	*Placement 15 cm deep*	*Broadcast and incorporated*	*Placement 15 cm deep*	*Complete incorporation at planting*	*Placement 10 cm deep*
Ammonium sulphate	6254	6202	1846	1800	6839	8567
Urea	5637	7002	2150	2121	6754	8330
No nitrogen	4154		2552		5924	
LSD, Method of application	408		1027		670	

Source. Rice Production Manual.

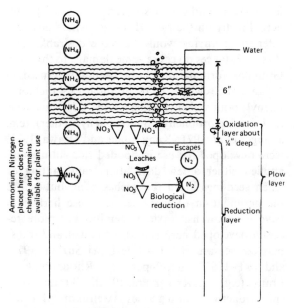

Fig. 5.16 Denitrification cycle
Source. After Mikkelsen *et al.* 1958.

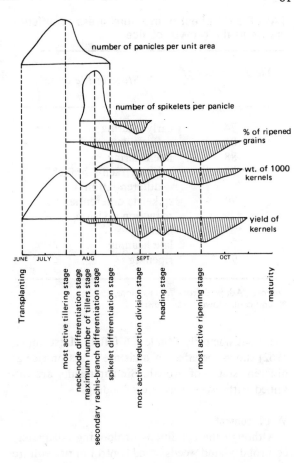

Fig. 5.17 Schematic presentation of yield-determining processes in rice, hatched areas negative, non-hatched positive
Source. After Matsushima, 1964, see text.

The timing of fertilizer application is important in rice, early applications being essential to high yields. Matsushima related yield to the various components and phases of growth and Fig. 5.17 shows the effect of nitrogen on these; clearly N is most needed in the tillering stage. Sometimes there is a second N utilization stage corresponding to the beginning of panicle formation and N top dressing is sometimes recommended at this stage.

Soil analysis is commonly used to assess fertilizer needs in rice, together with plot trials. Angladette (1964) has recently reviewed the assessment of nutritional status by plant analysis.

Irrigation. The quantities of water required to produce a crop of rice have already been indicated. This often exceeds the consumptive use of water in transpiration by a factor of about two.

There are two basic methods of irrigating rice; firstly (the more conventional) is to continuously flood the fields, and the second method is to employ intermittent irrigation keeping the fields near saturation. There is little consistent difference between the two methods.

In rain-fed irrigation little choice is available in regard to depth of flooding but where irrigation facilities are available the best all round depth is often stated to be about 5–10 cm. Comparisons of flooding depths in the range of 10–30 cm by various workers (e.g. Evatt, 1958; Have, 1967) have shown higher yields from shallower flooding. Water is led to the fields by a system of canals, and pumped onto the fields when the water level in the canals is below field level.

Sensitivity of the rice plant to moisture stress and to deep flooding varies with genotype, and in fact some varieties are adapted to growing under dryland conditions. In general, however, particular phases of growth are more sensitive to stress, particularly the stage of reduction division in the developing spikelets, and the early heading phase. The active tillering phase is less sensitive, presumably because compensatory growth can occur after a period of moisture stress or deep flooding which has killed off the weaker tillers. In the ripening phase the water must be drained slowly from the field some weeks before harvest to allow ripening to

TABLE 5.6 Effect of moisture stress at different stages in the growth of rice

Yield, per cent of control	Stage in life cycle
74	early tillering
86	active tillering
88	late tillering
75	primary rachis branch differentiation
79	spikelets differentiation
*29	reduction division
*37	heading
73	late heading
90	ripening

Source. Adapted from Matsushima, 1962.
*Significantly lower than control.

occur satisfactorily. Table 5.6 (after Matsushima, 1962) shows the effect of moisture stress on rice at different stages of growth. Similar results are obtained with excessively deep flooding.

Weed control

Although the conditions of flooding adequately control dry-land weeds, weed control in rice culture presents some special problems in tropical countries in which manual planting (which is necessarily slow) and the use of animal-drawn or small pedestrian cultivating implements are predominant. Thorough field preparation close to the time of planting is the most effective form of weed control. Straight row-planting also facilitates the use of inter-row cultivators. The hand-propelled rotary cultivator is commonly used in rice fields in Asia. The encouragement of the practice of planting in straight rows, however, is necessary in many tropical countries.

For broadcast sowing thorough field preparation is necessary. Grammoxone (paraquat) has been used for pre-killing of weeds without cultivation, apparently with some success, and, in the case of drill sowing, could be applied after planting but before emergence of the seedlings in clayey soils.

Control of broad-leafed weeds is effected by means of the hormonal herbicides 2:4-D, M.C.P.A. and others. 2:4-DP (2:4 dichloro phenoxyproprionic acid) is useful against sedges and Propanil is used as a contact herbicide against grasses. 2:4-D and M.C.P.A. are selective in action against broad-

leafed weeds and can be sprayed at rates of ½–1 lb/acre directly on the rice. However, where possible, inter-row spraying would be more desirable since these substances do cause some damage to the crop. Granular formulations have been much used in Japan and elsewhere. TOK granular (2:4 dichlorophenyl p-nitrophenyl ether) and CNP granular (2:4:6 trichlorophenyl p-nitrophenyl ether) (e.g. Jesinger et al. 1971; Kenji Noda and Ozawa, 1971) were developed for use in flooded rice culture and applied either a few days after transplanting or when seedlings are a few inches high, but before the germination of weed seeds. These herbicides have very low fish toxicity. Similar but more recently developed herbicides for flooded and transplanted rice are HE–314 (see Ippei Suzuki, 1971) and RP–17623, developed by Rhone-Poulene, France (e.g. Yuji Kawamura, 1971). All these herbicides are active against grasses as well as broad-leafed weeds. Various carbamates (e.g. Benthiocarb—see Kenji Noda and Ozawa, 1971) are used both in transplanted and row-sown rice for general weed control, and Chloramben has also recently been used for soil surface treatment in transplanted rice. Details of these various herbicides are given in the first part of this book.

MAJOR DISEASES AND PESTS

'Blast' caused by the fungus *Piricularia oryzae* is a widespread disease of rice causing brown spots on the leaf, stem and sheaths. High relative humidity and N nutrition favour its spread and silica (as discussed above) and potassium impart some degree of resistance to the disease. Being a perfect fungus numerous physiological strains exist and breeding for resistance is often only of local effectiveness.

Chemical measures are also used to control 'blast', Some of these are effective, but frequent rain in humid regions necessitates costly multiple spraying. Treatment of seedlings by dipping in such compounds prior to planting out is sometimes practical, and rogueing out of diseased seedlings is also helpful.

Brown spot or *Helminthosporium* blight caused by *H. oryzae* (*Cochlibolus miyabeanus*) is also a widespread rice disease particularly characteristic of poor soils and K-deficient conditions, and poor water control (deep flooding and drying out). The fungus causes sharply lined spots on the leaves. Better husbandry and resistant varieties effect control.

Helminthosporium sigmoideum (=*Sclerotium oryzae, Leptosphaeria salvinii*) is also widespread, particularly in South-East Asia, where it is almost endemic. It persists throughout the crop and causes reduced yields under poor conditions of husbandry. It produces small black lesions on the leaf sheaths and culm. Burning the straw effectively kills the sclerotia. The use of resistant varieties is the best control measure.

Cercospora leaf spot (*Cercospora oryzae*) produces elongate leaf lesions. It spreads rapidly if good screening for diseased seedlings is not practised. Resistant varieties and hygiene measures are used for control.

Various grain diseases are important, black smut (*Neovossia horrida*) and green smut (*Ustilaginiodea virens*) cause losses of grain. Hygiene measures of destroying by burning the infected panicles control these diseases.

A bacterial leaf streak (*Xanthomonas translucens*) causing interveinal streaks and water-logged appearance of the tissues which later exude a yellowish bacterical ooze, and another disease caused by *X. orzyae* can be serious in rice. They are controlled by good seed-bed hygiene and prevention of deep flooding. Copper and mercury compounds can be used on the nursery.

A number of virus diseases (tungro, yellowish dwarf, orange leaf, and grassy stunt) also occur in rice, being transmitted by leaf hoppers. Resistant varieties apparently exist and control can also be affected by rogueing out diseased plants and controlling the leaf-hoppers.

Various physiological diseases occur. A disease in South-East Asia known as 'penyakit merah' (red disease) has been attributed to nutritional deficiency (Lockard, 1959), although recent evidence suggests that it is due to virus. (Ou *et al.* 1965.) Ponnamperuma (1955) described a 'browning disease' from Ceylon which could be due to a nutritional imbalance. 'Akiochi', caused by sulphide poisoning, has already been mentioned.

The major insect pests of rice are various stem borers which are controlled by insecticides (chiefly BHC and systemic insecticides—see Rice Production Manual, of the International Rice Research Inst., Philippines) and by breeding. The rice bug *Leptocorisa acuta* attacks rice particularly in humid regions. It is controlled by light trapping, good weed control and by insecticides. Leafhoppers are controlled by insecticides.

Sometimes rats and birds are also troublesome in rice cultivation. The former are controlled by regional baiting programmes (chiefly using anticoagulants such as 'Warfarin') and the latter by scares of one kind or another.

6 COFFEE [Coffea spp.]

INTRODUCTION, SPECIES AND VARIETIES

COFFEE has been used as a beverage plant since at least the fifteenth century. There are four definite species of economic interest:

> *Coffea arabica*
> *Coffea canephora*=(*C. robusta*)
> *Coffea liberica*
> *Coffea excelsa*

All are small trees of up to about 20 ft under natural conditions, though marked differences in branching occur. The first two, *arabica* and *robusta*, are the important varieties in the coffee trade, with *arabica* being more favoured for its superior quality. *Arabica* coffee originated probably in the mountain forests of Ethiopia, and there are two varieties: *var. arabica* and *var. bourbon*. The latter is now considered to be of the best quality. The successful cultivation of the *arabicas* is carried out in high elevation sites in both the tropics and the sub-tropics, and the South American countries, of which Brazil is the most important, are the main centres of supply although *arabica* coffee is also an important crop in East Africa and other parts of the world. The humid lowlands of the equatorial regions are unsuitable for *arabica* coffee.

In the humid tropics, the second species, *C. canephora* (=*C. robusta*) is commonly grown (or alternatively hybrids of *C. canephora* and *C. arabica*). The name *robusta* is more widely known and will be used here. This species and its hybrids are widely grown in Asia and West and Central Africa. It is vigorous with good qualities of disease resistance, but the flavour is inferior to *arabica* coffee. *Robusta* coffee originates and is found wild in West African forests.

The two other species, *C. liberica* and *C. excelsa*, are used in hybridization to produce more suitable and disease-resistant types for particular localities. *C. liberica* is grown commercially in West Africa. In Malaya also, *liberica* is commonly grown on acid peat soils. *Excelsa* coffee is adapted to a more arid climate and is grown in North Africa as well as in other parts of the world, and is thought to originate in the Lake Chad area. A little is grown in Malaya. Both are poor quality coffees. Within the *C. arabica* varieties and *C. robusta*, there are numerous cultivars developed and propagated in various parts of the world. *Arabica* coffee is self-fertile and often self-pollinated, but *robustas* are largely out pollinated. Consequently, the variety of forms among *robusta* is much greater.

BOTANY

The vegetative plant

There are a number of different types of branching habit in coffee species and varieties; and further, the way in which pruning is carried out (see further on) also affects the structure of the tree.

Fig. 6.1 The growth characteristics of *Coffea canephora*

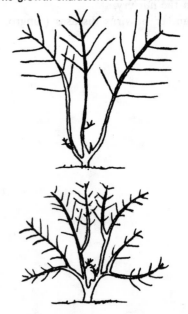

Source. After Haarer, 1963.
Notes. Above: *Var. robusta* trained to grow into a multiple-stemmed tree. Note the stiffer, more upright growth. It normally grows into a high tree. Below: *Var. nganda* assisted to grow into a multiple-stemmed tree. It normally sprawls.

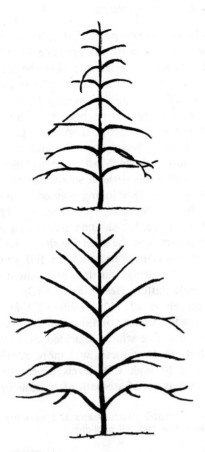

Fig. 6.2 The growth characteristics of *Coffea arabica*
Source. After Haarer, 1963.
Notes. Above: Branching habit of a young tree of *var. arabica*. Note the horizontal growth of the branches which tend to hang downwards. Below: Branching habit of a young tree of *var. bourbon*. Note the more upright branching habit and the tendency of the top of the tree to be flat.

Fig. 6.3 *Liberica* coffee in Malaya; branching similar to *Arabica*

Figs. 6.1 and 6.2 illustrate diagrammatically the basic differences between the two important species, *arabica* and *robusta*. Fig. 6.3 illustrates a stand of *liberica* coffee in Malaya planted from seed. Opposite branching and leaf insertion is characteristic of coffee species. Generally the species are evergreen in nature and leaf-shedding (as occurring in rubber) does not take place. Young leaves of *C. arabica* (*var. arabica*) are of a reddish-brown while others including bourbon are green. Stomata occur predominantly on the lower surface. They close readily indicating an adaptation to drought.(see p. 89 below).

The root system

The root system varies with soil conditions and depth of the soil. Feeding roots are largely super-ficial; deep roots (axial and tap roots) may penetrate up to 15 ft in friable soil. The depth of the rooting zone is important in relation to rainfall and drought (see below). A good description of the root system of *arabica* coffee is given by Fernie (1964). Various types of lateral roots are recognized; surface 'plate roots' which may branch and run parallel to the surface; and intermediate roots that ramify evenly and branch considerably. These are all located in the top two feet of soil; they bear smaller secondaries which bear feeder roots with root hairs.

Flowering

Many coffee varieties are photoperiodic (short-day plants), but in equatorial zones they do not generally show photoperiodic behaviour since the critical day-length seems to be about 13 hours. However, at higher latitude sites (see map Fig 6.5) flower initiation is often confined to the short-day season (see Alvim, 1958).

A seasonal dry spell or cool season is desirable for a good set of flowering buds in coffee (see 'Climate' below). Generally *robustas* will commence flowering and bearing at $2\frac{1}{2}$–3 years, and *arabicas* after 3–4 years. Flowers occur in clusters in the leaf axils of certain branches, generally only on older wood at the best climatic locations. They are scented and insect pollination is important in the out-pollinating *robustas*. Wind pollination in the absence of strong winds is restricted to the radius of one or two trees (Charrier, 1971). In *arabica* varieties out pollination occurs only of the order 7–9 per cent (Fernie, 1964). In regions with a single dry season, only one flowering and bearing period occurs and this is judged to be good in coffee. In regions of bi-modal rainfall and dry seasons, double bearing generally occurs (Wormer and Gituanja, 1970a).

From the blossom to fruit maturity may take between 8 to 11 months in a single-bearing region. The mature berries are bright red and must be harvested promptly to retain quality.

Growth quantities

Studies on the growth of coffee have mostly been concerned with flower and fruit development. At present, improvement of yield by the combination of desirable plant features (particularly height) and physiological efficiency is mainly carried out in Brazil (e.g. Monaco and Carvalho, 1969; Bicudo, 1969), where coffee is a crop of national importance. Flower initiation under field conditions occurs during a dormant season, after which the buds may remain unchanged until the break of the season (Frederico and Maestri, 1970). Conditions affecting assimilation but inhibiting vegetative bud expansion favour flowering (Robinson, 1964; Alink and Wolf, 1964). Removal of shade to allow more photosynthesis (Snoep, 1932; Dierendonck, 1959), and seasonal cold (Haarer, 1962) which selectively inhibits vegetative shoot expansion, promote flowering. (see also 'Climate' below and Salter and Goode, 1967). In biennial flowering, pruning during one period stimulates flowering in the next (Wormer and Gituanja, 1970).

Development of the fruit depends on conditions following the flowering period, including rainfall, sunshine, depth of soil and husbandry. Abortion of young fruits occurs most frequently at the pinhead stage (Huxley and Ismail, 1970). Subsequent to flower initiation, starch reserves accumulate in the flower parts and are mobilized during flower development (Crope et al. 1970; Maestri et al. 1970). Studies of growth of fruits from the pinhead stage by Ramaiah and Vasudeva (1969), indicated phasic development (bi-sigmoidal curves) in arabica and linear growth in robusta, possibly the former being due to environmental fluctuations. Fruit development has been promoted by 2:4-D and 2:4:5-T sprays at 10 ppm which reduced berry drop, presumably by increasing the sink activity of the fruits (Ann. Tech. Rep. Coffee Board Res. Dep. 1969–70, p. 87, 1971).

Cannell (1971a) reports that total net assimilation was increased by the presence of developing fruits. During a vegetative period the dry matter assimilated went to the leaves (61 per cent), new roots (10 per cent) and the remainder to the trunk and roots. During fruiting, both leaf and root increments were greatly reduced. Assimilation appears to be a limiting factor in coffee and is related to the question of shade. Most modern coffee production is carried out without shade, although reduction in stomatal functioning may occur in fully-exposed foliage (Wormer, 1965, Fig. 6.4). However, in mature coffee, self-shading occurs to a sufficient extent to eliminate the need for shade. Rather different cultivation is necessary in full sunlight to prevent overbearing and the development of a biennial yield pattern (see Haarer, 1963).

Bierhuizen et al. (1969) showed that stomata closed readily under conditions of moisture stress in arabica coffee which indicated drought resistance, but photosynthesis was more adversely affected due to high CO_2 levels developed in the leaves. Recovery of stomata after moisture stress

Fig. 6.4 Stomatal aperture and soil moisture level in coffee in shade and full sunlight

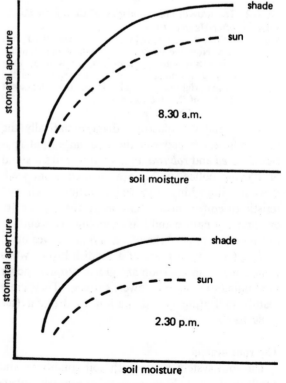

Source. After Wormer, 1965.

took at least five days. Nunes *et al.* (1969) report that light saturation was achieved at the rather low level of 0.11 cal/cm²/min (rather less than 10 per cent full sunlight) and net photosynthesis was highest at 20°C, decreasing above this in *arabica* coffee.

The situation in relation to foliage illumination appears similar to cocoa (Chapter 7). The correct illumination and presence of foliage is important for high berry yields (Cannell, 1969, 1971b; Phillips 1970; Janardhan *et al.* 1971). Cannell and Huxley (1970) showed that any fully-expanded leaf on the same branch or its shoots is active in growth of berries. The systems of pruning to develop the canopy and the bearing frame of the bush (see further on) have been questioned by Huxley (1970) and Huxley and Cannell (1970), particularly the multiple-stem system which gives incomplete ground cover, poor distribution of light in the canopy and a difficult-to-control leaf/fruit ratio. They suggest that a hedge-row system combined with pruning to develop a suitable leaf/crop ratio may be more satisfactory.

It seems likely that different genotypes will exist with better assimilation properties, a greater response to high light values and better foliage orientation as has been found in other crops (e.g. tea, Chapter 8) which may account for physiological superiority observed in some clones already known (e.g. Bicudo, 1969).

Osorio and Castillo (1969) showed that initial growth of germinated seedlings was related to seed weight, but these differences did not persist.

Stem girth (as with cocoa, Chapter 7) has been shown to be correlated with yield in both *robusta* and *arabica* coffee in India (Srinivasan, 1969).

CLIMATE AND SOILS
Temperature and moisture

The *arabica* coffee varieties are definitely adapted to a cooler climate than that of the equatorial lowlands, and the origin suggests they are most ideally suited to high elevation sites within the tropics which are free of frost; particularly for the favoured *arabica* types. Fig. 6.5 (after Haarer, 1962) indicates the main coffee-growing areas of the world. The average temperatures of the *arabica* regions are in the broad range 13°C to 27°C and according to Wellman (1961) the best average temperature for *arabica* is around 24°C. An elevation of 1,500 ft is often thought to be a lower limit for this species in equatorial regions. Nunes *et al.* (1969), in a laboratory study of both *arabicas* and *robustas* found net photosynthesis was highest at 20°C; it seems likely that reduced assimilation in fully-exposed leaves of coffee might (as with cocoa and tea) be a result of excessive leaf temperatures. Die-back of coffee shoots in hot weather has been attributed to lack of shade (D'Souza, 1971). Higher temperatures force rapid growth and early bearing in coffee, and premature exhaustion of the tree (Dierendonck, 1959). The size of the trunk and the accumulation of reserves are reduced. At cooler temperatures, flowering is confined to old wood and overbearing is not a problem. The following

Fig. 6.5 The coffee-growing countries of the world

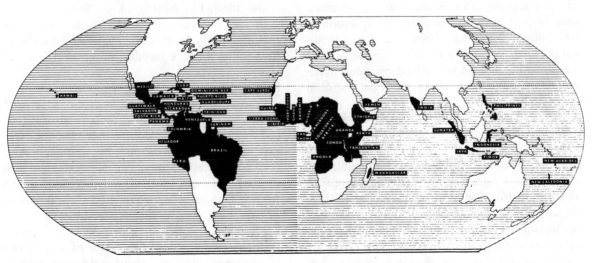

Source. After Haarer, 1962.

summary is from Salter and Goode (1967):

Cool moist climate—vigorous growth, flowers on old wood, no flowers on young wood.

Warmer drier climate—moderate growth, flowers on both old and young wood.

Hot and dry—poor growth, defoliation on older wood, flowers on young wood only.

Hot and with water available—very vigorous growth, abundant flowers.

The success of *arabica* coffee at high elevations in the tropics is also attributed to the generally favourable rainfall and atmospheric humidity of such regions.

Other species of coffee are more adapted to lowland conditions. In the humid lowlands, vegetative proliferation may limit flowering and require the pruning of vegetative buds to obtain adequate fruit set.

The best rainfall regime for coffee is one of even distribution over the major part of the year with a short dry season of one or two months in which rainfall is less than 2″ to facilitate flower initiation. Dagg (1971) estimated water use by conventionally spaced coffee in Kenya as 0.57E unirrigated and 0.69E under irrigation which indicates some ability to survive moisture stress. In terms of rainfall this amounts to an optimal rainfall of about 50″/annum, adequatley distributed. Closer spacing, as is the modern trend, may require more rainfall and/or irrigation. Some of the most successful coffee areas (as in Kenya) receive only about 40″ per year, and success here is linked to the deep-profiled, friable soils which permit minimal runoff and good storage and to the common practice of mulching to conserve soil moisture. In such marginal rainfall areas, great benefits have been achieved by irrigation in experimental trials (e.g. Alink and Wolf, 1964). Boyer (1969) in a study of soil moisture showed that growth could be maintained provided available water in the upper soil horizons did not drop below 25 per cent.

The effect of a periodic dry season or, alternatively, a cold (but above freezing) period which, as with many tropical crop species, favours flowering in coffee, is related to obtaining a favourable balance between vegetative and reproductive growth and the distribution of assimilates. Related to the availability of assimilates, conditions affecting photosynthesis such as moisture availability, sunlight and shading also affect flowering. Moisture stress promotes flowering because it inhibits vegetative growth and makes available more assimilates for both flower and fruit development (e.g. Robinson, 1964), and in cases where moisture supply such as in irrigation promotes flowering and fruit set, this is presumably because it has allowed greater photosynthesis and assimilation (e.g. Alink and Wolf, 1964).

Studies conducted in Kenya (Cannell, 1971c) indicate that moisture supply is most critical during a rapid phase of berry expansion (see Ramaiah and Vasudeva, *loc. cit.*), about 10 to 17 weeks after flowering. This is also when the best benefit from irrigation may be expected. In areas of bi-modal rainfall, as in Kenya, the pruning of bearing branches should be undertaken to encourage fruiting during the most favourable season; that in which the most rainfall occurs, and to avoid biennial bearing (Cannell, 1971d). Changing of the main bearing period by pruning has been discussed by Wormer and Gituanja (1970b).

Heavy rainfall during harvesting is not desirable (Cleves, 1970).

Frost and wind

Coffee production is severely limited by frost (Chousay, 1963), although coffee is grown in sites in which frost occurs in South America. Only slight frosts, however, are tolerated and severe losses have been experienced in Brazil following occasional heavy frosts. Frost damage in Brazil, for example, was estimated to reduce 1970/71 coffee production by about 65 per cent. About 20 per cent of producing trees were so severely damaged that they were expected to give no harvest at all. Quality is also affected by frost (see List, 1969).

Coffee is sensitive to strong winds and it is a common estate practice to retain wind breaks of trees in fields (see Vijaendraswamy, 1971).

Soils and drainage

Coffee is grown on a great variety of soils from deeply-weathered tropical latosols to coastal peat and swamp soils. The overall requirements, however, are that the soil should be reasonably deep (1–2 m) so that the roots can penetrate, and well drained. On both lowland and peat soils in the tropics, *arabica* coffee is not grown. With *arabica* in particular, good drainage is essential.

Much of the coffee of Central America is grown on rich volcanic soils. The coffee of East Africa is produced on deep well-structured but intrinsically infertile latosols. Soil requirements are closely tied up with climate, a poorer or shallower soil being satisfactory in regions of good even rainfall, and a superior soil being required in regions of marginal rainfall. Coffee has high nutrient requirements so that soils low in nutrients, as it is with many of the inland tropical soils, require fertilizer applications.

Soil reaction for coffee should be in the range moderately acid to near neutral. Amorim *et al.* (1968) showed that *arabica* seedlings did best in the range pH4–6. At high pH, roots showed grey discoloration, foliage was chlorotic and leaf levels of nutrients decreased. Poor performance at high pH may have been attributable to iron deficiency.

PLANTING
Land preparation, spacing and yield

Land preparation for coffee has been considered in a number of excellent publications (e.g. Haarer, 1962; Mitchell, 1964). Most handbooks, including the above, recommend jungle clearing by felling and burning, with prior killing of the trees if this is necessary to effect a good burn. However (Deuss, 1969) indicated that long-term yield and condition of coffee were better on land cleared without burning, though initial problems in establishment were more difficult. Stump removal and prior cultivation of the land are also recommended. If it is considered necessary that the coffee should be shaded, suitable jungle trees may be left. Leguminous species with a spreading canopy are most desirable.

Conventional field spacing for coffee varies from about $7' \times 7'$ for compact *arabica* plants at high elevations, to $10' \times 10'$ for *robusta* varieties and up to $12'$ for the larger *liberica* and *excelsa* species. Closer spacing will give better early yields but wide spacing will give better prolonged yields. An example of a spacing trial of *arabica* is given by Tremlett (1952). Treatments were $8' \times 7'$, $7' \times 7'$, $6' \times 6'$, and $4' \times 4'$. The closer spacing was superior up to the first 12 years but by 16 years the $6' \times 6'$ spacing had caught up and become superior to the others. It is possible to extrapolate and infer that a yet wider spacing would be superior for a long plantation life. The technique of thinning from say a $4' \times 4'$ spacing to an $8' \times 8'$ spacing by removing alternate trees is worth considering, although the added yield must be weighed against the additional work and planting costs.

In Sao Paulo Brazil, traditional spacing has been of the order 4×2.5 m (*c.* $13 \times 8'$—Ettori, 1970) which is now considered much too wide; yields in the 3rd, 4th and 5th years in one trial were 480, 900, and 1900 kg/ha made coffee respectively, compared with 1200, 2460 and 4200 kg/ha at a spacing of 2.5×1.5 m (*c.* $8' \times 4.8'$). Venezuelan coffee is traditionally grown at *c.* 3 m spacing with 1 or 2 plants per point; improved yields have been obtained by 3×1 m spacing (Garcia, 1969). World average yields are, however, much lower than the figures given above. Deuss (1971) reports yields of smallholder coffee as 160 kg under poor conditions of husbandry and 500 kg/ha under improved conditions.

Nursery and seed production

In *arabica* coffee varieties which are usually self-pollinated, seedlings may be obtained from beneath superior trees growing in the field. In other varieties of coffee, open pollination is more usual, so that progeny collected in this way may be genetically very variable and therefore unsuitable for planting. Consequently, the seed of other species, as well as *arabica* seed in many cases, is specially produced by crossing appropriate parent trees. The supply of superior coffee seeds is often undertaken by government agencies. Storage of seeds for periods of up to a year can be achieved in cannisters at high relative humidity (Bouharmont, 1971).

In producing seedlings for planting it is usual to prepare nurseries for the proper care of the seedlings until they are considered large enough to plant out to the field. It is generally considered cheaper to maintain seedlings in a nursery than in the field and this is the primary reason for establishing nurseries. Even so, in Brazil, the planting of seeds directly in the field is commonly practised and special techniques have been developed for this purpose (see Wellman, 1961). Pits about 18" deep are dug at the planting points and some of the surrounding top-soil is placed in them. Groups of seeds are planted at each corner of the pit and the best-selected seedling after a certain stage is left to reach maturity. The seedlings in the pit are protected by arranging a pyramid of sticks over each pit. For this technique, a friable soil is required to prevent waterlogging.

Frequently coffee is planted to the field using seedlings prepared in a nursery, and there are usually two phases of the nursery stage. First the

TABLE 6.1   Effect of transplanting method on subsequent growth of arabica seedlings in the field

		Out of 21 trees in each treatment				
		Died		2nd year		Maiden
Age	Treatment	1st year	2nd year	Height (inches)	Trees bearing	yield (quarts/tree)
Young	Ball-of-earth	2	4	62	6	14.8
,,	Bare, cut back	6	9	65	4	12.7
,,	Bare, not cut	3	3	61	5	16.9
Over 1 year	Large earth ball	0	0	72	5	17.0
,,	Bare, cut back	1	4	64	0	8.4
,,	Bare, cut back	1	2	58	0	8.4

seeds are germinated in a pre-nursery bed, in rows a few inches apart. A planting depth of about 1″ in the pre-nursery bed is satisfactory. Beds are usually of about 4-foot width, which is suitable for weeding, etc. with paths between them, and of any convenient length. A good soil mixture is used or in some cases clean sand, and the beds are raised to facilitate drainage. Sometimes soil fumigants such as methylbromide are used to combat pests and diseases in the pre-nursery beds. Such beds are usually shaded but this is not necessary with adequate watering, or in a wet season. Germination takes place quicker in warm climates, the range of germination time being from approximately 3 to about 8 weeks. Removal of the outer pericarp ('pergamino') hastens germination. Sometimes a mulch of dry grass or other material is used over the nursery beds.

Transplanting from the pre-nursery to the main nursery is carried out at the cotyledonary stage or even earlier. Traditionally nurseries have been established in areas of flat ground free of flooding, a soil of medium clayey consistency being preferred. In a ground nursery of this type 8 to 10 inch triangular spacing is used and transplanting is carried out by digging out 'balls-of-earth' surrounding each of the seedling roots (a clayey soil aids in preparing these ball-of-earth seedlings). Alternatively they are dug out at transplanting time as 'bare root' seedlings. The survival and growth of ball-of-earth seedlings is better particularly when transplanting at a late stage, as is indicated by Table 6.1 from McClelland (1917).

In recent years, polythene bags have been increasingly used for raising coffee seedlings. Seeds are direct-planted to bags of size about 8″ × 10″ laid flat, filled with a good quality soil. Such container seedlings grew 66–162 per cent better in one trial (Vine and Mitchell, 1969).

Coffee seedlings should be well fertilized in the nursery with a complete fertilizer. Nitrogen is an important fertilizer in such plants (Carvajal et al. 1969).

Transplanting should be done in a season with adequate moisture, and an age of six months is generally considered satisfactory for the above spacing or size of polythene bag.

Cuttings have not been successfully used in arabica and robusta coffee planting. Rooting is difficult and they must be taken from orthotropic shoots.

Graftage has been used on some occasions where a resistant or well-adapted root stock is required. In Guatamala, for example, graftage of arabica coffee on nematode-resistant and well-adapted robusta stocks has been a success (Gutiérrez and Jiménez, 1970).

Manuring and irrigation

Coffee is generally considered to be a fairly demanding crop in terms of fertilizer requirements and lack of response to fertilizers in some cases may be attributed to the frequent use of good soils in coffee growing. Nutrient removal by a crop of 1,000 lb/acre made coffee is of the order:

N—34 kg P_2O_5—6.1 kg and K_2O—51 kg

Much of the K is contained in the pericarp of the berry and can be returned to the soil.

Nitrogen is the most important nutrient from the point of view of yield, and overbearing is associated with depletion of N (e.g. Benac, 1969; Robinson, 1969). Coussement et al. (1970) recommend a ratio of 3:0.5:1 for N:P:K for arabica coffee in

Burundi, at a rate of about 700g per tree. Generally recommended rates are, however, lower than this.

The type of N fertilizer has been the subject of much investigation. Chilean nitrate ($NaNO_3$) is not a satisfactory source of N for coffee because of the high Na level (Subramaniam *et al.* 1969; Espinosa, 1970). Subramaniam *et al.* (1969) report that both calcium ammonium nitrate and urea raised yields. The best responses are generally reported from ammonium sulphate (e.g. Espinosa, 1969). Robinson (1964) recommends ammonium sulphate in soils of pH above 6.5; ammonium sulphate nitrate at pH 6.5–5.2 and calcium ammonium nitrate in soil of pH less than 5.2.

The degree of shade and exposure strongly affect the response of coffee to fertilizers. In mature shaded coffee most response has been obtained from a combination of N and K. Phosphorus is important in the early stages of growth but in mature stages there are fewer reports of a marked response to P fertilizer. The importance of K seems to be related to obtaining a favourable balance of N and K, the former often being present in the soil at relatively high levels under shade conditions, both because organic matter in the soil is maintained under these conditions, and because the shade trees commonly used also fix nitrogen. N and K responses have been reported by numerous workers (e.g. McClelland, 1926; Coste, 1956, etc.). When balanced nutrition is achieved, however, the response of shaded coffee to chemical fertilizer is relatively low. Some typical results of the effects of N, P and K on

TABLE 6.2   Results of manurial trial on shaded coffee at Balehonnur

N: 22 kg/ha	Yields kg ripe cherries per hectare				
	N	P₂O₅: 33 kg/ha	NK	K₂O: 44 kg/ha	CONTROL
	N	*NP*	*NK*	*NPK*	*CONTROL*
1941	870	729	1010	1010	608
1942	1379	1328	1468	1496	814
1943	66	98	124	93	47
1944	247	264	336	290	224
1945	1270	1245	1671	1918	809
1946	440	584	365	545	516
1947	515	767	785	767	646
1948	309	336	336	280	299
1949	497	720	531	837	741
Average	629	671	737	816	528
Control %	119	127	140	154	100

TABLE 6.3   Yields of two *Arabica* coffee varieties in kg ripe cherries as influenced by shade and fertilizer

Fertilizers first year: 100 g superphosphate+100 g amm. sulphate per plant
second year: 250 g 10:5:20/plant/half year
third year: 500 g 10:5:20/plant/half year

	Bourbon			Typica		
	+shade	−shade		+shade	−shade	
−Fertilizer	176	449		163	250	1038
+Fertilizer	223	666		162	464	1515
	339	1116		325	715	

+shade   664 ⎫
−shade   1831 ⎬ shade effect

shaded coffee are given by van Dierendonck (1959) and a sample is shown in the Table 6.2. The most striking yield differences occur from year to year, which are probably due to variations in climate.

Fertilizer responses of unshaded coffee are often much higher, and under conditions where soil fertility is maintained, absolute yields are also often considerably higher. An example of the effects of fertilizer on shaded and unshaded coffee is given by Triana (1957), shown as in Table 6.3

Deficiencies of micro-nutrients such as iron, boron and zinc have been reported in coffee (e.g. Franco and Mardes, 1953; Gallo *et al.* 1970), which have been treated by foliar spray. Iron, however, is apparently best supplied to the soil in the chelated form. Manganese toxicity, in manganese-rich soils, has also been suspected in coffee. Ananth and Hanumantha Rao (1970) report that foliar applications of Zn oxysulphate at 0.2 per cent and ZnSO at 0.25 per cent gave satisfactory responses. Copper will often be supplied in fungicidal preparations used to control coffee diseases.

Foliage composition is often used to determine coffee nutrient requirements. Table 6.4 (after Dierendonck, 1959) indicates the range of mineral composition of coffee leaves.

TABLE 6.4   Mineral composition of coffee leaves (% dry matter)

	Normal	*Range of Variation*
N	2.4–3.0	1.9–4.0
P_2O_5	0.2–0.3	0.1–0.45
K_2O	2.3–2.8	0.3–4.0
CaO	1.5	0.5–2.0
MgO	0.5	0.2–0.9

Values for other nutrients, particularly micro-nutrients, are not clearly established. (See also Carvajal, 1969; Bould *et al.* 1971 etc.) Colonna (1970) reports that samples of two leaves per tree from forty trees taken at random give sufficient information on the nutritional status of a homogeneous planting in *excelsa* coffee.

Fig. 6.6   Single-stem pruning of *Arabica* coffee

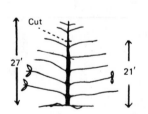

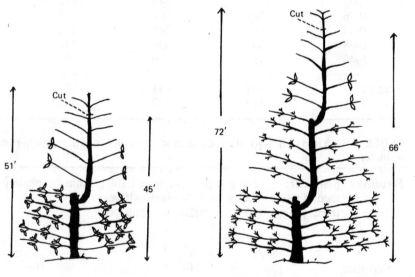

STAGE 1   When tree reaches height of 27″ cap at 21″.

STAGE 2   One sucker allowed to develop from below first capping. When it attains height of 51″ cap back to 45″.

STAGE 3   One sucker allowed to develop from below second capping. When it attains height of 72″ cap back to 66″.

Source. After Fernie, 1964.

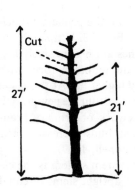

Cut

27′

21′

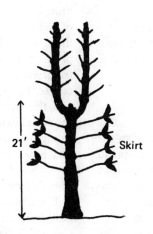

21′

Skirt

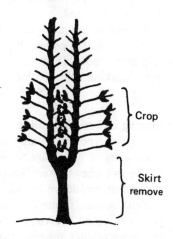

Crop

Skirt remove

STAGE 1 When tree reaches height of 27″, cap at 21″.

STAGE 2 Two stems developing from top node–branches immediately above them removed. Small crop on skirt.

STAGE 3 Further development of two stems: skirt removed and crop on lower branches.

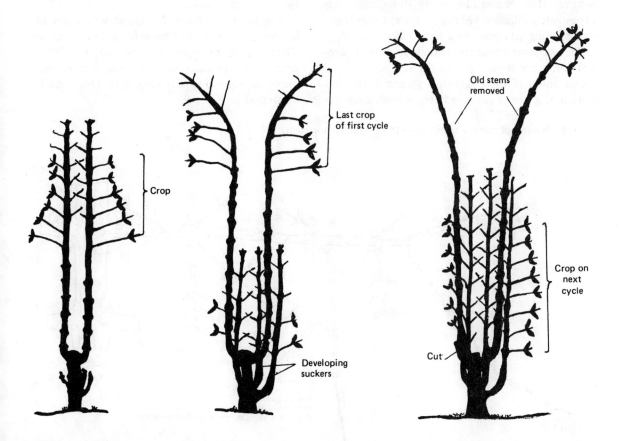

Crop

Last crop of first cycle

Developing suckers

Old stems removed

Crop on next cycle

Cut

STAGE 4 The optimum–bearing stage.

Fig. 6.7 Multiple-stem pruning of *Arabica* coffee
Source, After Fernie, 1964.

The best time to apply nutrients is at the beginning of the rainy season or when the dry season is expected to end.

Because of practical difficulties, irrigation is not often practised in coffee, though considerable yield response is possible in regions of marginal rainfall (e.g. Alink and Wolf, 1964).

Pruning and tree formation

Various types of pruning are used in coffee, the main aims of which are to limit indefinite spread of the branches to keep harvesting lines open, to remove unproductive branches, to produce a convenient bearing frame for harvesting and to limit production and prevent overbearing. A review of the conventional pruning techniques is given by Fernie (1970). The most usual method at present is probably the 'multiple-stem' method with which a 2.7×2.7 m spacing is desirable. Recent evidence, however, (Huxley and Cannell, 1970) indicates that a completely different hedge-row system, combined with pruning may produce a more efficient canopy.

The two most common methods of pruning *arabica* coffee are shown in Figs 6.6 and 6.7.

The object of the single-stem system is the formation of a tree of restricted height, with a permanent frame of primary branches. Once established, the crop is borne on both secondary and tertiary branches arising from the primaries. Maintenance of the primaries is of central importance so that these may last as long as possible. The type of pruning of shoots arising from the primary branch framework is illustrated in Fig. 6.8. Ultimately, of course, the primary branches are progressively lost and the tree will assume an umbrella form with the crop borne increasingly higher up. At this stage stumping may be desirable to encourage new growth and build up a new framework of primary branches. Notching of dormant buds increases their growth from the stump (Cannell, 1969).

The object of the multiple-stem system is the establishment of a number of upright stems of equal size which bear the crop, and the cutting away of old ones as they become unproductive and replacing them with new shoots originating as suckers. Generally one upright stem will bear crops for two or more seasons.

Another type of tree formation which has similarities to a method of bush formation in tea (see Chapter 8) is the 'Agobiado' method. This is much used in the production of *robusta* coffee, particularly the *nganda* variety (Fig. 6.1). This method is illustrated in Fig. 6.9.

Fig. 6.8 Pruning of primary branches in the single-stem system

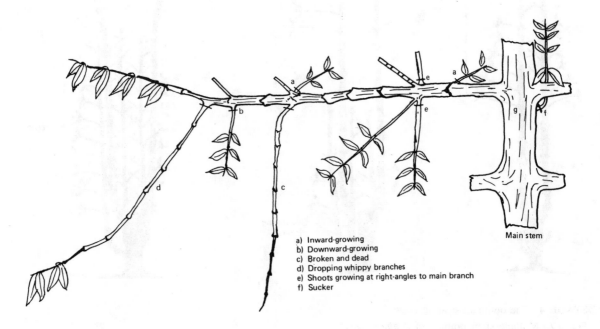

a) Inward-growing
b) Downward-growing
c) Broken and dead
d) Dropping whippy branches
e) Shoots growing at right-angles to main branch
f) Sucker

Main stem

Source. After Fernie, 1964.

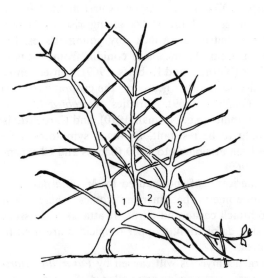

Fig. 6.9 The 'Agobiado' method of pruning coffee
Source. After Haarer, 1963.
Note. Three strong-growing uprights are encouraged to grow; the laterals from these will form the main permanents.

Harvesting

Harvesting is one of the most labour intensive tasks in coffee growing. The traditional method is to pick periodically or when necessary, collecting loose fruit from the ground at the same time. Recently a technique has been developed of harvesting fallen fruits by means of a plastic net spread over the ground (Silva *et al*. 1969). Motorized harvesting machines that shake the branches have also been investigated (Phillips and Rest, 1969). An artificial ripening agent (2-chloroethylphosphoric acid) which releases ethylene has been used to accelerate ripening.

Field maintenance and weed control

It has long been established that grass weeds are particularly bad for coffee (e.g. Pereira and Jones, 1950), probably because of the known effect of grass in immobilizing nitrogen. Clean tillage, however, has bad effects on the soil and often causes subsequent yield decline (e.g. Gilbert, 1938; Martin, 1936, etc.). The growth of legume covers has been recommended between the rows of coffee, but here competition for water seriously affects production in dry climates and in dry seasons unless the covers are controlled. The following Table 6.5 illustrates the effects of different cultural treatments on coffee in Uganda.

TABLE 6.5 Effects of cultural treatments on production of coffee in Uganda (1932)

	1930	1932	1931	Mean
Clean-weeded	101	113	83	100
Green manure, 2 covers annually	98	89	106	98
Banana-leaf mulch	90	110	122	107
Ordinary weed cover	109	72	62	81
Legume cover in wet season (cut back in dry season)	102	114	128	115

Lavabre (1971) reports that effective ground covers in West Africa are *Fleminga congesta*, *Leucaena leucocephala* (=*L. glauca*), *Pueraria javanica*, *Mimosa invisa*, *Stylosanthes gracilis*. Balasubramaniam (1970) also reports that *Mimosa invisa* has desirable characteristics for coffee ground cover. It has procumbent growth and shallow roots, provides good coverage and reduces weeding cost.

More recently, mulching has been recommended for coffee growing, using heavy-yielding straw grasses as mulch material. It seems likely that mulching is a superior field treatment, particularly in situations where moisture stress is likely to affect production to any great extent. Vincente-Chandler *et al*. (1969) report that mulching with coffee pulp increased yields in an area with a marked dry season.

Yield increases in unshaded coffee could probably be stimulated with proper erosion control and soil care and in this connexion, the effect of mulch on unshaded coffee should be noted. The dramatic effect of mulch on unshaded coffee yields are shown by Table 6.6 from van Dierendonck. Similar results have been obtained by other workers (e.g. Pereira and Jones, 1954).

TABLE 6.6 Effect of mulch on the yield of coffee

	Bourbon	Nova Mundo
Light mulch 37 tons/ha guinea grass	1521	1188
Medium mulch 77 tons/ha	1532	1405
Heavy mulch 120 tons/ha	1765	1506
Control	1309	1013

Chemical weed control is widely used in coffee at the present time. Aminotriazole and dalapon have been used to control *Pennisetum clandestinum* in Africa (Gestin, 1971). Grammoxone and simazine have been used in India. Mitchell (1969) reports that diuron, fluometuron, linuron, simazine and atrazine can be safely used in coffee.

The herbicide 2:4:5-T ester at 700 g/litre has been used to kill shade trees (*L. glauca*) which have become overgrown or have re-seeded (Snoeck, 1971) by painting the lower trunks.

MAJOR DISEASES

By far the most important disease in coffee is caused by the rust *Hemileia vastatrix*, which is distributed throughout Africa and Asia, including Papua-New Guinea. Certain coffee genes offer resistance to this rust (e.g. Goujon, 1971). Vishveshwara and Govindarajan (1970) report, for example, that a natural *C. arabica* × *C. canephora* hybrid in India is known for its resistance to all physiological races of coffee rust. In Brazil, a breeding programme is underway to combat the possible spread of the disease (e.g. Bettencourt and Carvalho, 1968). The disease is evidenced by lesions on the lower surface of the leaves giving off orange-coloured spores. On the upper side, the lesion appears as a translucent 'greasy' spot, becoming darker with age and finally turning brown. Nutman and Roberts (1971) report that the spread of the disease is mainly through uredospores which are dispersed by water. Apart from breeding, control of rust is based on copper sprays (Mulinge, 1971), and quarantine measures. Recently Rayner (1956) has advocated periodic prophylactic spraying for the control of leaf diseases in coffee, but this has been thought to adversely affect the biotic flora of the coffee plantation.

Another coffee disease which has assumed epidemic proportions, particularly in Africa, is coffee-berry disease (CBD) caused by *Colletotrichum coffeanum var. virulans*. It attacks *arabica* coffee varieties, and particularly under wet conditions (e.g. Woodhead, 1969) and conditions of cold (Mu-thappa, 1970). It often appears first at the base of the berries and the stalks giving rise to brown patches, but it can attack any young tissues. Infection occurs by means of conidia from diseased regions (Nutman and Roberts, 1969); consequently the disease is often geared to the cropping pattern (e.g. Griffiths, 1969). Varieties of *arabica* differ in susceptibility (e.g. Nutman, 1970), and *C. robusta* is thought to be resistant. Die-back symptoms have also been attributed to CBD (Saccas and Charpentier, 1969).

American leaf spot caused by *Mycena flavida* is troublesome in the American tropics, where it will also attack cacao and citrus. It attacks leaf tissues causing defoliation. Copper fungicides are used in control.

Tracheomycosis wilt caused by *Fusarium xylarioides* causes yellowing and collapse of infected trees of both *arabica* and *robusta* varieties of coffee, particularly under dry conditions and without shade. Infection is often *via* the root system. Mulching and shade are used in its control.

Various root diseases affect coffee, often arising from old shade trees which form centres of infection. *Armillaria mellea* is of widespread occurrence; it causes foliage wilting and yellowing, followed by death of the tree. *Rosellinia bunodes* is another root-rot fungus which occurs on newly-planted forest land; it causes a collar rot. Other root rots of minor importance also occur.

'Pink' disease caused by *Corticium salmonicolor* has also been reported in coffee (see also 'Rubber'). It is controlled by pruning diseased branches and with copper sprays. A coffee-bark disease caused by *Fusarium stilboides* has been reported by Baker (1970). It is controlled by removing and burning affected trees.

A recent review of coffee diseases is given by Wellman (1961). Coffee does not suffer from insect pests that cause widespread or catastrophic losses. Thrips cause defoliation, sometimes quite seriously, and various bugs (*Antestiopsis spp.*) attack fruit. They are controlled by insecticides.

7 COCOA [Theobroma cacao]

INTRODUCTION, VARIETIES

Cocoa (*Theobroma cacao* L.) in its natural habitat is an understory tree of the South American and Central American jungles in regions of high rainfall and temperature. It has apparently been cultivated or gathered from ancient times (see Baker, 1961) and entered into trade extensively with the opening of the Americas. Trinidad became a cocoa producer following the introduction of the tree in 1525 but until 1830, Venezuela was the major cocoa producer. Subsequently other countries in South and Central America became important producers. Cocoa reached West Africa in 1879 when a distinct type from Brazil or Surinam was introduced via Fernando Po, and subsequently Ghana and Nigeria became major producers. Cocoa also reached Malaya, Sumatra, Ceylon, Madagascar and Samoa in early times but the types grown in these countries today are of a different origin from the original introductions, being more vigorous and hardy. Cocoa has also been introduced to New Guinea where it is an important crop. Basically two groups of cocoa varieties with their origins respectively in Central and South America can be recognized. The first is the 'Criollo' type which is of superior quality, possessing generally elongated, strongly-ridged, red or yellow pods with a relatively thin woody skin, and plump, pale-coloured seeds. The second is the Amazonian 'Forasteros' type of cocoa which possesses greater hardiness, vigour and yield and was traditionally grown in Brazil and West Africa. The 'Forasteros' possess thick-walled rounder pods, yellow when ripe and with flattened, violet-coloured seeds. The quality of the Forastero cocoas is distinctly lower than that of the Criollo but in present times these are widely grown because of their hardiness and high yield. Amongst the former, however, the Amelonado cocoas of West Africa and the Bahia region of Brazil are particularly suitable for the manufacture of milk chocolate. The particular Amelonado cocoa of West Africa is considered to be very uniform genetically.

A third group of cocoa varieties, sometimes considered to have varietal status, is the 'Trinitarios' from Trinidad. This, however, is a complex group composed of genes from the original Spanish Criollos introduced there and subsequent introductions from eastern Venezuela and elsewhere which have Amelonado features. The Trinitarios exhibit the characteristics of a hybrid population and are probably the richest source of material for the improvement of cocoa by breeding. Early Trinidad types introduced to Ghana were hybridized with local Amelonados and are now considered to breed true.

More recent additions to the gene pool of cocoa were certain upper Amazon types which are very vigorous (collected by Dr. F.J. Pound, see Bartley, 1968). They are small-seeded but this character has been improved in breeding programmes. The upper Amazon types are self-incompatible and therefore thought to be hybrids.

The chromosome number of cocoa is $2n=24$ and generally they are considered to represent only one species. Incompatibility occurs in cocoa (*vis.* the upper Amazon types mentioned above) and some trees will not set fruit with their own pollen. In some clones this can be overcome by hand pollination, which is important for producing in-bred lines for breeding purposes. Mixing in a little pollen from compatible strains carrying a recognizable marker gene increases fertilization between incompatible types in some cases (e.g. Glendenning, 1967). Pollen can be stored by freeze-drying (Cabanilla and King, 1966). Reviews concerning cocoa varieties are by Baker *loc. cit.*, Wood, 1963; Soria, 1970 and others.

BOTANY
The vegetative plant

Cocoa is a small tree, reaching a height of about 30 ft if unpruned. The habit varies somewhat but a very characteristic feature is the mode of branching. A seedling produces a straight main stem to

Fig. 7.1a   Cocoa seedling showing fan branching following cessation of growth of the main stem

a height of about 3 to 5 ft which then ceases to
grow, branching to give three to five, usually evenly-
spaced side branches known in the cocoa world as
a 'fan' or 'jorquette' (Fig. 7.1). These fan branches
are more or less horizontally orientated. The terminal bud is usually determinate and ceases development after formation of the 'fan'. Continued
growth in height takes place by the formation of
a side shoot, usually on the stem beneath the fan
branches, which is called a 'chupon' and forms a
new main stem, again for several feet in height before it forms a second series of fan branches. If unpruned this form of growth may continue to give
four or more series of chupons and fan branches,
but in estate practice the tree is usually pruned to
at the first or second layer (see 'Planting' below).
The fan branches of the lower story become enlarged and woody and these, along with the enlarged chupon main trunk, form the main fruit-
bearing surfaces of the tree (see below). Generally
chupon and fan branches remain distinct, the former giving rise only to branches of the chupon type
except for the terminal determinate fan layer, and
the latter giving rise mainly to further fan branches,
only occasionally to a branch of the chupon type.
This has significance in establishing trees from cuttings which are usually obtained from fan branches.
There are some cocoa varieties which lack the chupon type of branch habit and possess only fan-

Fig. 7.1b   Young cocoa trees under coconuts, pruned at the first jorquette or fan

type branching. The leaf arrangement differs on the two types of branches being a $\frac{3}{8}$ arrangement on the chupons, and alternate on the fan branches. Leaves on the fan branches are produced periodically in 'flushes' (see 'Climate' below) at successive nodes. The length of internodes increases with age and then decreases as a branch matures. Leaves persist for about two flushes.

The root system

Seedling cocoa (grown from seed) produces a central tap root and ramifying branch roots. When grown from cuttings only adventitious-type roots are produced; but since cocoa is a small-statured tree it does not suffer from wind damage to the same extent as for example, rubber. Lateral roots are largely superficial in the top 6 inches of soil and the tap-root seldom penetrates more than about 5 ft (McKelvie, 1962).

Flowers and fruit

The flowers of cocoa are formed on the mature wood on 'cushions' which are morphologically compacted side branches (Lent, 1966), each flower being formed in the position which would have been a leaf axil on a normal branch (Fig. 7.2). Flowers occur on the wood of both chupon and fan branches that have reached an adequate size. Wastage of flowers through inadequate pollination is a feature of cocoa. The construction of the flower is designed to prevent self-pollination (Fig. 7.2) and cross-pollination by crawling and flying insects

is thought to be the main mechanism affecting pollination. Midges (*Ceratopogonidae*) are possibly the main pollination vectors (Posnette and Entwistle, 1957). Pollination occurs in the early part of the day, mainly due to the fact that pollen release occurs early, and stigma receptivity slightly later in the day (Sampayan, 1966). The use of insecticides to control insect pests could also affect pollination. However, Soerjobroto (1967) found no adverse effect of 'Endrin' used to control the cocoa moth *Acrocercops cramerella* and Ventocilla *et al.* (1969) report that BHC caused slight reduction in midge population. Aerial dusting is considered to be less harmful than spray applications (Ventocilla *et al.* 1969). The weather conditions (sunny as opposed to rainy) particularly early in the day affect the main pollinatiors in Indonesia. Pollen viability is affected by insecticides and copper fungicides (Smith and Soria, 1968).

Fruits (Forastero type) are shown in Fig. 7.2c.

Growth quantities, light and shade

Growth and yield in cocoa are closely related to environmental conditions, particularly in regard to shade and light. There seems little doubt that young seedlings have a shade requirement for optimal growth, but in mature cocoa there is now considerable evidence that optimal yields can only be obtained under full sunlight and with adequate fertilization (see p. 109); except under adverse physical conditions where shade trees may act as a buffer between the tree and the environment, or where

Fig. 7.2 a&b   Diagrammatic sections of inflorescence cushion and flower

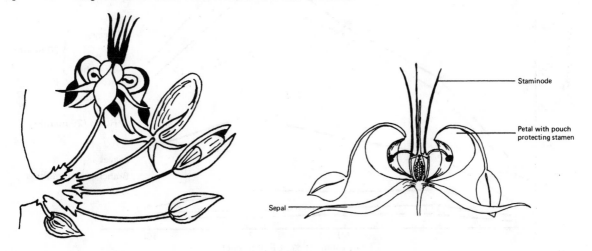

Source. After Cobley, 1956.

Fig. 7.2c Pods developing on the main 'chupon' stem of young cocoa trees

shade trees have been shown to affect soil improvement through root penetration and organic matter accumulation (e.g. Zevallos and Alvim, 1966). Even in young seedlings, varietal differences exist in shade requirement (Kentin and Asante, 1969).

Himme and Petit (1957) have studied the growth and development of cocoa under different light regimes from 5 per cent to 100 per cent sunlight in the Congo. There was a strong indication of a progressive change in the optimal light conditions with development (up to 20 months' age). Changes in plant weight and leaf area with age in relation to light intensity are shown in Figs. 7.3 and 7.4. Assimilation rates (approximately) assuming a mean leaf area between that at 10 and 20 months are as follows (from data of Himme & Petit):

Light level	Net Assimilation $g/dm^2/week$
5%	0.025
25%	0.040
50%	0.039
75%	0.050
100%	0.044

McKelvie (1962) reports NAR values of a similar order ($0.06g/dm^2/week$).

Fig. 7.3 Dry weights of leaves, branches and roots of cocoa in relation to light intensity, at 10 and 20 months after planting under field conditions

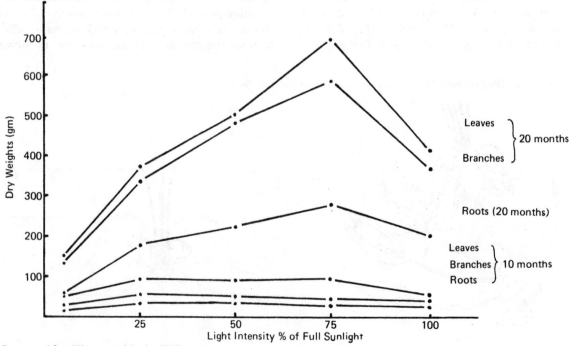

Source. After Himme and Petit, 1957.

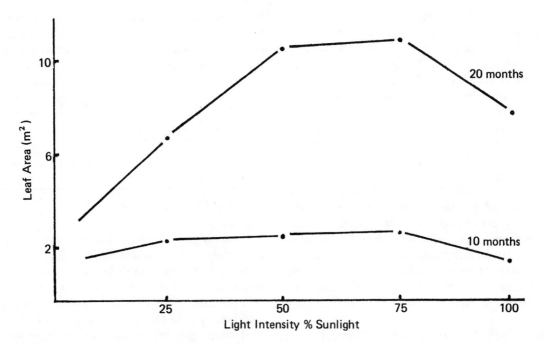

Fig. 7.4   Leaf area of cocoa in relation to light
Source.   After Himme and Petit, 1957.

These are of the order to be expected from non-deciduous tree species of the tropics. Cocoa possesses small sunken stomata on lower leaf surfaces. The main reduction in growth as light levels exceed 75 per cent can be attributed to both a lowered net assimilation rate (Wormer, 1965 has shown that stomatal aperture decreases in full sunlight in coffee) and reduction of leaf area, both through reduced number, reduced area of individual leaves and a faster rate of defoliation. However even in this relatively young cocoa (up to 20 months) shade in excess of 25 per cent is seen to reduce growth under the conditions of these experiments. Prof. G. Lemee (cited by Poncin and de Bellefroid, 1957) has apparently carried out assimilation studies on cocoa and observes that under damp and cloudy conditions assimilation continues to increase to full sunlight, but under sunny conditions with adequate rainfall, assimilation varies with sunlight as indicated in Fig. 7.5, with a threshold being approached at lower levels of illumination.

Vernon (1967) has shown with cocoa in Ghana that yield (Y) can be related to light (L), up to 60 per cent full sunlight by the quadratic equation

$$Y = 9.6 + 2.5L - 0.013L^2$$

At levels above this the relationship did not hold.

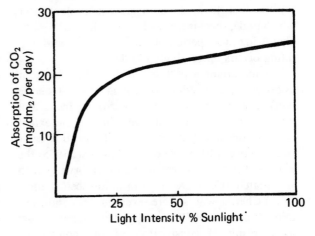

Fig. 7.5   Assimilation in relation to full sunlight with adequate rainfall
Source.   Data from Prof. G. Lemee, cited by Poncin & Bellefroid, 1957.

In the experiment of Himme and Petit cited above, the degree of branching was increased by light (as might be expected) and precosity of flowering and fruiting shifted from maximum at 50 per cent sunlight at 10 months to a maximum at 75 per cent sunlight at 20 months from planting.

Growth, of course, varies between different varieties and Ascenso and Bartley (1966) report that most hybrids grow at a faster rate than inbred lines.

In mature cocoa, the degree of self-shading is

much higher and with adequate fertilization (see below) it is now generally recognized that growth and yield are potentially much higher without shade. Temporary shade however is necessary in the establishment phase, and in adverse physical environments (such as a truncated soil profile or locations with a severe dry season) some degree of shading appears necessary even in mature cocoa.

Yield of pods has often been related to trunk girth (e.g. Sajoso, 1965; Are and Afonja, 1966; Glendenning 1966) particularly on the ability to sustain a good girth increment following the commencement of bearing. Under adequate physical conditions, girthing is usually better without shade. Mabey (1967) has found a relationship between trunk collar diameter (D) and total plant weight (W) as satisfied by the equation $W = a \times D^b$ where b is a value which ranged between 2.66 and 3.06 for different groups of trees and with increasing shade.

McKelvie (1955) has shown that, under West African conditions, pod growth occurs in phases; for the first 40 or so days growth is slow after which pod growth increases between 40 and 85 days. Following this a period of embryo growth and pod filling occurs up to about 140 days.

An important aspect of yield is the extent of abortion and shedding of young fruits (cherille wilt); this occurs particularly in young fruits and once the pod has reached a certain size the extent of abortion rapidly decreases. Peak abortion rates occur between 40 and 85 days (McKelvie, 1955) the period of rapid pod growth before embryo growth is initiated. (Compare oil palm inflorescence abortion.) Cherille wilt is more severe in dry conditions and Shimoya (1967) has suggested that it occurs as a result of constriction of the conducting elements to the fruit during drought. However Nichols and Walmsley (1965) have shown that phosphorus uptake (P^{32}) continued for several days following the cessation of growth which does not support this idea. The work of Nichols (1964, 1965b) indicates that changes in the ovary and production of auxins control wilting and bring about a cessation in susceptibility to wilting; and this is supported by experiments with antiauxin (tri-iodobenzoic acid) applications (Krekule, 1969) which increase wilting. Fruit thinning reduces the incidence of cherille wilt and Nichols regards this phenomenon as a natural fruit-thinning process. (For review see Nichols, 1965b.) Generally the number of mature pods produced is only a minute fraction of the number of flowers formed because of low fruit set and high abortion rates as shown in the following table (from McKelvie, 1962):

Tree	Flowers	Sets	Pods wilted	Pods harvested
1	15,144	173	123	50
2	18,308	178	115	63
3	700	26	20	6
4	9,196	75	57	18

CLIMATE AND SOILS
Climate

Cocoa is grown largely within the humid tropics. According to Papadakis (1965) the best cocoa areas in West Africa have a rainfall surplus over evaporation of less than 1000 mm, but it seems likely that this criterion is related more to soil formation (see below) than to rainfall. Murray (1964) states that cocoa growing is not possible in regions in which average temperature is lower than 21°C and in which the average for the coldest month is below 15°C and the absolute minimum is about 10°C. Seed viability has been found to be completely lost if a temperature of 4°C is maintained for as little as 15 minutes. Upper temperature limits are considered to be about 30°C as a continuous temperature although fluctuations up to 38°C are tolerated (Voelcker, 1955). In Ghana, Greenwood and Posnette (1950) found that flushing was reduced at mean maximum temperatures below about 29°C. Sale (1965, 1966) reports that leaf size and duration were reduced at a constant temperature of 30°C in cocoa cuttings grown under controlled conditions. The fact that leaf development was affected in much the same way by sustained temperatures above 30°C as leaves of young cocoa under full sunlight conditions suggests that the adverse effect of full sunlight may be caused by high leaf temperatures, particularly in young red leaves with inadequate conducting tissue. (Compare Tea.)

The production of flowers and fruit development is particularly sensitive to low temperatures. Fig. 7.6 (after Alvim, 1966) indicates the effect of temperature (and rainfall) on the extent of flowering and leaf flushing in cocoa in Bahia (Brazil) where seasonal temperature fluctuations are marked. Leaf flushing continued periodically at lower temperatures (which agrees with the observations of Sale above) but flowering was extremely sensitive to

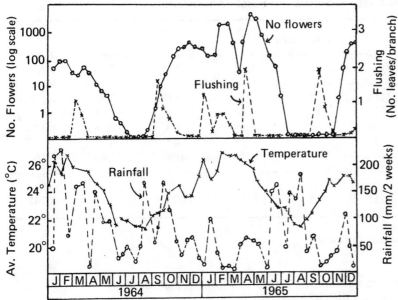

Fig. 7.6 Bi-weekly data on flowering and flushing of cocoa trees in Bahia, together with temperature and rainfall data

Source. After Alvim, 1966.

average temperatures below 23°C (rather higher than the 21°C indicated above as minimal for cocoa).

Cocoa is generally considered to do best under condition of little or no dry season. In regions with uniform climatic conditions, a relatively even distribution of fruit production occurs, but in countries with distinct wet and dry seasons the main crop is usually produced during the wet season with a decline during and after the dry season. Ali (1969) in Ghana, has shown a positive regression of rainfall on yield in the drier periods and a negative regression in the wet season, indicating the importance of sunlight under rainy conditions.

Alvim (1966) regards 100 mm rainfall per month (c. 4″) as minimal for cocoa (c. 50″ per annum). Optimal rainfall is probably of the order of 80″ to 100″ evenly distributed, being approximately equal to evaporation under humid tropical conditions. Thorold (1955) reports that a rainfall of over 100″ depresses yield in Fernando Po, off the coast of West Africa. Generally cocoa will show reduced flowering and yield with relatively slight moisture depletion below field capacity (e.g. 35 per cent depletion, Smith, 1964; 15 per cent depletion, Sale, 1967a). Murray (1966) has shown in watering experiments that the low humidities of a dry season reduce growth, even with adequate soil water, but excessively high humidities (90–100 per cent) have been shown to reduce leaf growth (Sale, 1967b). Moisture stress reduces yield through its effect on bean size

(Toxopeus and Wessell, 1970) and, as indicated before, through fruit abortion. There is much evidence for varietal differences in sensitivity to climatic conditions (e.g. Opake and Toxopeus, 1967; Cruickshank and Murray, 1966). Potentially high-yielding varieties are likely to be more sensitive to moisture stress.

Soils

Cocoa is unusual among tropical crops in that until very recently, it was largely grown without fertilization. Consequently, yield was often closely associated with soil type. Cocoa is nevertheless grown on a wide range of soils. In West Africa the soils predominately used for cocoa are inland latosols, and because of the fact that most cocoa was grown by peasant smallholders, without the use of fertilizers in the past, only relatively nutrient-rich soil was considered suitable for the crop. Crosbie and Brammer (1957) consider that only ochrosols,* derived under intermediate rainfalls (45″ to 65″ per annum) from a variety of parent materials are sufficiently rich in nutrients for good cocoa yield. Oxysols,* derived under higher rainfall conditions, are more leached and consequently poorer in nutrients. Formerly these were not used for cocoa in Ghana. Tables 7.1 and 7.2 indicate the nutrient status of good and bad cocoa soils based on this reasoning. Smyth (1966a) has reviewed literature relating to soil suitability for cocoa.

*Soil classification of C.F. Charter.

TABLE 7.1 Available nutrients in a representative 'ochrosol' in lb/acre/inch

Horizon layer in inches	Exchangeable bases per inch				Total P	Organic C	Total N
	Ca	Mg	K	Mn			
2	1,271	241	27	27	141	15,555	1,156
9	114	51	10	3	138	4,165	442
10	14	5	4	1	63	1,026	123
12	10	4	3	1	49	437	67
3	18	16	2	2	108	848	164
Total lb per acre for 36 inches	3,875	1,095	235	117	3,073	86,671	8,747

Source. After Crosbie and Brammer, 1957.
Note. Parent rock: Phyllite; annual rainfall: 60 inches.

TABLE 7.2 Available nutrients in a representative 'oxysol', in lb/acre/inch depth

Horizon layer in inches	Exchangeable bases per inch				Total P	Organic C	Total N
	Ca	Mg	K	Mn			
2	177	71	41	7.0	79	11,323	895
3	29	22	6	0.3	52	3,090	282
7	19	11	5	0.4	55	1,797	178
16	7	12	3	0.6	70	1,611	200
Total lb per acre for 28 inches	611	474	189	26	1,812	70,274	7,078

Source. After Crosbie and Brammer, 1957.
Note. Parent rock: Phyllite; annual rainfall: 80–85 inches.

Ochrosols (Table 7.1) are usually only slightly acid, compared to oxysols (Table 7.2) in which the pH may be close to the finial pH of fully-leached kayolinitic clays (pH 4–4.6).

The data in Tables 7.1 and 7.2 take into account the assumed dry density and the amount of oven-dry fine earth.

Because rainfall is strongly seasonally distributed over most of West African cocoa areas, soils need to possess adequate moisture storage capacity in the root zone for optimal yield. Such soils are often deeply friable, permitting deep root penetration. Crosbie and Brammer (1957) in Ghana, and Wessell (1970) in Nigeria, consider that fertilization is only economical on soils of good physical characteristics and when cocoa is already yielding relatively well (see further on).

In Trinidad, only soils of relatively good characteristics are considered as first class for cocoa (Havord and Hardy, 1957) and, contrary to the procedure in West Africa, fertilizers are recommended for poorer soils.

In East Malaysia, some cocoa is established on soils derived from basalt which are consequently richer in nutrients. These soils have a pH of up to 6 in the surface horizons and down to 5.1 in the lower horizons. Some soils derived from granitic rocks, which are poor in nutrients, are also planted with cocoa to a limited extent, but these give rise to various leaf symptoms; light yellow mottling of older leaves, marginal and intraveinal necrosis of young leaves (Wyrley-Birch, cited by Brown *et al.* 1967). Soils of basaltic origin also show similar symptoms if they are too acid (Brown *et al.* 1968).

Allen (1966) states that Mn and Al toxicity problems occur on these acidic soils (causing leaf necrosis and die-back). Brown and co-workers consider that cocoa benefits from fertilizers on such soils (see further on). Cocoa is also grown on volcanic soils in Grenada (Cruickshank, 1970).

Cocoa is planted with success on heavy coastal clays under coconuts in Malaya, and in Africa on some heavy riverine clays (e.g. Poncin and Bellefroid, 1957) with impeded drainage, gleyed in the surface horizons and with concretions of latrite. In these soils the tap root penetrates only 90 cm, and shade is considered necessary for optimal production. (Murray, 1967 comes to similar conclusions about shade on soils of restricted root volume in Trinidad.) Analyses of good, average and poor soils considered by Poncin and Bellefroid are as follows:
For good soil:
   pH generally 4.5 to 5.5, sometimes 6.5
   R_2O_3 1–3 M.E./100 g.
   P_2O_5 1.3 mg/100 g.
   Exchangeable bases 3.5 to 8 M.E./100 g.
   Fine earth 50 to 60 per cent (topsoil), 60–90 per cent (subsoil).
   Concretions of laterite 50–70 per cent subsoil.
Areas of average soil have the following characteristics:
   pH 4.2 to 4.5.
   Exchangeable bases 1.5–3 M.E./100g.
   R_2O_3 1 to 3.5–M.E./100 g.
   P_2O_5 0.6 to 1.5 mg/100 g.
   Fine earth 45 to 60 per cent (topsoil), 60 to 70 per cent (subsoil).
   Concretions 60 to 70 per cent (subsoil).
The areas of poor soil, which nearly always border on the swamps, give the following:
   pH 3.9 to 4.4.
   Exchangeable bases 0.7 to 2.5 M.E./100 g.
   R_2O_3 2 to 3.7 M.E./100 g.
   P_2O_5 0.5 to 1.4 mg/100g.
   Fine earth 50 to 65 per cent (topsoil), 60 to 75 per cent (subsoil).

In New Guinea and in Ecuador, some immature volcanic ash soils of pH near or above 7 are used for cocoa and support good growth. However in soils of high pH iron nutrition may be a problem.

Drainage. According to Urquhart (1961), cocoa may be unaffected by flooding of several weeks duration in a riverine environment but is adversely affected by stagnant water. The construction of field drains will be required on heavy soils. The physical conditions for cocoa grown under shade are considered to be less exacting than when grown in full sunlight or reduced shade, the shade acting as in 'buffer' under less than optimal conditions. Fennah and Murray (1957) illustrate the relationships between soil aeration and light, and soil reaction as shown in Fig. 7.7. Gavande (1968) studied aeration in a pot trial on 18-month-old cocoa and considered that a critical oxygen diffusion rate is 20×10^{-8} g/cm²/min. At field capacity silty and clayey soils tended towards inadequate aeration, but not sandy soils. Alvim (1968) has attributed lowered yields of cocoa in Bahia, Brazil to reduced root growth and foliar renovation following excessively wet seasons and soil flooding.

Fig. 7.7a Relationship between optimal growth and light at different proportions of soil water to soil air

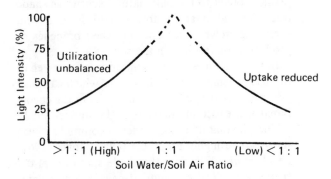

Fig. 7.7b Relationship between optimal growth and soil water/soil air ratio at different levels of soil acidity

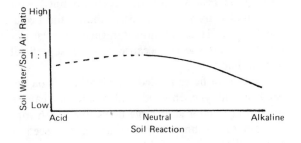

Source. After Fennah & Murray, 1957.

PLANTING
Land preparation, spacing and yield
Since in most circumstances young cocoa requires shade for initial development, the clearing of new jungle areas for planting of cocoa often involves the retaining of desirable natural trees for

shade. Removing undergrowth and small saplings is the first step after which the land may be lined and holed. Very large trees are also usually removed. Sometimes leguminous shade species are planted along with the cocoa to eventually replace the natural shade (e.g. Blow, 1968). If natural trees are not used as initial shade, the planting of leguminous trees must precede the planting of cocoa or temporary shade trees of crop plants such as bananas must be used. Shade trees may be progressively thinned by poisoning with development of the cocoa (e.g. Liefstingh, 1966).

There is great variation of opinion among planters as to the best species to use for shade. *Erythrina spp.*, *Albizzia falcata*, *Laucaena glauca*, and *Parkia spp.* are commonly planted leguminous species. *Cassia seamea* has been reported to be harmful to cocoa (Poncin and Bellefroid, 1957). Some planters prefer to use only natural species as shade trees. The ideal shade tree should have a thin spreading crown which does not shed branches or produce toxic residues in the soil.

In modern times more attention is being given to the use of temporary shade species which can also give a crop in the period of immaturity of the cocoa. Bananas are used in Ghana (e.g. Hammond, 1962).

The planting of cocoa under coconuts has been shown to be highly profitable (Owen Jones, 1967; Leach *et al.* 1971) since the shade species is also a crop species. Tall coconuts at conventional planting provide about 50 per cent shade which is close to ideal for cocoa in some circumstances. In India good results have also been obtained with cocoa under *Areca catecha* (Bhat-Shama and Leela, 1968). Some planters, however, regard palms as unsuitable shade species because of their superficial rooting habit.

The type of holes needed for planting cocoa depend on soil type. In good friable soil they can be small, but in poor soils large holes filled with topsoil aid establishment. Sometimes cocoa seed is planted directly to the field without a nursery phase (e.g. Blow, 1968) and in suitable soils the planting point is simply worked to remove weeds and roots. A number of seeds (usually three) are planted and later thinned, leaving the most vigorous seedling. In good-structured soils and with polybag seedlings holes of appropriate sizes are usually made at the time of planting (e.g. Leach *et al.* 1971).

Spacing for cocoa has been extremely variable. Close spacing (*c.* 7′×7′, 800 trees/acre) gives initially higher yields but cumulative yields are less different from wider (e.g. 12′×12′) spacing (Wood, 1964). In cultivation without shade, closer spacing (above 800 trees/acre) is generally used but with more vigorous upper Amazon clones and their hybrids a slightly wider (*c.* 10′×10′) has been recommended (Wood, 1964). Under coconuts in Malaya a spacing of two rows 10′ apart in the coconut avenues (*c.* 30′), and 6½′ apart in the rows has been used for cocoa (Leach *et al.* 1971).

Yields of peasant cocoa grown under shade and without fertilization are of the order of 500 lb dry beans per acre. With good husbandry and planting material, yields of 1,500 lb or more are possible in shaded cocoa. With close-planted unshaded cocoa, yields of 4,000 lb or more may be obtained.

Seed production and nursery practice

Cocoa is today planted, by preference, from hybrid seed from known crosses of superior high-yielding trees. This results also in a well-structured tree from the point of view of access and harvesting (see 'Pruning' below), which also has a substantial tap root giving resistance to wind damage and a better exploration of the soil. In making crosses, random pollination is prevented by pinning an insect mesh (at least 22 meshes/cm) to exclude natural pollinators over the trunk. Using forceps pollination is done by applying a ripe anther to the style (see Toxopeus, 1969). A one acre seed garden containing some 400 productive trees is sufficient to supply about 400 acres annually in the 4th to the 6th year, rising to 1,200 acres in the 11th to the 15th year from planting. Seeds may be planted directly to the field, as explained before, or raised in small polythene bags prior to planting out (Fig. 7.8). Older methods are described by Quartley-Papafio (1964). For the production of seeds on a larger scale, hand pollination may be too labour intensive and hybrid seed gardens may be established using naturally self-incompatible trees as parents, planted in alternating rows (see Edwards, 1969). This relies on natural pollinatiors and must be isolated from other cocoa. A certain amount of self-pollinated seed is also produced. A similar method of seed production has also been used in rubber. Seeds of cocoa remain viable for only about 10 days from maturation of the pod, but may be stored for several weeks in charcoal powder at 30 per cent moisture.

Cocoa is also grown commercially from cuttings, particularly when the large-scale production of

Fig. 7.8   Cocoa seedlings in polythene bags in a shaded nursery
Note.   Small-sized perforated polythene bags (about 6″ × 4″ laid flat) are suitable for
            cocoa seedlings.

trees resistant to diseases is necessary. Both cuttings and budding techniques are used, the former being more common. Cuttings are usually obtained from recently-matured flushes of fan branches and may be single-leaf or several-leaf cuttings (Fig. 7.9). Traditionally cuttings have been produced in concrete bins filled with sawdust or other medium (see Braudeau, 1957; Urquhart, 1961) but at the present time they are almost invariably raised in polythene bags under a close-fitting polythene cover to maintain humidity, which gives a high rate of success (e.g. see McKelvie, 1957; Burle, 1957).

Sawdust mixed with compost and sand is commonly used as a rooting medium. It should drain and aerate adequately. Newhall and Diaz (1967) report that cocoa-bean endocarp (parchment) is also a good medium for seedlings, being less prone to waterlogging and to nitrogen de-mineralization than sawdust, and sand enriched with superphosphate, bone meal and crushed hoof and horn flakes has also been used as a medium. For seedlings in polythene bags jungle topsoil is commonly used. The pH of potting soil for cocoa should be in the range pH 5 to 6 (Urquhart, 1961). Iron deficiency has been reported in cocoa nurseries resulting from excessive organic matter, Praedo and Miranda (1967), and this can be combated by foliar spray.

Cuttings are usually dipped with rooting hormone prior to planting.

Buddings have been used to a lesser extent for cocoa, and chiefly in Indonesia (see Burg, 1969). A technique which theoretically gets the best of both worlds was developed by Topper (1957) which uses patch budding (similar to rubber, see Chapter 9), thus providing the tree with a tap root; and after the development of the bud-shoot, the production of adventitious bud-wood roots close to the stock/scion junction is induced by incising into the bud-wood and inserting rooting compound, after which the area is earthed up to encourage rooting (Fig. 7.9d).

Usually fan branches are used to obtain material for bud-wood and cuttings because these are more numerous. This has the disadvantage of producing a tree without the main chupon stem and is more difficult to harvest. However, Ascenso (1968) reports that success has been obtained with wood from cut-back chupon branches which produce side shoots.

The use of marcotts for propagation of cocoa has not been adopted commercially although Toxopeus and Okoloko (1967) have described a technique for raising marcotts which gives a success of above 60 per cent.

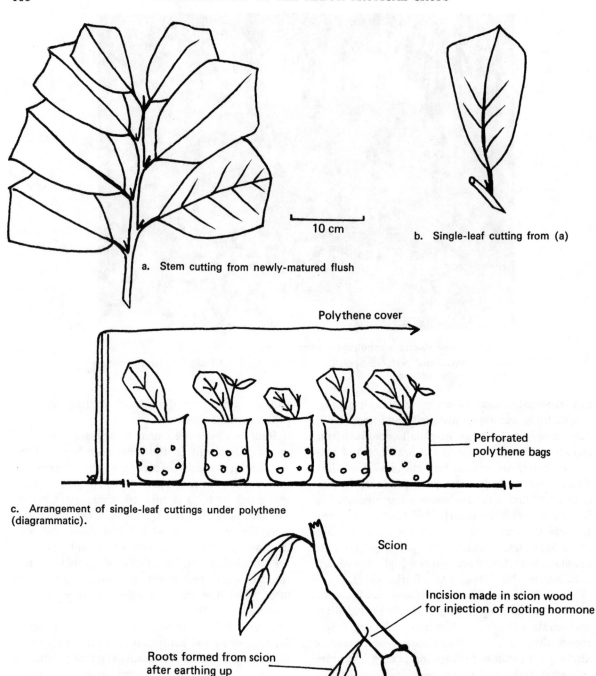

10 cm

a. Stem cutting from newly-matured flush

b. Single-leaf cutting from (a)

Polythene cover

Perforated
polythene bags

c. Arrangement of single-leaf cuttings under polythene
(diagrammatic).

Scion

Incision made in scion wood
for injection of rooting hormone

Roots formed from scion
after earthing up

Stock

Stock tap root

d. A budding with scion roots developed

Fig. 7.9 Propagation material in cocoa

Sources.   a and b. After Urquhart, 1961.
              d. After Topper, see text.

Manuring

The state of knowledge of fertilizer use in cocoa is not very satisfactory and this is attributed to genetic variability in cocoa plantings, soil heterogenity and pest occurrence (Urquhart, 1961). Variable results from fertilizer experiments may also be due to variation in pollination and fruit set. Fertilizer response is also complicated by the use of shade.

Homes (1957) used sand cultures to assess the best combination of cationic and anionic nutrients using his 'systematic variables' method in which factorial analysis of nitrogen, sulphur and phosphorus combinations with the cationic nutrients potassium, calcium and magnesium is the main feature.

Results of plant weight of 1 year-old seedlings are as follows:

Control	N K	N Ca	N Mg	S K	S Ca	S Mg	P K	P Ca	P Mg
12	52	89	108	41	62	98	7	74	80

This draws attention to the importance of Mg in the cation balance, and this is generally agreed to be an important nutrient in cocoa. Homes gives the following composition of the major cations and anions as optimal.

NO_3: 37 $\frac{1}{2}SO_4$: 29 $\frac{1}{3}PO_4$: 34
K : 21 $\frac{1}{2}Ca$: 35 $\frac{1}{2}Mg$: 44

Generally nitrogen and phosphorus are also recognized as important nutrients (e.g. Wessell et al. 1967) the former particularly when cocoa is grown in the absence of shade and the latter in phosphate-sorbing soils common in the tropics (see Ahenkorah, 1968). Significant K fertilizer responses have also been reported (e.g. Verliere, 1967), and on leached sedimentary soils with good physical properties considerable increases have been reported from complete fertilizer, as well as lime and Mg (Silva and Cabala, 1966). Fennah and Murray

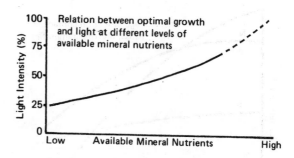

Fig. 7.10 Relationship between light and nutrient supply
Source. After Fennah and Murray, 1957.

(1957) give the following relationship (Fig. 7.10) between light (100 per cent = no shade) and mineral nutrients.

Urquhart (1961—data from R.D. Cunningham) gives the following table which illustrates the interaction of fertilizers ($N:P_2O_5:K_2O:MgO$—13.4: 90.7:90.4:45.8 lb/acre—followed by additional N at 100 lb/acre and P_2O_5 at 30 lb/acre) and shade.

TABLE 7.3 Yields of cocoa (in Ghana) in relation to shade and fertilizers

Treatment	Yield (2 years)
Shade, no fertilizers	958
No shade, no fertilizer	2,348
Shade, fertilizer	1,211
No shade, fertilizer	3,091

Source. Adapted from Urquhart, 1961.

In some of the major cocoa-producing countries, fertilizers have been little used in the past, but the value of fertilizers in raising yields has become appreciated in recent years, particularly in unshaded, close-planted pure stands of cocoa. Nutrients removed by a crop of 500 lb of cocoa/acre have been

TABLE 7.4 Effect of cocoa on soil properties (Nigeria 0–4')

	pH	%C	%N	Capacity m.eq./100 gm	Bases m.eq./100 gm	% Base Saturation
Forest soil	6.40	2.83	0.270	11.4	10.3	90
Cocoa 5'×5'	6.45	2.11	0.222	7.7	6.6	86
,, 7½'×7½'	6.20	2.06	0.196	7.3	5.7	78
,, 15'×15'	6.00	2.05	0.155	6.8	4.7	69

Source. After Kowal, 1959.

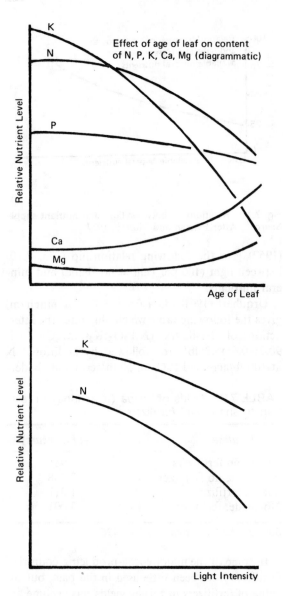

Fig. 7.11 Relative changes in leaf levels of nutrients with age and light intensity
Source. After Murray, 1967.

calculated at about N—12 lb, P—6 lb and K—9½ lb by Urquhart (1961) to about N—32 lb, P—4 lb and K—10 lb (Adams, 1962). The effect of cocoa on soil impoverishment is indicated in Table 7.4 (after Kowal, 1959) compared with forest soils. The table emphasizes the importance of adequate ground cover on maintaining fertility, and nutrient removal by the crop.

Wessell (1970) reports yield response in Nigeria to high N (120–160 lb/acre) and P (30–60 lb/acre) to levels of over 1200 lb/acre dry cocoa (normal

yields being about half of this). Cunningham *et al.* (1961) consider that yields of up to 2000 lb/acre are possible with existing Armelonado cocoa in Ghana with adequate fertilization on close-planted unshaded trees, and up to 3,500 lb/acre with vigorous hybrids. Urea has been shown to be a satisfactory source of N in cocoa. On leached basaltic soils in Sabah, Brown *et al.* (1968) report beneficial results from magnesium limestone, rock phosphate and N:P:K fertilizers at a ratio of 3:1:2. Lime at a rate of 1200 kg/ha has been recommended to raise soil pH in Bahia, Brazil (Cabala *et al.* 1967).

Leguminous shade trees probably fix nitrogen in most circumstances but there is a lack of quantitative information on the subject. Generally fertilizer responses under leguminous shade trees are not economical. Organic manures have been used extensively in Grenada and Ceylon, and in the past in Trinidad. Mulching, particularly with cut bushes and leguminous shrubs, has also been reported as beneficial. Variable results have been obtained from planted leguminous covers (e.g. Jordan and Opoku, 1966).

Nutrient requirements for cocoa have been assessed by both soil and foliage analysis although Smyth (1966b) considers that soil analysis is not sufficiently evolved to enable accurate nutrient assessment. The levels of leaf nutrients vary greatly with age and also with light and shade. Fig. 7.11 (from Murray, 1967) shows relative changes in nutrient levels with these two factors.

Murray (1967) gives the following analyses (Table 7.5) for leaf nutrients in the second and third terminal leaves of newly-matured flushes. Verliere (1967) in West Africa found the highest correlation between leaf Ca and yield. N/P and Ca/K ratios were also related to yield. Interactions between fertilizers also affect leaf-nutrient levels. Both N and K have been reported to reduce P leaf content which was highly correlated with yield (Wessell, 1967).

TABLE 7.5 Nutrient levels (% dry matter) of cocoa leaves

	Deficient	*Low*	*Normal*
N	1.50	1.50–2.00	2.00
P	0.13	0.13–0.20	0.20
K	1.20	1.20–2.00	2.00
Ca	0.30	0.30–0.40	0.40
Mg	0.20	0.20–0.45	0.45

Source. From Murray, 1967.

With reference to phosphorus, Wessell (1970) regards 0.21 per cent as optimal for young leaves and 0.13 per cent for old leaves, and soil P levels as adequate above 12 ppm available P. In regard to boron nutrition Tollenaar (1967) considers that boron deficiency occurs at soil levels less than 0.2 ppm, and is treated with applications of borosilicate to the soil (20–80 lb/acre) or by foliar spray of soluble boron fertilizer. Boron deficiency is characterized by leaf curling and tip browning, branch defoliation and formation of dense 'broom' branching and tip die-back. Distorted and constricted pods and parthenocapfic development without seeds also occurs (Ascomaning and Kwakwa, 1967). Iron deficiency has been reported in nursery seedlings (see above) in high organic media, which can be corrected by foliar spray (Praedo and Miranda, 1967).

Response to fertilizers is considered by Wessell (1970) to be most economic on naturally high-yielding areas. In this connexion Jiménez Saénz (1967) has shown that nutrient uptake was higher in high-yielding clones that they gained a greater fresh weight per unit of nutrients absorbed. With low-yielding trees and under poor conditions of husbandry, fertilizers would be wasted. Ruinard (1966) considers that frequency of weeding is more important than fertilizer use in peasant cocoa in Ghana.

Pruning

There is very little scientific information on pruning and this practice is generally carried out as a rule-of-thumb procedure by experienced planters. The idea of pruning is to restrict height growth at the first or second chupon and induce a spreading type of canopy. If the first chupon (in seedling cocoa) reaches a height of 4 to 5 ft (see 'Botany' above) then pruning is generally maintained at this level; if shorter, then pruning may be maintained at the second level. The resultant tree has a vertical trunk of several feet height and a canopy of spreading branches. The fan branches are usually maintained at three or four which become stout laterals and also bear flowers and pods. Maintenance pruning of excess laterals and new chupons and restricting lateral spread to prevent excessive overlapping of trees, particularly at close spacing, is carried out once in two years or less often (see Cook, 1966).

With cuttings from fan branches, the straight upright trunk is initially lacking and trees may be pruned to give three or four vigorous branches; alternatively chupon growth, which often occurs later, is encouraged in an attempt to produce a tree resembling one grown from seed.

General husbandry and weed control

A mulch of vegetable matter is sometimes used to suppress weeds and reduce erosion, and has been reported to be better than cover legumes such as *Calopogonium* and *Pueraria* (Jordan and Opoku, 1966; Wessell, 1967–8). However, 'bagasse' mulch (from sugar-cane) reduced N availability in soils. Weed control in mature cocoa is much reduced because of the dense shade. In young cocoa traditional weeding has been manual circle weeding for about 3' diameter, and inter-row slashing.

Chemical weed control is much used at the present time and is considered to be more economical. Kasasian (1965) has reported that simazine, linuron and diuron are useful for long-term weed control and do not produce any marked toxicity symptoms in seedlings when used at twice the normal rates. Grammoxone has been used in conjunction with these herbicides as a contact weed killer. The leaves of the plants, however, must be protected against the spray. Dalapon at 5–10 lb/acre has also been recommended for grasses and 2:4-D for broad-leafed species but, again, foliage contact must be avoided. 2:4:5-T has been used for woody species but has been reported to be somewhat harmful to cocoa. In Malaya sodium arsenite has been used successfully for general weed control in cocoa, and more recently the arsenicals MSMA and DSMA, which are less toxic.

MAJOR DISEASES AND PESTS

Cocoa has a number of important diseases which cause widespread crop losses. The fungus *Phytopthora palmivora* causes a range of symptoms including pod rotting (black pod), seedling wilt and leaf fall (Hislop, 1964). The fungus, which is a virulent parasite, also effects other important crop species including rubber, citrus and coconut. In cocoa it can spread from infected pods to invade the flower cushion and stem where it gives rise to 'canker' (Vernon, 1971). The disease spreads by means of sporangia or zoospores in moist conditions, and Turner (1960) considers that different strains exist which may combine to produce new virulent strains capable of infecting resistant selections. However, much emphasis has been given to the selection of resistant types and this is likely to be the basis of control in the future (e.g. Dakwa, 1966/7; Weststeijn, 1967). Rocha and Medieros (1968) discuss the mechanism of resistance which may be related to the presence of fungitoxic phenolics in the fruits. Other methods of control are reduction of shade and periodical spraying with Bordeaux mixture or other copper fungicides.

Another serious disease is 'witches-broom' caused by the fungus *Marasmius perniciosus*. It is endemic to the South American region. The main symptom is the production of broom-like branching of the shoots with undeveloped leaves and enlarged, branched, flower cushions. The upper Amazon clones have been used as a source of resistance to this disease (e.g. Bartley and Amponsah, 1966).

A disease which has become serious in South America in recent years is trunk canker caused by the fungus *Ceratocystis fimbriata* (e.g. Reys *et al.* 1966; Scheiber, 1969). Some clones, particularly from the Amazon types are highly resistant, but the resistance has been reported to be a recessive factor (Soria and Esquivel, 1968).

A fungus disease which is serious in several South American countries is 'monilia pod rot' or 'watery pod rot' caused by *Monilia roreri*. The infection develops internally and the first indication of the disease is the appearance of spots which later turn brown and expand in size. Spores form on the spots and the pod interior is filled with watery liquid. Spraying with copper and other fungicide, particularly during wet weather and during flowering, is used for control. The possibility of resistant types also exists (see Ampuero, 1967).

'Cocoa die-back' is a condition which occurs universally in cocoa and is serious in some areas. Until recently a specific causal organism for die-back was not known and the disease was attributed to a variety of conditions including boron deficiency, and various micro-organisms of low pathogenicity. In 1971, however, Phillip Keane, working in New Guinea, demonstrated that the primary phatagon of die-back is a species of *Onchobasidium*. This has subsequently been confirmed in other countries. The symptoms are a browning of young twig tips which become desiccated and withered, followed by defoliation. Because of the role of boron in conferring resistance to disease (Rajaratnam, 1971) deficiency in this element may still be central in the die-back condition. Replanting with cuttings from resistant trees is used to combat the disease.

A virus disease which has caused widespread losses in cocoa in Africa is called 'swollen shoot'. The virus causes swellings of fan and chupon branches and initially a red-vein banding and later chlorosis alongside the leaf veins (see Longworth, 1963). Several strains have been isolated which are spread chiefly by mealy bugs. The best method of control, where feasible, is by phytosanitary procedures. The Amelonado type of cocoa is particularly susceptible to swollen-shoot. Amazon types possess more tolerance to the disease but unfortunately also constitute reservoirs of infection which are difficult to detect.

Capsids (chiefly *Sahlbergella singularis* and *Distantiella theobroma*) have caused widespread damage to cocoa in the past in West Africa, Java, Ceylon, New Guinea and South America. Capsids feed on the pods and young shoots. The punctures made by capsids in feeding often become infected with the fungus *Calonectria rigidiuscula*. Control is based on the use of insecticides.

8 TEA [Camellia sinensis]

INTRODUCTION, VARIETIES

TEA (*Camellia sinensis*) has been cultivated as a beverage plant in China for at least two thousand years. The wild progenitors of tea are not known but its origin is thought to be on the Asian continent, possibly near the source of the Irrawaddy river (see Eden, 1965). Various types of tea can be recognized which have been assigned varietal status. These are the *assamica*, *sinensis* (China) and *cambodia* types from Indo China. They all hybridize freely (Carpenter, 1950) and Sealy (1958) considers that there are only two undoubted varieties in the botanical sense: *sinensis* and *assamica*.

Tea is regarded as a tropical crop since it forms the basis of the economy of a number of tropical countries. It is, however, much more widely distributed than any other perennial tropical crop species. It has been cultivated at high elevations in equatorial regions of Ceylon, Africa, Java and Sumatra and in northern India, using teas of the *assamica* type, following the breakdown of the China tea trade of the last century. The technique of hand picking is used predominately in all the above areas. The mechanization of tea cultivation and plucking was developed in Russia which is now also one of the major tea-producing countries.

The production of *China* teas was traditionally more or less confined to China and Japan where it is, even today, produced mainly by peasant cultivation, but China-type teas (jats) are now used in other tea-producing countries of the world, largely because of their greater hardiness and yield. The term *jat* is used in the tea business to refer to any tea of specific genetic composition, particularly those multiplied by vegetative means to large populations.

In recent years tea growing has spread to numerous countries including Brazil, Peru, Mexico, Turkey, the USA., Zambia, Mauritius, New Guinea and even Australia (see Mann, 1970).

BOTANY

The vegetative plant and productivity

Left unpruned, tea grows into a small tree of about 30 feet in height which is roughly conical in shape. In cultivation it is pruned to form the characteristic hedge canopy needed for plucking (Fig. 8.1). The form of the aerial parts of the tea plant depends greatly on the method of bush formation and pruning (see later). Tea produces alternate leaves and branches which develop from the buds in the leaf axils.

Leaf morphology varies considerably. Normal leaves have a serrated margin. Jats of Assam origin (*assamicas*) often have light-coloured and large leaves. Others are narrow and rather darker-coloured and are usually thought of as China (*sinensis*) jats. Leaves produced in the early stages of growth and after pruning are generally larger than later-formed leaves (see further on). Tea leaves are glabrous due to a heavy cuticle and have a characteristic glossy appearance. Stomata occur predominately on the lower surface. Internally, scattered highly-thickened sclerids occur which are of significance because they can be used to determine the purity of made tea.

With the tea plant, leaf growth is of central importance because it is, in fact, the crop yield itself. Leaf yield is strongly affected by temperature (as indicated in the first chapter of this book) hence tea yield is generally greater (but with a lower quality) at lower elevations in the tropics than at the traditional high elevation sites (e.g. Nanayakkara, 1967—see later).

The traditional Assam type of jat, with large leaves more horizontally disposed, is considered to be essentially low-yielding because the canopy does not permit good light penetration and the development of a high leaf area index (Hadfield, 1968). (This aspect of canopy development was discussed in the first part of this book.) This canopy feature is also unresponsive to high fertilizer levels. Fur-

Fig. 8.1a   Hedge planting of tea in Malaya

Fig. 8.1b Normal shoot showing two terminal leaves and bud, subtended by maintenance leaves

Fig. 8.1c   Massive root of high-elevation tea

thermore, high insolation of the upper leaf layer due to its horizontal orientation causes overheating, hence shade becomes beneficial. Hadfield considers that this canopy characteristic has limited the response of tea in northern India to improvement in the level of husbandry, particularly in fertilizer use.

Jats with smaller, more erect leaves permit better light penetration, are more efficient dry-matter producers and respond in terms of leaf yield to higher fertilizer (particularly N) levels. Furthermore, since inclined leaves are not overheated, shade is unnecessary and overall photosynthesis higher.

Studies with labelled CO_2 (Sanderson and Sivapalan, 1966) showed that carbon assimilated by mature leaves moves both to the developing shoot tips and to the roots. Carbon assimilated by young leaves does not move out of the leaf, consequently the illumination of the lower mature leaves is central to high production. In plucking, the terminal two immature leaves and the bud are removed (see further on).

The root system

The production of shoots for harvesting depends much on the assimilates of the mature leaves or maintenance foliage but regrowth following pruning is more dependent on the carbohydrate reserves of the root system (e.g. Barua, 1966). Consequently root development, whether the tap root of seed-grown tea or the adventitious roots of cuttings, and conditions of soil and disease and the type of pruning affecting root growth, are important in tea cultivation. Temperature also affects root growth, a higher root/shoot ratio developing under cooler conditions. Hence tea grown at high elevation in the tropics generally has better root development and is longer lasting and able to withstand more frequent pruning than lowland tea. Generally when roots reach a diameter of 1–2 mm starch reserves begin to be laid down (see Eden, 1965). Seed-grown tea has a dominant tap root (Fig. 8.1), but root distribution varies with clone or jat. Some bushes develop deep lateral roots while others produce largely superficial roots. In general, however, the absorptive root system is predominately superficial. Eden (1940) found that less than 5 per cent of the total roots could be found at a depth of two feet. Various workers (e.g. Park, 1928; Webster, 1953) have found mycorhyza associated with feeding roots but the function is not known.

Shade

As with cocoa and coffee, there is an interaction between shade in tea and fertilizer requirements and yield. For example Thung (1966) reports that removal of shade in Java required the use of additional fertilizers. Fully-exposed leaves may become small and yellow particularly without fertilizers. Ripley (1967) studied the micrometrology of shade on tea and showed that the diurnal range of temperature was reduced by shade. As mentioned before, Assam-type jats with broad horizontal leaves suffer from heating and consequently the effect of shade seems more related to temperature control than to a direct light effect (compare cocoa). Sharma (1968) states that there is a rapid decrease in the rate of net photosynthesis if the temperature of the leaf exceeds 35°C (see further on). Shade, when required, therefore should be light and should reduce temperature without reducing photosynthesis to a critical level.

Flowers

Flowers are produced in the axils of older leaves. They open when mature to release three large seeds of about 1.5 cm diameter. In planting, floatation is used to separate good and bad seeds.

Periodicity of shoot growth

Shoot and leaf production of tea shows a periodicity (flushing) which is apparently endogenous in origin and unrelated to climatic fluctuations. It is likely that most plants show periodic growth of this nature but this fact has become accentuated in tea because of the relationship to plucking and yield. Bond (1942, 1945) has described the sequence of events in flushing and dormancy. During dormancy the terminal bud develops slowly and unfolds to produce a much-reduced scale leaf or a smaller intermediate leaf known in the tea business as a 'janam'. The dormant bud is referred to as a 'banjhi'. The bud comes out of dormancy by producing normal leaves and elongated internodes (Fig. 8.1b), and then after a number of these have been produced, relapses into the state of 'banjhi'. Fig. 8.2 indicates the variation in yield with periods of flushing and dormancy.

The periodicity of tea flushing is also independent of plucking and a normal sequence of events is for the bud to produce two scale leaves, a 'janam' or fish-leaf, and a number of flush-leaves. The primordia of these various leaves are already pre-

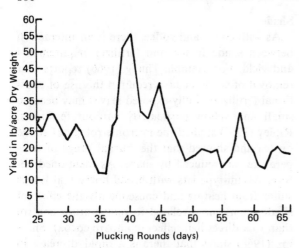

Fig. 8.2 Periodicity of tea flushing as shown by yield

Source.   After Eden, 1965.
Note.   Plucking rounds at 9-day intervals. Data from
            Ceylon.

formed at the time of the commencement of a
'banjhi'.

Periodicity appears to be related to nutrient supply. Under conditions of N deficiency an extended
series of reduced leaves will be produced, and under
conditions of high nutrition a continuous series of
flush-leaves may be produced. There is evidence
that control of flushing can be achieved by the use
of the plant growth regulator gibberellic acid and
various growth inhibitors (Yanase, 1970).

CLIMATE AND SOILS
Climate

Tea, as explained before, is grown successfully
in a great range of geographical situations and consequently in a great range of climates from humid
tropical to sub-humid and temperate. Carr (1972)
has recently reviewed the climatic requirements for
tea. Fig. 8.3 (after Eden) shows areas of traditional
China tea culture and regions of the plantation tea

Fig. 8.3   Map showing main tea-growing areas

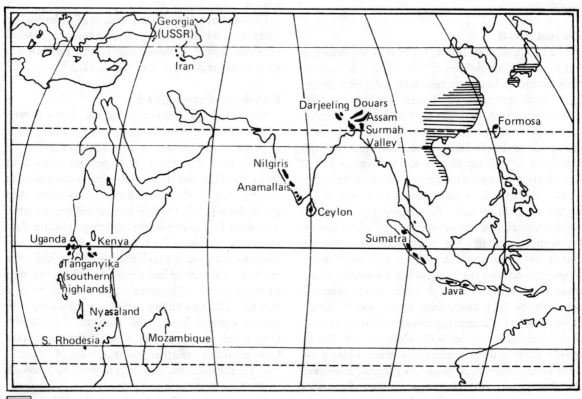

▤ Areas of traditional cultivation

■ Mainly areas of plantation industry

Source.   After Eden, 1965.

industry. Being a leaf crop, as opposed to the more complex situation of a fruit crop, tea is probably less critical than most tropical crop species in regard to climatic requirements. From the point of view of high yield, however, temperature, moisture supply and photoperiod are all important.

Minimum rainfall for tea is considered to be in the range 1150–1400 mm (c. 40″–55″) per annum (e.g. Eden, 1965). Distribution of rainfall is also important and Eden observed that if monthly rainfall remains below 50 mm (c. 2″) for several months, crop production declines severely. In Malawi tea consumed water at an average rate of about 0.9E, where E is the Penman estimated open-water evaporation (Dagg, 1970; Willatt, 1971). Weather factors affected evaporation in the range 40–115 mm per month (c. 2″ to 5″) and water use has been expressed by the relationship:

$$Et = Eo[0.9a + (1-a)0.9n]$$

where a is the fraction of soil covered and n is the fractional number of rainy days per month.

The relationship between yield and moisture has been studied by a number of workers. An irrigation trial with watering frequencies ranging from twice weekly (no soil moisture stress) to watering at 15 atm. soil tension, showed a pronounced effect of water up to the highest level (Fordham, 1971). However response to moisture is also affected by clone or genotype (e.g. Dutta and Sharma, 1967). Portmouth (1957) found that preceding rainfall some two months earlier was most strongly correlated with yield emphasizing the importance of soil water in tea growth, and Laycock (1958) found annual yield in Malawi could be expressed by the equation:

$$Y = 0.0909E + 0.0472M - 0.0600D + 1.7932$$

where Y = yield in 100 kg/ha, E = rainfall from early rains (Nov.–Dec.) in cm, M the main rains (Jan.–May) and D rainfall in the dry season (June–Oct.). The absence of strong response to dry-season rainfall could be related to the photoperiodic requirement of tea since this is the short-day period in Malawi (see below).

There appears to be no upper limit to rainfall for tea, bearing in the mind the type of soil and its tendency to waterlogging, and the associated effect of rainfall on sunshine and temperature. Tea appears to be adequately supplied with moisture via the soil and therefore high rainfall associated with

high sunshine, such as in areas predominated by convectional rain falling in the late afternoon, might provide the best growing conditions. However, Petinov (1961) reports that aerial irrigation water increases yield in southern Russia, although in East Africa (see Annual Reports, Tea Research Institute) this was found to reduce yield, but the reduction was attributed to lowered temperatures. The work of Williams (1971) supports the view that yield is reduced under conditions of low humidity below about 80 per cent under tropical conditions and in certain areas mist is considered to be an important source of moisture. On the whole, however, tea seems possessed of certain xerophytic features (vis. the thick cuticle—see p. 113—and plant water use lower than E or open surface evaporation). Probably near optimal moisture conditions would be obtained by evenly distributed rainfall of the order of 6″ per month under tropical conditions.

Temperature is very important in relation to the production of leaf in most plant species. Hadfield (1968) and Sharma (1968) found that assimilation increased up to about 35°C in tea and then declined. Differences in the temperatures developed with the horizontal and vertical leaf form have already been mentioned in relation to shade. Horizontally disposed leaves can reach levels 10°C higher than ambient air temperature; less so with vertically orientated leaves which reached 6–8°C higher than ambient air temperatures.

Green (1970) in Malawi has shown that leaf expansion occurs only above 21°C. In Malaya, under humid tropical conditions, higher tea yield in the lowlands (mean max. temperature about 32°C) than in the highlands (mean max. temperature c. 25°C) is attributed to temperature, since conditions of husbandry, soil and clonal material used are otherwise similar. In these locations however, rainfall is generally adequate and leaf heating probably minimal.

Soil temperature also affects leaf expansion in tea, 20°C being a critical level. Optimal soil temperature (at 0.3 m) has been found to be about 25°C (see Carr, 1972).

Tea appears to go into a state of leaf dormancy under short-day conditions, (e.g. Schoorel, 1949; Barua, 1969) and this is of considerable importance in high latitude areas such as northern India, Malawi, Russia and so on. Gibberellins and kinins have been reported to release dormancy in tea buds (Kulasegaram and Kathiravetpillai, 1971).

Frost and hail

Tea is adversely affected by frost but nevertheless is not killed, and frost occurs in many of the important tea-growing areas.

In the East African highlands hail sometimes causes major crop losses. Attempts have been made to suppress hail using silver iodide to seed hail-bearing clouds (Henderson *et al.* 1970).

Wind

Wind breaks of shelter trees are often retained in tea planting and have been reported to increase yields provided sufficient soil moisture is present, but under dry conditions trees compete for moisture with the tea.

Soils and drainage

Tea is grown on a fairly wide range of soil types, but because of the requirement for an acidic soil reaction (see below) is more restricted than some of the other pan-tropical crop species. In the high rainfall areas required for tea, and the location of much tea in inland areas, latosols derived from a variety of materials are commonly used. These include latosols of granitic origin and those from various sedimentary and other parent materials. In Sumatra and East Africa, soils from basic volcanic rocks are used for tea but because of the high rate of leaching these are also of an acidic reaction.

Ikegaya (1970) reports that yield is associated with exchangeable base saturation in Japan. However, the most important soil chemical feature for tea soils is an acidic reaction. The upper limit is considered to be in the range pH 6.0 to 6.5 and moderately good tea can be grown on soils of pH 4.5 or even pH 4.0. The requirement for acidic conditions has been associated with aluminium accumulation (Chenery, 1955), and tea has been shown to contain up to 17,000 parts per million of aluminium. The available aluminium status has been a useful diagnostic characteristic for good tea soils, possibly because aluminium plays some regulatory role in uptake of ions such as manganese (see Eden, 1965) or, as has recently been suggested, is associated with P uptake through an inorganic Al/P complex molecule or ion (Sivasubramaniam and Talibudeen, 1972).

Tea soils must be free-draining since tea is adversely affected by a high water table whether stable or fluctuating (e.g. Ghosh, 1968; Dutta, 1968). Heavy clays are therefore less suitable unless adequate drainage improvement is carried out (e.g. Silva, 1967, 1968).

PLANTING

Land preparation, spacing and yield

Tea requires more exacting land preparation than tall-statured tropical crops because of the low plucking table or hedge system developed, and the frequent need for access for harvesting. Furthermore, in the hilly locations often used for tea planting, careful contour planting and terracing are desirable on many soils (for example, Venkataramani (1968) and Eden (1965) state that the three most important aspects in land preparation for tea are: (1). adequate protection from soil erosion; (2). good preliminary cultivation to assist weed suppression and root development, and (3) shade to protect soil and the young developing plant. The desirability of shade, however, is dependent on climatic conditions and the clone of tea planted and has already been discussed.

Removal of stumps from the soil and large roots is normally carried out, followed by burning. However the spreading of ash to avoid high concentrations which would produce patches of high soil reaction (which would affect seedling development) must be carried out. Eden recommends the growing of cover plants after clearing to combat erosion. Maize or oats (e.g. Robson, 1971) might be planted initially, depending on the climate. In clearing unforested land burning is best avoided.

Narrow bench terraces are recommended on steep land and contour planting should be undertaken, particularly on land too steep for the use of machinery (see Eden, 1965). On easily-eroded and impermeable soils, careful drainage and terrace formation are desirable. Terrace drains of not more than 1:100 slope are recommended. Where possible, self-draining permeable soils such as granitic latosols are best used for steep land cultivation, and in fact many tea areas are established on stable soil systems where contour planting has been neglected (Fig. 8.4).

Spacing is important to yield in tea. In contour planting on hilly land contour distances of average $2\frac{1}{2}$ ft are recommended (e.g. Glover, 1949; Daniel, 1951). Traditional spacing of $4' \times 5'$ (2,720 plants per acre) has been far too wide for maximum yield. Eden has assembled various data indicating the relationship of density to yield which is shown in Fig. 8.5 (after Laycock, 1961). These curves do not

Fig. 8.4 Tea on steep land in Peninsular Malaysia
Note. This is a particularly stable granitic soil and contour planting has not been followed.

conform to those of plant species capable of adapting to a range of spacings, except at densities above about 6000 per acre, and this may be related to the pruning-back procedures used in tea cultivation (see p. 112).

The quantities of water transpired by tea (see above) do not suggest that dense spacing will place heavy demands on water.

Recent investigations indicate that optimal spacing for yield may be of the order of 12,000 plants per acre (Laycock, 1967; Dutta, 1968 and Annual Report of the Tea Research Institute, East Africa 1965). In cultivating tea at a 'conventional' spacing of 152 cm (*c.* 2500/acre), Barua and Dutta (1971) found that shoot density rapidly decreased from the centre to the periphery of each bush. They reported that optimal spacing (in dm) could be obtained by dividing square root of the total number of shoots by the shoot density of a central 2×2 dm square. For the tea under consideration this worked out at about 9 dm (approx. 5000/acre), rather lower than that indicated by workers in different locations and with other jats. It seems highly likely that clones with vertically orientated foliage will be found to have a higher optimal density than broad-leaved types.

Planting out should be done in a humid season and the branches of the young seedlings (see below) orientated along the planting lines to encourage spread of the bushes in this direction.

Yields of made tea per acre are indicated in Fig. 8.5. With newer clones, however, particularly those with the more productive canopy characteristics yields are much higher. Yield is also closely related

Fig. 8.5 Effect of plant density on yield in relation to age

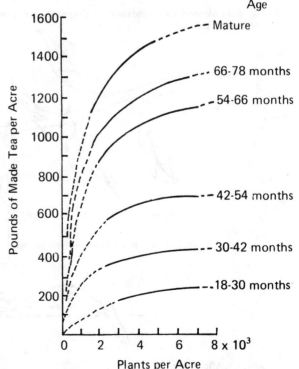

Source. From Eden, 1965; after Laycock, 1961.

to nitrogen fertilization (see further on). High-yielding clones in India are reported to produce average yields of 2500 kg/ha in the 6th and 7th year of planting (see *Two and a Bud* 17, p. 53) and in Ceylon the tea clone TRI 2032 has yielded an average of 2735 kg/ha in the first year and 4800 kg/ha in the second year of three pruning cycles, with heavy N fertilization (Fernando, 1969). This exceptional clone has also been reported to produce the phenomenal yield of 9000 kg/ha (Harler, 1966; Nanayakkara, 1968).

The higher temperatures of tropical lowlands, with adequate rainfall, also promote high leaf yield, but as explained before quality is reduced.

Nursery

Traditionally seed was used for planting tea and various attempts have been made to produce clonal seed from planted nurseries of superior trees of one or more types—from trees which combine favourably to produce superior progeny (e.g. Wright, 1942). Seed-bearers are planted at a density of 300/acre and are pruned to a height of 10–12 ft. At the present time most new commercial tea is established from vegetatively propagated material from superior mother bushes in the area. While the scope for improvement of tea by genetic manipulation is considerable, since tea is a highly hetrozygous plant (e.g. Richards, 1966), the vegetative method has the advantage of reproducing climatically adapted plantings of uniform genetic composition. Furthermore, the preservation of quality has been reported to be best achieved by vegetative multiplication (Magloblisvili and Arobelidze, 1970).

In addition to desirable production and disease resistance characters, such vegetative material must root easily from cutting. Variations exist in this character.

Single-leaf cuttings are most commonly used in tea, although Green (1970) has suggested that when excess cutting material is available, cuttings with two or three leaves may establish sooner. Figs. 8.6 and 8.7 illustrate the nature of single-leaf planting material used and the method of planting. Poedjowardojo and Darjadi (1969) report that slightly more active rooting occurred in cuttings obtained during a dormant stage. Flower-bearing nodes are not suitable for cuttings. Whether rooting hormone (napthalene acetic acid—NAA) is used depends on the natural rooting ability of the clone and is determined by experience. Figueroa (1970) reports that high levels of NAA (10,000 ppm) are needed in

Fig. 8.6   Method of preparation of single-node cuttings for vegetative propagation and method of insertion of cuttings in the soil

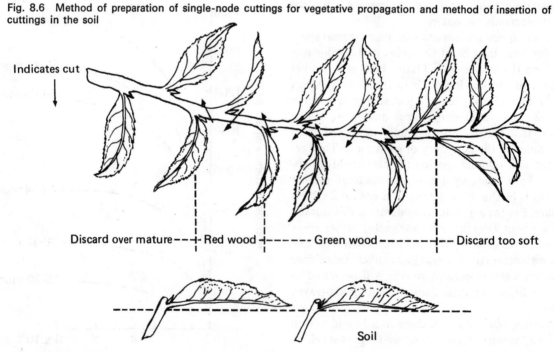

Indicates cut

Discard over mature --- Red wood --------- Green wood ------- Discard too soft

Soil

Source.   After Eden, 1965.
Note.   A sharp cut is needed for rooting to take place.

Fig. 8.7a A nursery bed for single-leaf cuttings under polythene cover

Fig. 8.7b Rooted cuttings transferred to open nursery

Peru, and that China jats respond better to rooting hormone than Assam types.

Fresh cuttings should be stored in shallow vessels of water or in polythene bags if not planted immediately. In any case they should be planted within 48 hours (e.g. Barua and Grice, 1968). Cuttings should be taken early in the day for best results.

A sandy rooting medium is required for tea, with a pH not higher than 5.5, and a low content of organic matter, at least in the upper layer which is in contact with the cutting. Subsoil is commonly used as opposed to topsoil (e.g. Pilot, 1968; Green, 1968), and addition of compost is avoided since this retards rooting although callus development occurs.

Leaf cuttings are rooted in perforated polythene bags or sleeves or in nursery beds as shown in Fig. 8.7. These are covered to maintain a high humidity but watering must be carefully controlled and not too heavy. Mist propagation has been shown to be successful in Kenya (Hartley, 1969), and fumigation of nursery soil against nematodes and soil pathogens is sometimes required. Phosphate or super phosphate (e.g. 0.5 kg/m³ Pilot, 1968) is commonly added to the nursery soil, and after establishment (about 1 month) a complete liquid fertilizer can be used but rather sparingly (e.g. 1 oz per 1000 cuttings). Nursery failure can often be attributed to over-zealous fertilizer use on the part of tea planters (see Pethiyagoda, 1970).

When rooting has occurred tea plants are removed to an open nursery planted in individual containers (usually polythene sleeves Fig. 8.7). Shade should be at first heavy (*c*. 80 per cent) but progressively removed in preparation for planting out. Seedlings in the open nursery may be pinched off to encourage branching before planting out, approximately eight months after cutting under favourable conditions. Branch shoots often grow faster than the original from the leaf cutting.

The selection of seedlings at planting out has been unsatisfactory in tea since vigour of nursery seedlings bears little relation to subsequent field performance (Green, 1971).

Grafting

Grafting is used in tea chiefly to propagate generative material either for breeding or vegetative propagation. Both cleft grafting and patch budding are used (e.g. Grice, 1968; Barua, 1968; Katsuo, 1969). Graftage produces vigorous material but is unsuitable for field planting because of the high number of plants required per acre.

Manuring and irrigation

Undoubtedly the most important fertilizer for tea is nitrogen, and this is related to the use of this element in leaf growth. Urea has been shown to be a suitable N source both in soil application and as a folia spray (e.g. Raghavan, 1970; Roy *et al.* 1970; Bhavanandan and Fernando, 1970). However losses of N from urea applied at heavy rates tend to offset the advantage of its use in terms of cost. Nitrogen is more effective in unshaded tea, as might be expected from experience with other crops, and with high-yield jats.

In the past sulphate of ammonia was used extensively in tea fertilization. In cases where sulphur deficiency occurs (e.g. Grant and Shaxson, 1970) it is advantageous to use this source of N periodi-

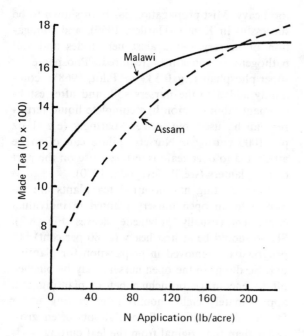

Fig. 8.8   Response of tea to N in Malawi and Assam
Source.   After Eden, 1965.

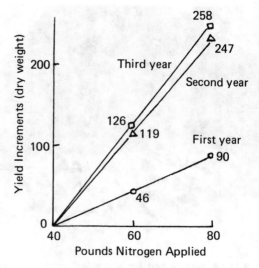

Fig. 8.9   Manurial response at different stages following pruning
Source.   After Eden, 1965.

cally. In Malawi, Grant and Shaxton consider the supply of sulphur once in three years as satisfactory. Calcium ammonium nitrate is not suitable for tea as a regular N source because of the build-up of a high Ca ratio in the base elements of the soil and the raising of pH to undesirable levels (e.g. Wilson, 1967, 1968).

Tea responds to high levels of N as shown in Fig. 8.8 (after Eden, 1965 from various data). However response to even higher levels occurs in more productive jats in full sunlight. Akhmetov and Bairamov (1967) for example report yield response to 400 kg/ha of N.

In some estates it is the practice to increase fertilization only on highly productive fields. The response of tea to high N levels increases following regeneration of the canopy after pruning as shown in Fig. 8.9.

Phosphorus nutrition is usually satisfied by annual application of phosphate of the order 20 kg/ha P_2O_5. Potassium is less often reported as a nutrient producing yield response in tea, although the crop removes significant quantities of K. Eden (1965) reports that in Ceylon a 12-year cessation of K fertilization is required before K responses become evident in mature tea. Young tea, however, with limited root systems often responds to K.

Micronutrient deficiencies occur in tea on the leached soils frequently used for its cultivation. Zinc deficiency is most common and this is usually supplied as a foliar spray, often mixed with fungicides or pesticides for convenience of application (e.g. Venkatarami, 1968). Boron, magnesium and copper deficiencies also occur (see Harler, 1971).

Foliar diagnosis is usually used for assessment of tea fertilizer needs, using the two terminal leaves and bud as sample material. Optimal N levels recommended are in the range 3.5–4.8 per cent (Lin, 1965; Akhmetov and Bairamov, 1967); P of the order 0.25–0.35 and K of the order 1.25–2.0 (Lin, 1965; Tolhurst, 1971).

With high-yielding clones it seems likely that irrigation could be highly profitable in tea; chiefly sprinkler irrigation has been tried (Othieno, 1968; Winter, 1969). Willatt (1970 and 1971) has shown that irrigation increases plant survival and root and shoot growth. The effect of aerial water application on yield as opposed to water supplied *via* the roots has been discussed before under 'Climate'.

Pruning, plucking and bush formation

This book is concerned mainly with tropical regions and therefore the methods of mechanical plucking developed in Russia and other countries are not considered here. Since plucking is also a light type of pruning these two operations are considered together. The main aims of pruning as outlined by Eden (1965) are as follows:

1. To maintain the plant permanently in the vegetative phase.

2 To stimulate, in particular, the young shoots that constitute the cropped portion of the bush.

3. To form as extensive a frame as possible in two dimensions on which in the third dimension, the 'flush' will rapidly and continuously regenerate.

4. To keep the height of the bush within the bounds of easy and efficient plucking.

5. To renew the actively growing branches so that replacement of healthy wood and foliage keeps pace with the ravages of death or damage; to maintain a sufficient volume of mature foliage to meet the physiological needs of the plant and to promote rapid renewal of flush suitable for manufacture into tea of quality.

In the process of plucking tea which is carried out periodically (see below) it will be apparent that the leaf area index is being continuously reduced. From the production point of view it is highly desirable that this process tends to maintain the tea in a condition close to optimal leaf area index which, as discussed in the first part of this book, is the most productive condition of the canopy. It may be said that the operation of pruning should be undertaken when regeneration of the lower regions of the canopy to achieve a more productive leaf area has become necessary.

In India the practice of continuous top pruning at the time of plucking or the removal of shoots projecting above the plucking table keeps the table low and also opens the lower horizons of the canopy to more light for generation of new shoots from below. For this reason severe pruning, or cutting back the canopy to about 18″, is carried out only once in twelve or more years. When this top pruning is not practised, severe pruning needs to be carried out more frequently (from three to six years) as is largely done in Ceylon, particularly in the past.

The various types of pruning are illustrated in Fig. 8.10. Pruning requires the mobilization of starch reserves from the roots for regeneration of the canopy and is much more effective when reserves are high. In equatorial locations elevation greatly affects the accumulation of reserves from a level of about 11 per cent in the lowlands to about 24 per cent at 6500 ft (Tubbs, 1936). The occurrence of a dormant season (short days—see above) inhibits flushing and causes a build-up of reserves and this is usually the season in which pruning is carried out (e.g. Dutta and Grice, 1966). In tropical lowland situations pruning has to be greatly curtailed and life of a tea bush is generally much shorter. In lowland situations pruning is best done in stages and the lighter type of rim-lung pruning

Fig. 8.10 Diagrammatic illustration of types of pruning used in tea cultivation

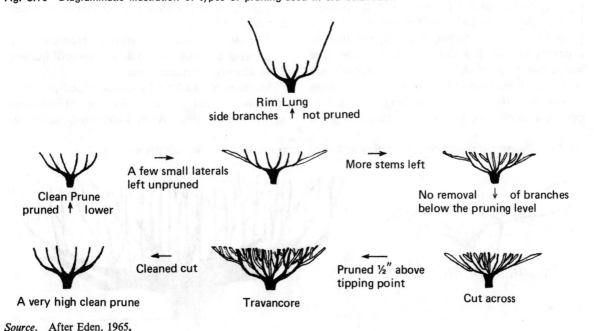

Source. After Eden, 1965.

employed (Venkataramani and Venkata Ram, 1968). Prior to pruning, fertilization with P and K as well as the normal N is recommended and resting for 8 weeks aids build-up of reserves (Sakar, 1970). Pruning after a short dry spell, in the absence of a winter season, should be advantageous in terms of reserves. Pruning during rains is not recommended.

A light form of pruning, used to level the plucking table when it cannot be conveniently improved by breaking off individual shoots by hand, is known as 'skiffing'. The value of skiffing in stimulating yield has been questioned (Cooper, 1931). Skiffing of tea has been considered in the *Encyclopedia of Tea* of the *Tocklai Experiment Station*.

An extreme type of pruning is known as collar-pruning and this is carried out only in exceptional circumstances, usually to combat diseases. In collar-pruning the bush is cut back at ground level and should be covered with a thin layer of soil to reduce die-back of the stump.

Bush formation. Bush formation in the establishment phase also involves pruning procedures to reduce apical dominance and encourage the growth of side shoots. The pinching off of seedlings in the nursery has already been mentioned. In the past, three main methods of bush formation were used. In the first the tea bush was allowed to grow for a period of up to three years before cutting back to 18″ and then allowing it to regenerate to a plucking height of about 27″. In the second, the cutting-back height is about 6″ from the ground after the plant has reached a height of 3–5 ft, and this encourages the development of a low frame. Tea which has been established by this method and subsequently medium-pruned at about 18″ will have a two-tiered structure of branching, and is still quite commonly encountered in estates. A third type of pruning is the cutting back of young bushes when the branches are no more than pencil thickness on two occasions, first at 6″–10″ and later at 12″–18″.

A fourth method of bush formation much used in modern times is 'pegging' (e.g. Visser, 1969; Manipura, 1971). The plant is pruned back and after about 6 months in the field, lateral branches are pegged with wire hooks at about 25° from the horizontal; after a further six months the branches will be fixed in this position. Vertical shoot production on the pegged branches is encouraged by this process and these may be pinched back to encourage a dense stand as shown in Fig. 8.11.

Plucking. The frequency between plucking rounds is a management problem and depends on many factors including the cost of labour. Obviously too frequent rounds will produce too few shoots to be economical, and too infrequent plucking will produce overgrown shoots that may need to be discarded. The question of periodicity of flushing and effects of temperature and fertilizers also need to be taken into account. In practice, frequencies vary from several days to about two weeks during periods of active leaf growth.

Generally the best plucking level is considered to be two leaves and a bud (Fig. 8.1) from newly developed shoots. Plucking too low, back to the unexpanded scale and small leaves of the flushing cycle, greatly reduces subsequent yield since it is important to leave sufficient maintenance foliage for photosynthesis.

The use of growth regulators to promote flushing during dormant period and to control flushing have already been mentioned.

Mechanical plucking by means of self-propelled machines is not practised in tropical countries at the present time. A back-mounted motor tea

Fig. 8.11 Diagrammatic illustration of 'pegging' and frame formation after pinching off laterals

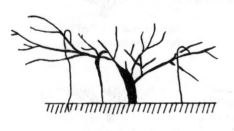

plucker has been used in Japan (Torii, 1966), and pruning shears fitted with a collecting bag have also been used with some success (e.g. Whitlam, 1966).

Weed control and mulching

In the past, weed control has been by hand cultivation. Chemical weeding is now widely used and has the advantage of not disturbing the soil (e.g. Hainsworth, 1969). Dalapon (for control of grasses) and paraquat have been extensively used in tea (e.g. Venkataramani, 1968; Newing, 1971), but spraying must be confined to the soil and weed surfaces avoiding the tea foliage by using a droplet-generating spray head as opposed to a mist sprayer. Diuron, simazine and atrazine have been used for long-term control (e.g. Wettasinghe, 1968; Dutta, 1967). The arsenicals DSMA and MSMA have also been used in tea (Roy *et.al.* 1970).

Mulching with grass, leguminous and other materials has been much used in tea and generally promotes yield (Barbosa, 1970; Tolhurst, 1970, etc.). Furthermore, prunings should be left on the field to act as a mulch, unless they harbour disease organisms (e.g. Newing, 1971).

MAJOR DISEASES AND PESTS

Tea is affected by a number of serious diseases of which 'blister-blight' caused by *Exobasidium vexans* is probably the most important in the Asian region. The disease causes small translucent spots which expand and become depressed concave regions giving the impression of blisters on the reverse side of the leaf. These become white and release spores of the fungus. Young leaves primarily are attacked and chiefly during wet weather (Kerr, 1966). Control is by the use of less susceptible clones, by confining pruning operations to dry weather and by the use of copper fungicides (e.g. Silva, 1967). Recently nickel sprays have also been shown to be effective (Venkata Ram, 1967).

Various root diseases affect tea, sometimes seriously. In Asia the fungus *Poria hypolateritia* causes death to tea bushes, and in Africa *Armilaria mellea* causes similar 'sudden death' of bushes. These fungi proliferate on old stumps of felled shade trees and are controlled by careful excavation of stumps. Soil fumigation procedures are expensive and have not generally proved satisfactory. In lowland situations, *Ganoderma pseudoferreum* may be troublesome in the regions of old tree stumps. *Ustilina deusta* occurs in high elevation and high latitude areas. A disease known as 'violet root rot' caused by *Sphaerostilbe repens* occurs under waterlogged conditions.

Various stem and canker disease conditions also occur in tea. These are caused by *Phomopsis theae* which results in collar and branch cankers. Resistant clones offer the best control measure (e.g. Shanmuganathan, 1969), and improvement of soil conditions (e.g. Silva and Fernando, 1968).

A virus disease causing phloem necrosis occurs in tea (see Bond, 1944), and Eden (1965) describes a physiological condition in seedlings referred to as 'bitten-off disease'. This latter is marked by root rot and leaf etiolation followed by shedding. It is characteristic of poorly-drained soils and high soil reaction.

Insect pests can be troublesome because of the danger of using insecticides on a leaf crop. A good example of biological control of an insect pests occurs in tea. The tea tortrix caterpillar *Homona coffearia* attacks tea in Ceylon and has been controlled by annual pruning and by introduction of the wasp *Macrocentrus homonae* which is a natural predator in Indonesia (see Eden, 1965).

9 RUBBER [Hevea brasiliensis]

INTRODUCTION, SPECIES

RUBBER (*Hevea brasiliensis*) is a dicotyledonous tree which originated on the South American continent and first gained commercial importance in the mid-eighteenth century in Europe because of the unique elastic and water proofing properties of its precipitated latex. The invention of vulcanization by Goodyear in 1839 (which utilizes sulphur to improve the elasticity and durability of rubber) greatly stimulated the use of rubber and eventually led to the establishment of the rubber industries of South-East Asia which are, today, the main centres for the supply of natural rubber to the world. In spite of the development of synthetic rubber, natural rubber still retains a prominent role in industry and seems likely to continue to do so.

Prior to 1900, all the rubber came from naturally-growing wild trees in Brazil and neighbouring countries. Rubber was introduced to South-East Asia following the collection of seeds in Brazil by Wickham (see McFadyean, 1944). There are actually several species of *Hevea* (*H. spruceana, H. guianensis, H. callina, H. pauciflora, H. nitida* and others), and latex-bearing trees (e.g. *Ficus elastica, Castilloa elastica, Manihot glaziovii, Parthenium argentatum*, etc.) but only *H. brasiliensis*, or those clones which are recognized as *H. brasiliensis* has become established as the major source of natural rubber. The Asian rubber tree is generally recognized as *H. brasiliensis* and has a chromosome number of $2n = 36$ (see Dijkman, 1951) but it seems possible that it also contains germ-plasm of other species since Siebert (1947) and Baldwin (1947) have shown that the species readily hybridize, and that a wide range of geographic races exist. Furthermore, some Asian clones (now discontinued) have characteristics of other species; for example, clone PB84 from Malaysia has an overlapping-leaf characteristic of *H. guianensis*, and others have latex characteristics (deep yellow and black discoloured latex) of *H. nitida* or *H. pauciflora*. Yet another clones have flowering characteristics (all-year-round flowering) of *H. pauciflora*. The vigour and high yield of the successful Asian rubber tree clones is, however, attributed to the species *H. brasiliensis*

After 1920, rubber returned to South America as well as to West Africa as a plantation crop, but because of various problems, notably the occurrence of certain diseases, less than optimal climatic conditions and the absence of good technology and extension facilities, thriving industries have never really been established outside of the South-East Asian region except in Liberia.

Through selection and breeding, and improvement of planting and tapping techniques, the yield of raw rubber has been greatly increased from the average of about 500 lb/acre/year obtained from early unselected plantings to over 2,000 lb/acre from modern clonal rubber (see later).

The techniques of propagation and bud grafting used in the South-East Asian rubber industries have led to a high uniformity of planting material and an improved tapping panel. A newly-revived technique of crown budding, the production of new genetic strains and improvements in tapping methods are expected to save the natural industry from extinction as a result of competition with its synthetic counterpart. In the long-term, the prognosis for natural rubber is good, since the raw material used in the production of synthetic rubber, namely crude oil, has a definite limit of supply.

BOTANY

Vegetative and bark characteristics

Rubber is a tree crop and Figs. 9.1 and 9.2 show various aspects of the external morphology and tapping process of the crop.

From the point of view of yield, the nature of both the bark and the girth of the tree is of central importance. High-yielding clones have well-developed lactiferous tissue which gives a high latex yield per unit of cut length, and also a relatively high girth size and increment under conditions of

Fig. 9.1a Weeding in young rubber

Fig. 9.1b Mature rubber, clone RRIM 501

tapping. Fernando and Tambiah (1970) report that high-yielding clones have longer scieve tubes in the phloem. Features of the latex itself, its rubber content and the tendency for flow to stop due to plugging or for other reasons, are also important to yield, and generally the properties of the bark and latex are considered to be of central importance. The technique of crown budding to produce a three-part tree with a vigorous stock, a trunk possessing the desirable bark properties, and a crown resistant to wind damage (e.g. Yoon, 1967, 1971) and diseases (e.g. Radjino, 1969) offers considerable promise in improving yield. However, at present, most plantings consist of budded seedlings comprising a trunk and crown of a superior clone and a root stock from a suitable seedling (see 'Nursery Practice' below). Fig. 9.3 indicates the nature of the bark of the rubber tree (diagrammatic). The anatomical structure consists of a thin cork layer, a zone of hard bark containing numerous stone cells which decrease towards the inside, and the lactiferous tissue which is made up of parenchyma and groups of latex vessels, abutting on the cambium. The concentration of latex vessels increases towards the cambium and thus it is important to tap as close to the cambium as possible without

Source. Photographs by S. Subramaniam.
Note.   This clone is discontinued because of susceptibility to wind damages.

Fig. 9.2a Straight-trunked grafted rubber trees and old trees grown from seedlings showing tapering trunk

Fig. 9.2b Half-spiral tapping cut and collecting cup

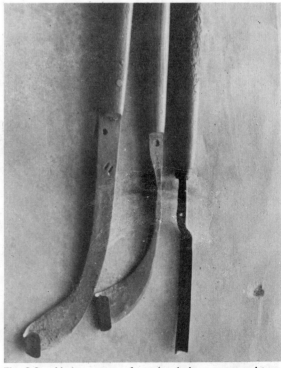

Fig. 9.2c Various types of tapping knives; recurved type for tapping high panels above chest level approximately

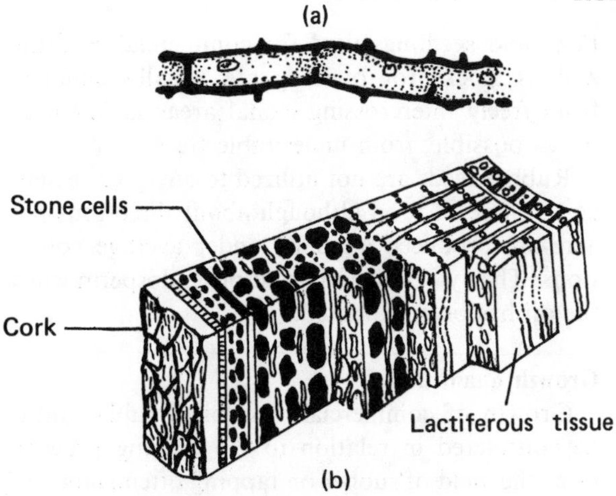

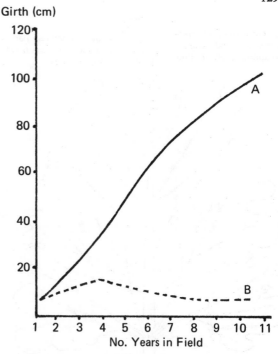

Fig. 9.3   Latex vessel and bark of rubber tree

Source.   After Bobiloff, 1923.
Notes.   a. Dissolution of cell walls to form part of the
            lactiferous system. b. Bark showing cork layer,
            stone cells and lactiferous tissue (from Dijkman,
            1951).

Fig. 9.4   Changes in girth (A) and girth increment (B)
in rubber

Source.   After Vollema and Dijkman, 1939. Clone LCB 501
            buddings.

actually touching it, an operation which calls for considerable skill on the part of the tapper.

The anatomy of the bark has been investigated by Arisz (1918, 1919–20), Bobiloff (1918, 1923), Schweizer (1949) and others. Latex vessels tend to run in concentric circles, are often laterally interconnected and have a spiral orientation with reference to the trunk as a whole (see 'Tapping' below). Apart from genetic factors determining the nature of the bark, it is also affected by climate and soil factors, and under dry conditions and in poor soil, the size of the stone cell zone increases.

The development of the trunk and quantitative aspects of growth are of considerable importance. Trunk shape varies between trees grown from seed and those from buddings, the former being more tapered (Fig. 9.2). Seedling trees are consequently opened for tapping at a lower height than budded trees. In mature trees grown from seedlings, the bark mantle is thicker in the lower regions (below about 4 ft) and thinner above. In budded trees the trunk and bark are very uniform in diameter and thickness.

Girthing and branching

Initially in the young tree the trunk gains rapidly in height, but once branching occurs, terminal growth slows down and the girth increment is accelerated (see Fig. 9.4).

Girth increment is also important from the point of view of resistance to wind breakage. Trees with

large crowns and a poor girth increment are more susceptible to wind damage.

The branching habit is also very important from the point of view of wind damage. Trees with few large branches joined at an acute angle (e.g. Fig. 9.1, clone RRIM 501) are more susceptible to wind damage. Trees with fairly light branches inserted at an obtuse angle are better from this point of view (see Wycherley *et. al.* 1964 and Fig. 9.5, after Sekhar and Subramaniam, 1972). Corrective pruning has been tried for achieving a better branching habit (e.g. *Ann. Reps. R.R.I.M.* 1960 and 1961) but crown budding in which the trunk and crown of the tree can be selected for desirable properties promises to be most useful in controlling susceptibility to wind damage. In addition to the type of branching and the strength and girth of the trunk, strength properties of the wood and trunk height are also important in determining resistance to wind damage. A dwarf rubber tree resulting from a cross between the Malayan clone RRIM 605 and Ford 351, which measured only about half the normal height has been considered to have potential for reducing wind damage (*Planters Bull. Malaya* 119, p. 53–1972).

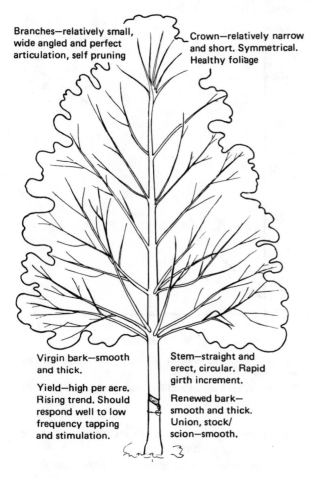

Branches—relatively small, wide angled and perfect articulation, self pruning

Crown—relatively narrow and short. Symmetrical. Healthy foliage

Virgin bark—smooth and thick.

Yield—high per acre. Rising trend. Should respond well to low frequency tapping and stimulation.

Stem—straight and erect, circular. Rapid girth increment.

Renewed bark— smooth and thick. Union, stock/ scion—smooth.

Fig. 9.5   Characteristics of the 'ideal' rubber tree
Source.   After Sekhar and Subramaniam, 1972.

The root system

Rubber trees grown from seedlings or buddings generally have a substantial tap root and extensive lateral root system. In hedge planting (see below) the root system extends laterally into the inter-row space for considerable distances (e.g. 18 metres, Schweizer, 1939). Root systems developed from cuttings or marcotts lack a central tap root and are very wind susceptible. Therefore they are not used in the commercial planting of rubber.

Flowering

Rubber is monecious, producing small panicles of flowers irregularly, or more usually, in certain seasons.

Under uniform humid conditions rubber sometimes flowers lightly all year round. Synchronous flowering is necessary for breeding work. Cross pollination normally occurs under natural condi-

tions and seedlings used for commercial planting and as stocks for budding are usually obtained from freely intercrossing clonal areas isolated, as far as possible, from undesirable trees.

Rubber seeds are not utilized to any great extent as a source of food although about three tons per acre are produced annually under average conditions. They yield oil and an edible endosperm when the oil has been extracted.

Growth quantities

Growth of commercially-exploited rubber must be considered in relation to the tapping process, since the yield of rubber on tapping often shows an inverse relationship to the growth increment of the trunk. Studies on growth of rubber trees under Malayan conditions have been made by Templeton (1968). An example of a growth curve for a rubber stand (Clone 501 from Malaya), and a curve for the development of leaf area are shown in Fig. 9.6. At face value these curves suggest that leaf area

Fig. 9.6   Growth in dry weight (crop growth rate) and leaf area development of rubber clone No. 501 from Malaya

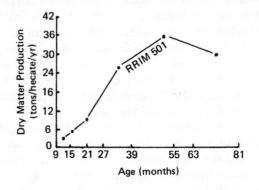

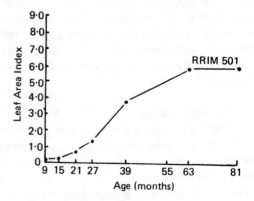

Source.   After Templeton, 1968.

development for clone RRIM 501 (and the tendency is even more pronounced in most other important clones) may be above optimal over the major portion of the life of a rubber tree. Crop growth rate, after four years, is showing a decline while LAI in this clone is levelling off at a value of about 5.8. Clone 501* is not a particularly leafy clone and others reach LAI values of 14 or more. Generally, optimal LAI for productivity in broad-leaved species is somewhere in the range 4–6, and it seems likely that the question of canopy development in relation to productivity might bear more detailed investigation in rubber. Studies on the photosynthesis of high, medium and low-yielding clones in Indonesia have been inconclusive except that photosynthesis was highest when the old leaves were present and new whorls were developing and lowest at the stage of young leaves development following defoliation (Adijuwana & Soerianegara 1970). Senanayake & Samaranayake (1970) were unable to demonstrate any relationship between stomatal number and yield in twenty-five clones; however, it seems possible that yield limitations imposed by photosynthesis might only appear in a situation where very high yields are obtained and with the use of stimulants.

*This clone is a precocious high-yielding clone but is no longer recommended for planting because of susceptibility to wind damage.

Templeton considers that the supply of assimilates is not the primary limitation to rubber yield, at least in the clones studied and under standard conditions of planting. This is supported by the studies of Chuah (1967) on carbohydrates in dry trees. The yield of rubber itself sometimes accounts for only a relatively small fraction of the dry-matter increment (between 10 per cent and 75 per cent of the growth loss of the trunk and crown in the clones studied, occurring as a result of the tapping process). Table 9.1 (after Templeton, 1969) indicates the effect of tapping on the production of dry matter in the form of trunk and crown. The effect of tapping on the increment of the crown dry weight amounts to values ranging between about 50 per cent reduction to virtually no effect. Table 9.2 shows the effect of heavy tapping on the content of certain carbohydrates in the latex of rubber trees.

This work has led to the view that the loss of some other factors during tapping (possibly proteins, RNA or ATP) is the primary limitation to girth increment and yield. However, it has been suggested (Bealing, 1969) that normal sugars and starch may not be the primary carbohydrates for the synthesis of rubber, since latex contains a high level of the carbohydrate quebratchitol.

Dijkman (1951) considers that the pattern of trunk development, which is central to yield, is

TABLE 9.1   Effect of tapping on dry-weight increment of trunk and crown of rubber clones

| | | Dry weight increment/tree | | | | |
		Trunk	Crown	Total Shoot	Kg Rubber	Rubber % weight loss
Clone 501 (Malaya)	Untapped	16.2	71.0	87.2		
	Tapped	4.3	37.4	41.7	4.63	10
	Loss	11.9	33.6	45.5		
Clone 612 (Malaya)	Untapped	22.8	89.4	112.2		
	Tapped	18.4	84.5	102.9	3.09	33
	Loss	4.4	4.9	9.3		
Clone 618 (Malaya)	Untapped	16.3	74.0	90.3		
	Tapped	9.9	73.4	83.3	5.40	77
	Loss	6.4	0.6	7.0		

Source.   After Templeton, 1969.

TABLE 9.2 Effect of tapping on latex constituents of rubber bark

		Normal Tapping S_2/d_2	Heavy Tapping 400% S/d
ug/g	Sugars	20.5	20.9
fresh bark	Starch	136.9	133.9
	N%	0.82	0.68
	Soluble protein mg/ml fresh bark extract	5.15	4.26

Source. After Chuah, 1967.

Fig. 9.7 Changing NAR and RGR of rubber clones

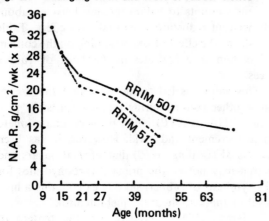

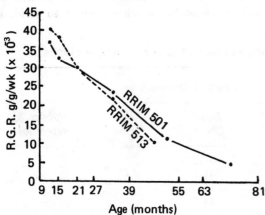

Source. After Templeton, 1968.

similar for all clones, the growth increment being higher in the early pre-tapping phase, as shown in Fig. 9.4. However it is now known that some clones retain a good girth increment in spite of regular tapping, indicating that this may no longer be the primary limitation to yield; and altogether the last word has not been said on the question of optimal canopy development in rubber. Comparisons of net assimilation rates (NAR) and relative growth rates (RGR) of various clones also supports this view. Fig. 9.7 illustrates changing values of these quantities in two clones. In both cases the values decrease rapidly over the first seven years, which in rubber represents virtually only the establishment phase. Also, in the more leafy Clone 513, decline is more rapid. Further, evidence from planting density and planting arrangement experiments show that these factors greatly affect girth increment (Fig. 9.8), and experiments on avenue planting (see below), which permits better light penetration, also support the view that optimal canopy development is generally exceeded in commercial plantings of rubber. A tentative conclusion might be that light and assimilates appear not to be the primary limiting factors in some modern rubber clones. However, in others in which girth increment is not strongly affected by the tapping process, the question of optimal canopy development needs careful investigation. Some modern clones are particularly leafy, producing foliage stands of LAI = 14. The ques-

Fig. 9.8 Effect of planting distance on girth increment in rubber

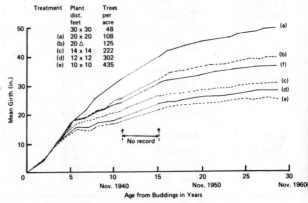

Source. After Westgarth and Buttery, 1965.

tion of achieving optimal LAI has a bearing on pruning investigations (mainly concerned with the prevention of wind damage) clone leafiness and on planting arrangement and planting density.

Studies on the growth of seedlings have shown that rapidity of germination is positively correlated with subsequent growth rate, and that plant height, and leaf area and stem diameter are all positively correlated. Jayasekera and Senanayake (1971) suggest the latter as a measure of vigour of rootstock seedlings. Fernando and Silva (1971) report that growth of seedlings was negatively correlated with plant latex content as determined by 'microtapping'.

TAPPING

In the early days of the rubber industry when rubber was collected from wild trees in the South American jungles, tapping was a very crude affair and the latex was usually obtained by slashing the trees to release it. Scientific tapping was developed initially by Ridley, who was one of the founders of the rubber industry in Malaya, and subsequently by other workers. Recent reviews have been published by Ng et al. (1969) de Jonge (1969) and other workers. Ridley's early work established a system of progressively cutting off a thin ribbon of bark to expose the latex vessels in such a way that the cambium is not disturbed and the bark regenerates after a period of time. Because the latex vessels are inclined spirally in one direction the slope of cut which samples the most vessels is downwards from left to right at an angle of about 30° (Fig. 9.2). Probably the most widely-used tapping system is that in which half the trunk is tapped on every second day, opening the cut when the tree has reached a girth of 20″ at a height of 60″ from the ground, in the case of bud-grafted trees, and at 50″ height in the case of trees grown from seedlings. Good tapping should consume about ¾″ of bark per month with this alternate daily system which consumes the first panel of bark in about five years. (It should be said that, apart from the introduction of scientific tapping by Ridley, the success of the rubber industry in the Asian countries has been dependent on the skill of the tapper, particularly the Tamil tapper.) After this the panel of the bark on the opposite side is tapped, and then the renewed bark of the first panel and so on. Finally, after about twenty or more years of tapping and re-tapping of the two lower panels, new panels may

be opened higher up and tapped with the aid of a ladder. This is usually done on an intensive system (see below) prior to replanting. The alternate daily ½ spiral system of tapping which is the most widespread in use, has been referred to by the formula s/2 d/2. A full spiral system is more intensive and is given the formula s/1, and tapping more frequently such as daily d/1, or less frequently, three-daily d/3 and so on.

Rubber latex contains in the range 25–45 per cent rubber solids, 10–20 per cent leutoids which are membrane-enclosed bodies thought to be similar to lysosomes of animal cells and various other constituents. The metabolic pathway for the formation of rubber is uncertain, but Bealing (1969—see above) suggests it is formed from the carbohydrates inositol and quebrachitol, both of which occur commonly in latex-producing species. The basic molecular unit of rubber is isoprene which polymerizes in the *cis* form with up to 1 million units to form rubber.

$$CH_3$$
$$CH_2 = C-CH = CH_2$$
ISOPRENE

Polymerization in the *trans* form gives rise to 'guttapercha' which is non-elastic. Once rubber is formed, it is resistant to enzymatic action and although energy rich, does not appear to be metabolized by the plant. Rubber is extracted from the latex by precipitating with acid, rolling to express water and then drying or processing in various other ways.

Research on tapping has involved the investigation of other tapping systems (reviews cited above), with the aim of obtaining higher yields per tapper to combat rising labour costs and lowered rubber prices. It has involved frequency of tapping, size of tapping cut, the effect of tapping systems on girth increment, the use of stimulating chemicals which prolong latex flow and tree spacing.

Longer tapping cuts (for example a full spiral as opposed to the standard half spiral) give initially a higher yield per tapping occasion but this occurs at the expense of girth increment and is therefore not practised normally, except towards the end of the life of the tree in the few years before replanting is due to take place.

Lower frequencies of tapping give more latex per tapping occasion, particularly with certain clones, and are also consequently less labour intensive. Girth increment is also favourably effected by lower tapping frequencies but overall yield declines.

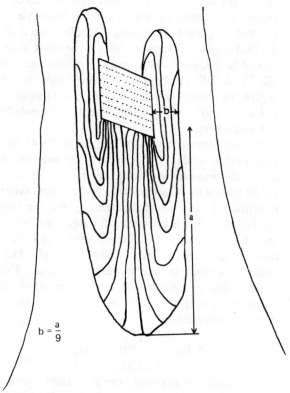

Fig. 9.9 Latex flow area of the tapping panel
Source. After Frey-Wyssling, 1932. Drawn from Dijkman, 1951.
Note. a=vertical flow influence, b=horizontal flow influence.

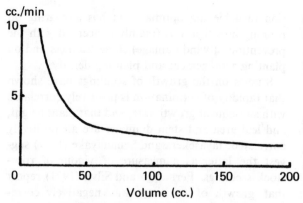

Fig. 9.10 Changing flow rate of latex with time-generalized form of flow curve
Source. After Frey-Wyssling, 1933.
Note. The horizontal axis may also be considered to be time.

Probably the most promising results from the point of view of the rubber industry have been obtained by the use of chemical stimulants, some of which are widely used at the present time, and this subject is considered in more detail below.

Latex flows chiefly from the region of the bark below the tapping cut, but also to some extent, from the side and extending above the cut (Fig. 9.9). The area of bark which is sampled is generally more extensive with high-yielding clones and may extend 5 ft or more from the cut (Frey-Wyssling, 1932). Flow rate is at first rapid and then falls off in the manner illustrated in Fig. 9.10 and is motivated by cell turgor. Thus the water balance of the tree is important to latex flow and production. Tapping is carried out in the early morning when the period of high cell turgor is essential for uninhibited latex flow (e.g. Nizwar & Soerianegara, 1970).

Plugging of the latex vessels of the cut normally limits latex flow (see below) and this has been shown to be under genetic control. The rate of plugging

has been expressed in terms of a plugging index (P=flow per minute in the first 5 minutes in relation to the total flow) and Waidyanatha & Pathiratne (1971) have shown that P ranged from 1.40 to 6.68 in a study of 19 *Hevea* clones in Ceylon. Plugging can however be very much influenced by certain chemical substances used in 'stimulation'.

Stimulation and latex vessel plugging

The stimulation of latex flow by use of chemicals is now widely practised and generally this yields more rubber per tapping occasion by inhibiting the process of plugging of the ends of the latex vessels, which ordinarily limits latex yield. The most common substances in general use are 2:4-D and 2:4:5-T, which are applied to an area of scraped bark below the cut at a concentration of about 1 per cent mixed with a neutral carrier such as palm oil. Stimulation is usually carried out twice yearly. Many other substances and procedures also stimulate flow including copper salts (Compagnon and Tixier, 1950) and the actual bark-scraping process itself (see Abraham and Tayler, 1967).

Plugging is thought to occur as a result of rupture of certain latex bodies called leutoids (see above, p. 133) which release a serum capable of precipitating latex, and evidence suggests that the rupture of leutoids is caused by shear forces in the narrow vessel openings at the exposed cut, where flow rate becomes most rapid. The experiments of Southorn (1969) on the rate of plugging and latex flow in relation to the size of the opening cut (which affects flow rate and therefore the shear forces acting on the leutoids), and of Milford *et al.* (1969)

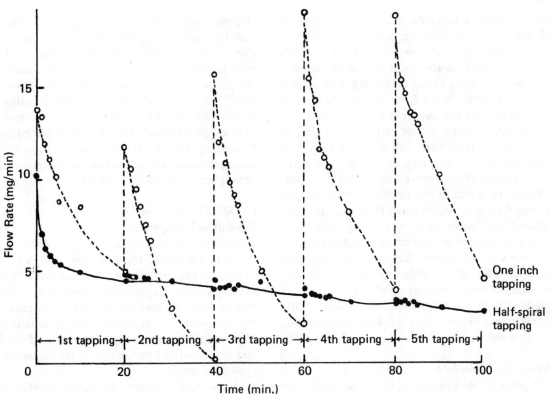

Fig. 9.11 Flow curves for repeated opening of half spiral and one-inch cuts expressed as flow per unit of cut surface
Source. After Southorn and Gomez, 1968.

on different clones indicate that plugging is the main factor limiting latex flow and that the rupture of the leutoids near the cut most likely occurs as a result of shear forces imposed by the rapidly-flowing latex. Fig. 9.11 indicates the effect of cut size on the rate of flow and the speed of plugging in rubber.

Stimulation with the growth regulators 2:4-D and 2:4:5-T reduces plugging and induces a more rapid and more prolonged rate of flow. Stimulation has also been reported by the use of 'pichloram' both alone and in combination with 2:4:5-T (Abraham *et al.*, 1968). The action is thought to be due to a reduction in viscosity of the latex, since turgor pressure is apparently not increased (e.g. Buttery and Boatman, 1964) by application of these stimulants, and this would be in accord with the observations of Waidyanatha & Pathiratne (loc. cit.) in which clones with a low plugging index had latex with a lower dry rubber content. The primary action of the growth regulators is not known but on evidence from other plants, it seems likely that they induce a greater water uptake by affecting osmotic regulants of the latex serum or by promoting mem-

brane integrity (e.g. Commoner *et al.* 1943; Glasziou *et al.* 1960). Thus, with the less viscous latex coagulated rubber particles can be swept out of the narrow channels of the vessels near the cut. Leutoid rupture is not prevented by the use of stimulants since the latex still contains numerous particles of ruptured leutoids. Stimulation also causes an increase in the area of bark tapped (cf. Fig. 9.9).

Stimulation results in an increase of up to twice the latex flow per tapping occasion but it is not recommended for young trees on the normal alternate daily half-spiral tapping rotation, since it reduces girth excessively. It is however generally recommended on older trees and may also possibly have application on young trees with reduced frequency of tapping in the future. At the present time it is probably safer to regard stimulation not as a means of increasing yield but simply as a means of obtaining the same yield with a lower tapping intensity, either by reduced frequency of tapping or by reducing the area of bark tapped (e.g. Abraham, 1971c; Samosorn & Ratana, 1972).

New and promising stimulants in the experimental stage at the present time are the gases acetylene and ethylene (see Abraham *et al.* 1971a, b and c). Ethylene oxide was first reported to stimulate flow by Taysum (1961), but now ethylene (in the form of 2-chloroethyl-phosphoric acid, which releases ethylene gas) appears most promising. It is applied either as a liquid in the form of a poultice stuck to the trunk near the tapping cut, or more recently, mixed with a neutral carrier such as palm oil and applied to scraped bark below the cut as with 2:4-D and 2:4:5-T. Initial results suggest that ethylene is superior to these two conventional stimulants and may be useful in conjunction with a reduced tapping frequency, which could be advantageous in reducing the depression in girthing commonly caused by the tapping procedure. Furthermore, recent work at the Rubber Research Institute, Malaysia, indicates that satisfactory latex yields may be obtained from extremely small cuts of a few cm length with the use of ethylene, which would completely eliminate problems of bark consumption and renewal.

Dryness or brown-bast

Dryness sometimes occurs in rubber and is accompanied by characteristic changes in the nature of the renewed bark which becomes dry and lumpy, with disorganized meristematic activity and disorganization of the lactiferous tissues and phloem. The inner phloem may assume a grey/brown discoloration. In normal estate practice it occurs only to the extent of a few per cent but with heavy

tapping (increased frequency particularly, but also with the full-spiral system) it becomes more prevalent (Chuah, 1967), especially when combined with the use of stimulants. Chuah considers that the primary cause is loss of proteins, RNA and possibly enzymes concerned with the building and functioning of the lactiferous tissues and the phloem. Mechanical disturbance of the cambium by tapping too deeply may also be causal in inducing dryness as might the action of micro organisms infecting the tapping panel.

CLIMATE AND SOILS
Rainfall and temperature

From the climatic point of view, rubber is best suited to the humid tropical zones and is unlikely to be successful in drier or colder regions with present-day clones. Experience in Indonesia indicates that in regions where the dry season exceeds three months, rubber is not successful (Dijkman, 1951) and rainfalls are generally in the range 2700–5000 mm (*c.* 100″–180″) per annum in the major rubber-producing areas.

Figure 9.12 indicates the rubber-producing regions of the world. Rubber, unlike some other tropical crop species (e.g. bananas, sugar-cane and pineapples), is very much more restricted in its latitudinal distribution and is not represented in the sub-tropics. Even in humid zones, latex yield is closely linked with variation in climate, particularly rainfall. In regions where even a short dry season

Fig. 9.12 Rubber-producing areas of the world

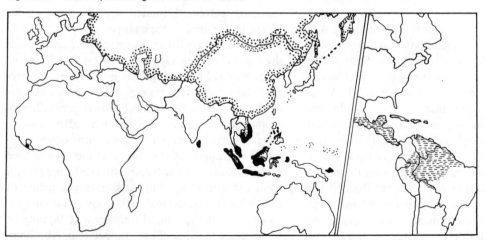

Source. After Dijkman, 1951.
Key. Present production area of *Hevea* in the Pacific and Liberia.
The Russo–Asiatic Bloc for the absorption of Pacific rubber.
Prospective area for *Hevea* in Central and South America.

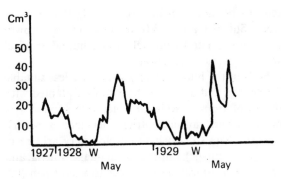

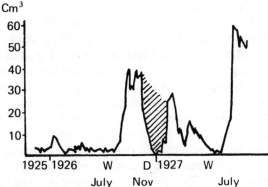

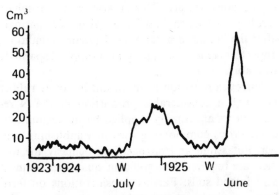

Fig. 9.13 Examples of latex flow in relation to the phenomenon of wintering (in Indonesia)

Source. Adapted from Schweizer, 1936; drawn from Dijkman, 1951.

Notes. W=natural wintering, D=artificial defoliation, the shaded area indicates estimated yield loss.

occurs (e.g. West Malaya where there are relatively dry seasons of only a month and even these experience about 4″ average rainfall) a phenomenon known as 'wintering' occurs in which leaf fall takes place and latex yields decline sharply. Fig. 9.13 (after Schweizer, 1936) illustrates the effects of wintering in Indonesia. The extent of the defoliation taking place during wintering varies greatly but there is a relationship between defoliation and latex yield since experimental defoliation of trees also causes loss of yield, corresponding to the period in which new leaves are forming. However, it seems likely that moisture stress is the primary cause of yield reduction during wintering. Wintering usually occurs once a year coinciding with a dry spell, but secondary wintering may also occur in regions marked by two dry seasons. Although moisture stress reduces yield, excessive rain interferes with tapping frequency (Wycherley, 1963) and increases the incidence of fungus diseases such a *Phytopthora, Oidium* and where it occurs, *Dothidella* (see p. 144). Fernando and Tambiah (1970) report that the latex vessel system has the ability to regulate moisture conditions within the tree somewhat so that potentially high-yielding clones with an extensive latex system would be better adapted to marginal climates.

Temperature also greatly affects rubber production in present-day clones, mainly by affecting the rate of growth of the tree. Experience in Indonesia indicates that a six-month increase occurs in the time prior to the beginning of the tapping stage for each 100 metre rise in altitude (about 1°F drop in the environmental temperature—Table 9.3).

Effect of wind on the crop

Rubber is a crop which has been very susceptible to wind damage as mentioned before, and a number of potentially high-yielding clones have suffered from this disadvantage (e.g. Clone RRIM 501

TABLE 9.3 Relationship between elevation and the beginning of tapping of rubber in Java

Height above sea-level (m)	No. of plantings sampled	Average age at tapping (years)	
		Before 1930	After 1930
0–200	361	5	4½–5
201–400	182	7	4⅔–5⅓
401–600	176	7½	5⅔–7
601–800	56	8½	7–8
801–1000	7	10	—

Source. Adapted from Dijkman, 1951.

from Malaysia—see Wycherley *et al.* 1964). High nitrogen nutrition, which induces massive crown development, enhances wind damage (e.g. *Planters Bull.* 43 (1959), p. 52), and in this regard the planting of leguminous cover crops can also enhance wind damage (*Planters Bull.* 57 (1961), p. 183). The best long-term approach to the wind problem is the breeding of clones with a good branching habit and low susceptibility to wind damage as indicated in Fig. 9.5. Pruning of trees to achieve a more satisfactory habit has also been considered (*Planters Bull.* 91 (1967), p. 147) but there is lack of experience in this practice. In general it is considered that corrective pruning should be carried out early ($1\frac{1}{2}$–3 years) removing potentially dangerous side branches or pruning to reduce trunk height by pinching off the terminal bud. Reducing the nitrogen fertilization of susceptible clones is also recommended.

Avenue planting tends to enhance wind damage because of the tendency for such trees to lean outwards into the inter-row space, particularly in the extreme forms of hedge planting.

Soils and drainage

Because of the fact that rubber cultivation is confined to the humid tropical zones, particularly South-East Asia, rubber is not grown on such a diverse range of soils as many other tropical crops. It is frequently grown on leached acidic inland soils of the humid tropics (latosols) which are often well structured and deep profiled but low in nutrients, or on coastal clay soils which are usually richer in nutrients (particularly those formed under a marine environment) but on which drainage problems are encountered. Wycherley (1963) reports that in Malaya the best-yielding areas are on inland soils because coastal areas, though fertile, experience higher disease incidence and heavy rains interfere with tapping. Rubber is also grown extensively on basaltic latosols which are deep, well-structured soils, higher in cationic nutrients than the latosols derived from other parent materials and with a higher reaction, but nevertheless still on the acidic side. Obviously nutrient requirements will vary with soil type. Some indications of nutrient requirements and fertilizer practice are given further on.

The optimal pH for rubber lies in the range pH 4 to pH 6.5 (*Planters Bull.* 50 (1960), p. 98) but the crop will tolerate pH in the range 3.8–8.0 (e.g. Kortleve, 1928; Vollema, 1949). Young seedlings

tend to be more sensitive to low pH than mature trees. Soil pH above 8.0 definitely causes growth retardation, but fortunately most humid tropical soils are acidic in reaction.

Soils of the humid tropics which are less suitable for rubber, but are nevertheless often planted with the crop, are peat soils (in which tree support is a problem and various nutrient deficiencies occur), acid sulphate soils (which develop very acid reactions on drainage), impoverished inland alluvia, and soils with a truncated profile due to a hard-pan clay layer or concretions of laterite.

Most coastal soils will require field drainage to at least 18″ (e.g. *Planters Bull.* 28 (1957), p. 20) and main drains of depth 2 metres. In low-lying coastal areas tidal gates are required. Most tropical soils require phosphate and this is generally supplied in the form of rock-phosphate during the immature years. Phosphate nutrition in mature rubber is done sparingly. Liming is not usually carried out unless soils are excessively acid (below pH 4) as calcium has a yield-depressing effect on rubber which is thought to be due to the enhancement of plugging in the latex vessels. This is also why phosphate fertilizers, which contain Ca, are also used sparingly after the initiation of tapping. The same conditions apply to magnesium which is also supplied sparingly to mature rubber.

Nitrogen nutrition is important in young rubber although a considerable proportion of the N requirements are usually supplied by the leguminous cover crops commonly grown in establishing rubber. Potassium and nitrogen, particularly the former, are required in significant quantities on most acid inland soils. Further considerations on fertilizer requirements of rubber are given under 'Manuring', p. 143–4 below.

PLANTING

Land preparation, spacing and yield

Being a tree crop, and often planted in newly-cleared jungle areas, land preparation is often fairly rough for rubber and is achieved by felling and burning, leaving large tree stumps to rot naturally during the establishment phase. Usually the planting of a leguminous cover crop between the rows is undertaken. Terracing is carried out on steep land with narrow-bench terraces constructed on the contour. In replanting, trees from old rubber areas are usually uprooted by the use of heavy tractors or bulldozers. This is advantageous in regard to root disease since the old root material is more or less completely removed from the soil. Sometimes felling and stump poisoning using sodium arsenite or 2:4:5-T is carried out instead of uprooting. (*Planters Bull.* 47 (1960), p. 39.)

The conventional spacing for rubber in estate practice in Malaysia is 20′ × 12′ (square planting) on flat land (*c.* 180/trees per acre) and 30′ × 8′ on hillside contours, 30′ being the average distance between contours—other distances used for contours are 25′, 45′ and 60′. In Indonesia the conventional distance for commercial planting is usually 6 × 6 metres. A theoretical consideration of spacing and yield in relation to age of the stand and to thinning out was given by Ham (1940) and is summarized by Dijkman (1951). Thinning out, to adjust optimal spacing with time, has however never been established in plantation practice.

Spacing affects not only girth increment (as discussed before) but also thickness and quality of the renewed bark. However, denser spacing is generally found to reduce wind damage (see Dijkman, 1951).

The results of a long-term density experiment carried out in Malaya over 30 years on clone AVROS 50 are reported by Westgarth and Buttery (1965) in relation to yield parameters, and in relation to profit (taking into account the effect of spacing on individual tree yield and yield per tapper—as discussed also under tapping systems) by Barlow and Lim (1967). The spacings used were as follows:

30′ × 30′ (square)	=	48 trees/acre
20′ × 20′ (,,)	=	108 ,,
20′ (triangular)	=	125 ,,
14′ × 14′ (square)	=	222 ,,
12′ × 12′ (,,)	=	302 ,,
10′ × 10′ (,,)	=	435 ,,

Girth increment, which is central to yield, was greatly affected as mentioned before (Fig. 9.8) with the lowest density trees reaching about twice the girth of the highest (*c.* 50″ compared to 25″). The percentage of trees which could be tapped also increased with reduced density from about 60 per cent to more than 95 per cent, and the yield per tree increased substantially with reduced density. The dry rubber content also increased but only slightly. Yield per acre, however, was, as might be expected, initially much higher (about twice as high) in the densest planting compared with the widest over the first eighteen years—but after this the difference narrowed and eventually an intermediate spacing gave the highest yield per acre. (Figs. 9.14 and 9.15 show respectively yield per tree and yield per acre in the Malayan experiment). Cumulative yield over

Fig. 9.14   Effect of density of planting on annual yield of rubber per tree

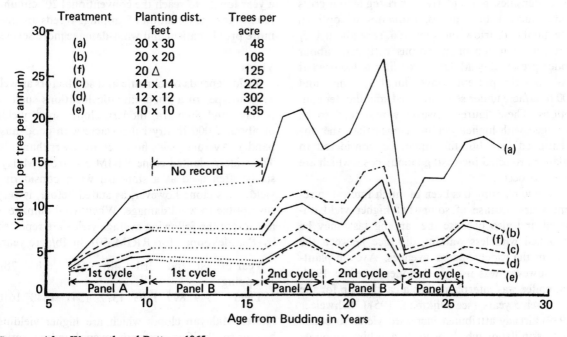

Source. After Westgarth and Buttery, 1965.

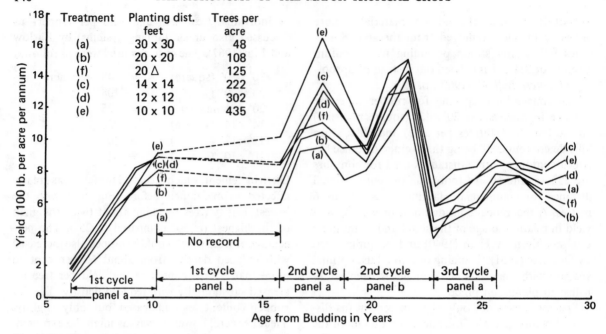

Fig. 9.15   Effect of density of planting on annual rubber yield per acre
Source.   After Westgarth and Buttery, 1965.

thirty years was higher at the denser spacings, but the yield of individual trees was much higher at the wider spacings. From the point of view of profit, however, the cost of tapping plays a central role in determining optimal density and the high individual tree yields or the yield per tapping occasion favours lower densities, particularly with rising labour costs and reduced rubber prices. Estimates of optimal tree number (Barlow and Lim, 1967) are about 125/ acre for estate-run organizations with paid labour under present-day Malayan conditions (somewhat less than the present conventional spacing), and 200 trees/acre under self-employed smallholder conditions. These figures however would tend to be modified with higher-yielding clones than the one mentioned here but this might be considered to balance projected increasing labour costs which are likely to occur.

Avenue planting has been of much interest in recent years because of somewhat higher yield obtained and also because the tapping task may be simplified with less walking distance between the trees in the widely-spaced avenues. Avenue planting, however, was used in Indonesia before 1934, and coffee was interplanted as a catch crop for the first twelve years (see Dijkman, 1951). Schweizer (1939) already attributed increased yields to better light assimilation which, according to him, amount-

ed to as much as a 37 per cent increase in foliage illumination. Avenue spacings tried in Malaya are $50' \times 4\frac{1}{2}'$, $65' \times 3\frac{1}{2}'$, $80' \times 2\frac{3}{4}'$, the latter being extreme. Evidence suggests that about a 65' avenue might be optimal, but in all avenue plantings early girth increment is delayed so that trees may take a year longer to reach the conventional 20″ circumference for the initiation of tapping. Weeds are also more problematical and wind-damage more severe.

Yields

Yield depends on climatic and soil factors as well as genotype. In a good climatic locations such as Malaya and Sumatra, modern clones show yields of about 2,000 lb dry rubber/acre with good husbandry. Average yields however are lower than this. The widely-planted clone RRIM 501 from Malaysia is often used as a standard when considering yield. This clone however, as stated before, is very susceptible to wind damage. When wind damage is not severe the following average yields are reported by Paardekooper for RRIM 501 in lb/acre/year:

Year of 1st tapping	1st	2nd	3rd	4th	5th	6th	7th
Yield	549	899	1193	1377	1547	1648	1610

Newer Malayan clones which are higher yielding belong to the five, six and seven hundred series of the Rubber Research Institute of Malaysia, (particularly RRIM 527, 600, 623 and 628) and a number of potentially very high-yielding clones are under trial at that institute. Some workers consider that about 5,000 lb/acre/year is the theoretical optimal yield which will be obtained from clones of the future.

Ceylon clones RRIC 36 and Nab 17 are reported to produce 2,400 lb per acre in the 4th year of tapping in a trial conducted in Malaya (Paardekooper, 1965).

The two most widely-planted clones in Asia are PR 107 and GT 1 which are planted from Ceylon to Vietnam and are tolerant of drier conditions and resistant to wind. Wintering is also less severe. These clones are capable of up to 2,000 lb/acre/year. Other clones which have performed well are PR 255 and 261, PB 5/51 and AVROS 2037 although 255 is susceptible to strong winds and PB 5/51 to attacks by *Oidium* at humid locations. PB 86, an old clone, is resistant to *Oidium*.

Nursery practice and crown budding

In the establishment of most tropical tree crops it is usual to have a nursery phase for propagation of the planting material and the maintenance of the young seedlings up to a certain age. Rubber seedlings established in a nursery can gain up to twelve months in time to maturity compared with buddings made directly on field-planted stock seedlings. The reasons for this have been discussed in the first part of this book.

In rubber, propagation and nursery procedures

are highly evolved. Cuttings and marcotts, as explained before, have proved unsuitable for rubber propagation because of susceptibility to wind damage of trees propagated from such material (due to the absence of a tap root). Buddings of one kind or another, consisting of buds from high-yielding clones on seedling stocks, are most widely used in rubber planting. Seedlings, particularly those from known high-yielding plantings, are also used for commercial planting but to a lesser extent. The latter require less attention in planting and establishment. In budding, production of stock seedlings must coincide with a period of seed fall; also a suitably wet season for establishment of the bud graft and later for planting out of the budded material is required and these factors vary from region to region. Seeds however can be stored for a few weeks in moist sawdust (Kelang and Haaren 1967) or for several months at reduced temperature (about 5°C) for the establishment of appropriately-timed stock seedlings.

Experience in Asia has shown that the stock has little effect on the development of the scion (Buttery, 1961) compared with other crops such as citrus (see Chapter 13) but the stock seedlings are nevertheless usually taken from clonal areas of good trees. Yahampath (1968) however, reports some distinct stock/scion interactions in Ceylon in which clone PB 86, showed better growth on RRIC 7 and 41 compared to the usual stock Tjl. The same clone on *Hevea spruceana* grew considerably slower.

Budding to propagate rubber has the advantage of producing more uniform stands of trees and was

Fig. 9.16 Closely-planted bud-wood nursery

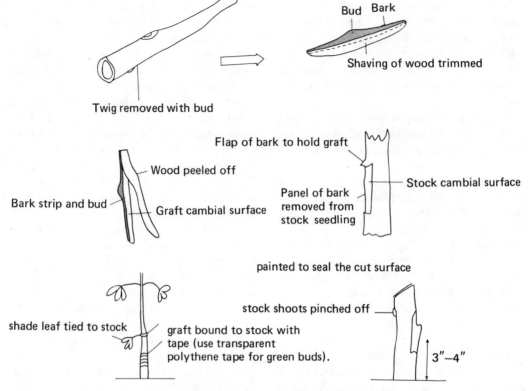

Fig· 9.17   Techniques of budding in rubber propagation (diagrammatic)

first used in the early part of the present century. Bud-wood, from superior clones, is produced in special nurseries (Fig. 9.16). Until 1962, fairly large seedlings were used for budding since younger material was difficult to establish by the techniques originally developed. This is known as 'brown-budding' because the stock and the bud wood have already developed brown bark. Establishing stock seedlings is mostly done in polythene bags at the present time but sometimes in well-prepared nursery beds. For brown budding, stock seedlings reach a suitable size after nine to fifteen months.

Green-budding is carried out on younger material and has the advantage that trees generally reach maturity somewhat earlier (about six months). The technique of budding is essentially the same for both green and brown-budding as illustrated in Fig. 9.17, but the tape used to bind the graft must be of transparent material in the case of green-budding, since the photosynthesis of the stem and graft tissues is important in the growth and establishment of the bud in these young graftings (see Hurov, 1960). In brown-budding, non-transparent tape can be used since the stock has considerable

reserves. The stock is cut back a few weeks after making the budding, which eliminates the dominance effect of the stock and allows the bud shoot to develop.

Planting of polythene bag material to the field is much simplified but stock seedlings established in the ground have to be lifted. Bare-root seedlings show high losses with rubber but raising a core of soil as illustrated in Fig. 9.18 has proved very successful in transplanting ground-nursery seedlings. This technique was developed at the Rubber Research Institute of Malaysia.

Mulching both with natural materials and with polythene has been shown to increase the growth of ground-nursery seedlings (Narayanan et al., 1968; Greenwood & Promdej, 1971).

Budded material is usually planted to the field when two or three whorls of leaves have developed on the bud shoot, by which time it has established a strong dominance and suppresses the development of other shoots on the stock.

One type of planting material used in rubber cultivation, which is useful for replacing points where the original budding has failed, is the

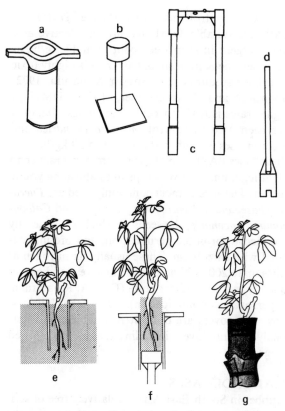

Fig. 9.18 Raising rubber seedlings for planting to the field (using a method of the Rubber Research Institute, Malaysia)

Notes. a. Metal core extractor for lifting core of soil (as in e and f). b. Plunger for extrusion of core (as in f). c. Double-headed hammer for driving in core extractor. d. Key for prizing core from soil. g. Extracted core wrapped in paper for transport to planting site.

'stumped budding'. This is essentially an overgrown budding (usually a brown budding which is more widely spaced in the nursery). A stumped budding may be produced from a budding of about 10' height which is topped, uprooted and trimmed to a club-like portion of the main tap root. Because of the considerable reserves in the stumped budding, and its large size, it usually establishes rapidly.

Recent investigations have been carried out on the cleft grafting of very young material, termed 'seed grafting', by Teoh (1972). In this, four-week-old seedlings established in small polythene bags are used as stocks and the apical portions of young growing shoots as scions. The method has given grafting success up to about 80 per cent and is considered to offer the possibility of reducing time to maturity. However, it has not yet developed to the stage where it can be recommended to the industry.

Crown budding. The idea of a three-part tree, which allows more flexibility in combining a wind or disease-resistant crown with a high-yielding trunk as tapping panel on a suitable root stock, originated with Cramer in 1926 (see Dijkman, 1951, p. 222). This procedure could be of considerable use in disease control in regions where leaf disease is a serious problem to the establishment of a rubber industry; for example, in South America where *Dothidella* (see p. 144) virtually prevents the production of rubber on a plantation scale. Problems of crown/trunk compatability arise and Radjino (1969), for example, reports that specific crowns reduced the yield of the high-yielding panel of clone LCB 1320. But even so, the technique offers great promise. In Malaya, effort has been directed to the grafting of wind-resistant crowns on the panels of high-yielding trees. Crown-budding experiments in Malaysia have been described by Yoon (1967, 1971, 1972a, b and c), and in Indonesia by Djikman (1951). Budding of the crown is best done when the normal-budded seedling has reached a height of 10 ft, which is the maximum height of the tapping panel (about 12 months after the initial budding) with the budding being made at about 7 ft. The crown bud tends to grow much faster than a normal bud, presumably due to the high level of reserves in the large seedlings, leading to the production of a top-heavy tree. However, this can be controlled by allowing one or two other buds to develop at the same time and later pruning them back.

Manuring

Considerable work has been done on the nutritional requirements of rubber in Malaya and Indonesia (see Shorrocks, 1964; Djikman, 1951 for reviews).

The rubber tree takes up some 82 lb N, 14 lb P, 62 lb K and 16 lb Mg per acre per annum (Shorrocks, 1965a and b). The nitrogen requirement is complicated by the establishment of cover legumes, but generally N is required during the establishment phase. On coastal soils, derived under a marine environment, nitrogen is the most important fertilizer. Angkapradipta *et al.* (1972) report that urea is a satisfactory source of N for young rubber but ammonium sulphate is more widely used. On inland soils both N and P are required and on soils, other than those volcanic soils derived from basic material, considerable quantities of K, and to a lesser extent Mg are also required. Phosphate is applied mostly in the establishment phase and reduced when tapping is commenced. Shorrocks (1965b) considers that inland soils with acid extractable K levels of below 100 ppm and Mg below 50 ppm require these fertilizers in rates equivalent to 150 kg/ha potassium chloride (muriate of potash) and 150 kg/ha kieserite. With soil levels of 300 ppm K and 100 ppm Mg, fertilizer dressings of one-third of these levels are recommended. Potassium and magnesium deficiencies are often reported in rubber plantings on inland soils.

Leaf-analysis procedures are often used in estate practice to determine nutrient requirements. The levels considered optimal vary with position of the sampled leaves and with season but generally lie in the range: N-3.0–3.5 per cent; P–0.20–0.28 per cent; K–1.0–1.6 per cent; Mg–0.17–0.26 per cent; Ca–0.38–0.82 per cent; S–0.14–0.26 per cent, all on on a dry basis (Shorrocks, 1964 and other sources). Micronutrient deficiencies are not common in rubber. Satisfactory levels in the foliage are considered to be as follows: Mn–11–38; Fe–66–86; B–20; Mo–1.7; Zn–21; Cu–13, in parts per million.

Irrigation is not practised in rubber cultivation at the present time.

Weed control and cover crops

Traditionally, sodium arsenite has been the herbicide for general weed control in rubber and is used for spraying along the rows of trees. In normally planted mature rubber the need for weeding is greatly reduced because of the dense canopy, but in avenue planting, weeding costs can be considerable. Inter-cropping with catch crops offers the best solution here, since the profit margins and the perennial nature of the rubber crop require that the expenditure on weeding be minimal. Cultivation of leguminous cover crops in the establishment phase also helps to control weed competition, but nevertheless there is a need to find a substitute herbicide for sodium arsenite which is considered to be dangerous because of its highly poisonous nature. Paraquat ('Gramoxone') has been used to some extent as well as MSMA (monosodium methyl arsonate) and DSMA (disodium methyl arsonate) but these are more costly than sodium arsenite. A promising herbicidal mixture for general weed control is 2:4-D amine/sodium chlorate at rates in the range 0.6–1.2 lb active ingredient 2:4-D amine in combination with 15–25 lb sodium chlorate per sprayed acre (*Planters Bull.* No. 91 (1967), p. 139). For longer-term weed control this urea herbicides DMU and CMU are suitable, DMU has proved effective as a pre-emergent in strip weed control at 1½ kg/ha as has ametryne & linuron, but atrazine is highly toxic to rubber seedlings (e.g. Siregar & Basuki, 1972; Boonsirat and Paardekooper, 1971). DMU at higher rates is toxic to rubber seedlings. Dalapon has been used to control *Imperata cylindrica* and other grasses in rubber at rates up to 20 kg/ha (e.g. *Agri. News* BASF 6, p. 9). It is common practice to plant leguminous cover crops in establishing young rubber. The species most commonly used are *Pueraria phaseoloides*, *Centrosema pubescens* and *Calopogonium mucunoids*, the seeds of which are generally scarified to promote germination, and pre-innoculated with rhizobium. These usually fix nitrogen at rates up to 200 lb N/acre in the first year and consequently reduce fertilizer costs (Dr. J.A. Rajaratnam —personal communication). On steep hillsides, however, covers are difficult to establish and normally natural covers including grasses are allowed to grow.

MAJOR DISEASES

Rubber in South-East Asia is relatively free of serious diseases but in South and Central America the establishment for a rubber industry is virtually prevented by the occurrence of South American leaf blight caused by the fungus *Dothidella ulei*. Attempts to obtain resistant clones have not been very successful because of the production of new pathogenic strains by the fungus and the long period required to produce a new rubber clone. Clones which have shown resistance are F×25, F×4037 and others. Crown budding of resistant crowns onto high-yielding panels is the standard procedure of establishing rubber in South America (Goncalves, 1968 also see *Coopercotia* 25, p. 26, 1968). There is considerable motivation for the establishment of successful rubber industry in tropical America (see map) and there is much research effort to discover ways of overcoming the *Dothidella* problem. The disease flourishes under humid conditions and Camargo *et al.* (1967) report that the fungus affects trees mainly when relative humidities are over 95 per cent for more than 10 consecutive hours and more than 12 times per month, which offers scope for strategic spraying of fungicides in localities which are not too humid. In planting in regions subject to *Dothidella* mixed clonal material should be used to reduce the build up of new pathogenic races of the fungus. This disease does not occur in South-East Asia at the present time but there is interest in the breeding of resistant clones as an insurance against its possible occurrence in the future.

Leaf blight caused by *Phytopthora* occurs in Asia, particularly India and certain regions of Java, Sumatra and southern Thailand, where it is sometimes serious. It has also occurred periodically in Malaya, in the drier north-western part of the country. It is controlled by prophylatic spraying with fungicides or dusting with sulphur to control its spread when an outbreak occurs. Its occurrence is often associated with a wet season following on a dry period in which 'wintering' and seed formation occur. Certain clones are more susceptible.

Another leaf disease is caused by *Oidium haveae* which causes leaf fall. It has been reported to be more serious at higher elevations and under excessively humid conditions. It is also controlled by spraying or dusting. (Fernando, 1971). PB 86, a widely-planted clone, and RRIC 102 prove resistant to this disease.

Various root diseases occur in rubber of which *Ganoderma spp.* and *Fomes spp.* are the most serious. These spread from stumps and large woody roots of both jungle and old rubber trees. It is not practical to remove jungle stumps during land preparation but in replanting old rubber areas, stumps are commonly removed by means of heavy machinery as mentioned earlier. Root diseases tend to be worse under wet conditions. Poisoning the old stumps with sodium arsenite has been used to control root diseases; also the innoculation of the stumps with saprophytic fungi to facilitate their breakdown has been considered. *Fomes lignosus* can be reduced by treating the planting holes of new rubber seedlings with sulphur (Satchuthananthavale and Halangoda, 1971). The fungicide 'drazoxolon' in grease had been reported to be a suitable collar protectant against *Ganoderma pseudoferreum* (*Planters Bull.* Malaysia 133, p. 78, 1971).

A stem disease known as 'pink disease' caused by the fungus *Corticium salmonicolor* attacks aerial branches in rubber, particularly at branch junctions and in regions of rough bark formation. It is also most serious under moist conditions. It is controlled by means of Bordeaux mixture or other copper fungicides.

Various panel diseases occur but usually not in epidemic proportions. Certain clones again are more susceptible. The most serious panel disease is probably 'black stripe' caused by *Phytopthora palmivora*. It is controlled by means of mercurial fungicides, however, Schreure (1971) reports that 'captafol' reduced bark renewal. Additions of 2:4D at 0.25–1.0 per cent to the fungicide increased yield but also increased bark proliferation.

There are no known virus diseases of rubber, and insect pests are usually not serious, possibly because of the presence of the latex. There are published reviews of rubber diseases by Sharples (1936) and Hilton (1963). The *Planters Bulletin Conference* No. 74 (1964) also deals with diseases of rubber.

10 TAPIOCA [Manihot utilissima]

INTRODUCTION, SPECIES AND VARIETIES

TAPIOCA (*Manihot utilissima* Pohl. = *M. esculanta* Crantz), otherwise known as cassava and manioc, is an important food crop of the tropics and also assumes importance in the world of international trade as a source of starch and as a constituent of animal feeds, particularly in the Economic European Community (e.g. *Far East Trade and Development*, 1969), where there is an expanding market at the present time.

There are over 128 species of *Manihot* (Pax, 1910) and the cultivated forms are thought to have arisen in the drier zones of Mexico and Brazil (Rogers, 1963), from where they have spread, following the discovery of the Americas, throughout the tropics.

There are many hundreds of clones of tapioca which are propagated vegetatively by means of stem cuttings. The chromosome number is 36, and some plant breeders have suggested that the cultivated tapiocas may be natural tetraploids since colchicine-produced 'polyploids' do not show superior vigour over ordinary varieties. However, 'triploids' produced in India (see Abraham, 1970) are thought to have superior vigour to their parent types. The species *M. utilissima*, *M. carthaginensis*, *M. aesculifolia*, *M. palmata*, *M. tweediana*, *M. saxicola* and others are thought to be closely related, and some may be natural hybrids. From the point of view of improving yield by breeding, incompatibility problems are encountered in making crosses (see Bolhuis, 1967; Jennings, 1970), but most plant breeders consider that these problems are not severe enough to impede the flow of genes in the genetic improvement of the species (e.g. Martin, 1970). However, tapioca is one of the least-studied of the major tropical crops, and it is clear that much basic genetic and cytological work needs to be done to characterize the genus. There is, however, great interest in tapioca as a supplier of starch and this crop should assume·increasing importance as an industrial crop in the future. Production studies on the crop (see 'Growth Quantities') indicate that considerable improvement in yield is possible, and for high yield the climate most likely to be best for tapioca is that of the humid-tropical zones.

Tapioca is largely grown for its starchy tubers, but the leaf is also valuable as a source of protein (see p. 154).

BOTANY

The vegetative plant

From the point of view of agricultural yield, only the vegetative plant is of interest, since the crop is propagated vegetatively and the agricultural product is the starch-filled tuberous roots. Fig. 10.1 illustrates the vegetative morphology of the crop.

Cuttings used for planting usually bear several nodes and buds, one or more of which may develop into shoots (Fig. 10.1a). The leaf canopy is formed after about 6 months (Fig. 10.1b). Tuberization of the roots commence after 6–8 weeks and continues for two years or more, although the crop is usually harvested after 10–18 months, depending on the location. It can, however, persist for a period of years and in some parts of the world (notably the more arid parts of Africa) it is sometimes maintained for several years as a reserve or famine. crop. The plant is essentially a perennial and some species (e.g. *M. glaziovi*) are trees. The stem continues growing for an extended period and constitutes a severe competitive sink with the tubers for assimilates. The stems are often unbranched but sometimes branch dichotomously after reaching a height of a few feet in field-plantings at conventional (*c.* 3 ft) spacing. The leaf canopy, however, usually does not increase in size after the first 6 months and will generally only maintain itself or decrease in size.

There is considerable variation in canopy morphology, as indicated by Fig. 10.1c, which illustrates three contrasting leaf types, but essentially the leaves have 5–8 lobes and are borne laterally on the aerial stem. Differences in leaf shape, size and

Fig. 10.1a   Germinating cutting of tapioca

Fig. 10.1b   Fully-developed tubers

Fig. 10.1c   Foliage characteristics of three tapioca varieties; fully-developed canopy and single leaf

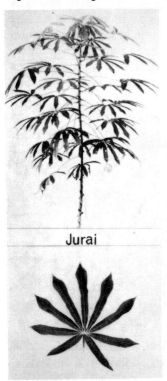

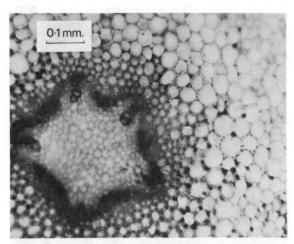

Fig. 10.2 Transverse section of an unexpanded tapioca root showing the pentarch stale

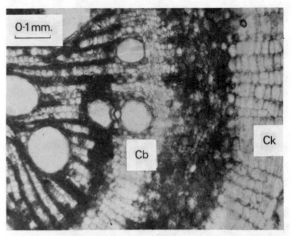

Fig. 10.3 A 2 mm diameter tapioca root in transverse section showing the commencement of cambial activity (Cb) and formation of a cork cambium (Ck)

orientation are associated with productivity (see 'Growth Quantities').

Stomata are located largely on the lower surface and leaf conductivities are relatively low compared with many other crops species (most annual species for example—see later), and it seems likely that this is the major bottle-neck to productivity. The low leaf conductivity may be related to the origin of the cultivated species in the drier tropics, and to the perennial habit which necessitates the survival of dry seasons.

The tuberous roots are formed as a result of activity of the root cambium, which produces expanded starch-filled parenchymatous cells in place of the normal lignified xylem cells of secondarily thickened roots. Thus, the process of tuberization, in

its simplest form, results from a change in the nature of cells differentiated by the cambium. Fig. 10.2 and 10.3 show respectively transverse sections of an undifferentiated tapioca root and one in which the cambium is just commencing activity. Fig. 10.3 also shows the formation of a cork cambium which persists and clearly has a protective function in the expanding tuber. Fig. 10.4 shows inner, middle and outer regions of a young tuber (4 mm diameter) in which the cambium can be seen to be relatively superficially located and to produce the massive zone of storage tissue which in matured roots may be 6″ or more in diameter in some clones. The extent of radical expansion of the storage parenchyma appears to be related to tuber size and yield (see later) at least in some clones.

Fig. 10.4 Outer, middle and inner transverse sections (left to right) of a 4 mm tapioca root

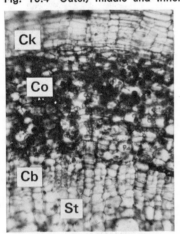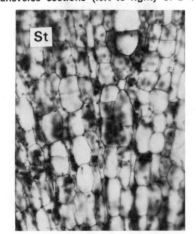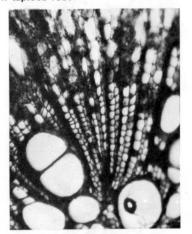

Notes. Note superficial location of the cambium (Cb) beneath the cortex (Co) and cork (Ck). Storage tissue (St) is shown in the left hand photograph and the centre.

TABLE 10.1   Relative conductivities of a range of crop species measured with a falling-pressure porometer at mid-morning under conditions of low moisture stress on four separate days (units arbitrary)

Crop	Conductivity (m. div./100 sec.)								
	Day	1	Day	2	Day	3	Day	4	Averages
Cucumber	25	20	22	17	31	18	37	25	24.4
Groundnut	36	31	42	50	28	33	18	25	32.9
Tomato	500	833	714	833	500	625	250	500	594.0
Sorghum	125	83	46	56	50	63	56	71	68.9
Ragi (millet)	416	500	333	500	333	500	294	417	401.0
Sugar-Cane	50	42	33	56	42	46	29	33	41.2
Rice	333	500	625	714	353	500	100	416	443.0
Oil Palm	25	18	17	26	19	31	17	26	22.4
Pineapple	15	15	33	38	56	42	46	39	35.5
Tapioca	7.6	8.6	7.5	10.0	6.7	10.0	8.3	8.3	8.4
Passion fruit (*Passiflora*)	5.9	6.3	8.3	11.1	7.4	10.2	11.1	10.0	8.8

The root system

The root system ramifies in the soil during the first few months, but when massive tuber growth occurs, the productoin of absorptive roots declines. The whole root system is generally fairly superficial and located in the top 2 ft of soil.

Flowering

Flowering occurs spasmodically or periodically although there is evidence that tapioca is a short-day plant (see Bolhuis, 1967). The seeds are not used agriculturally, but both control and synchronization of flowering are of importance in genetic work.

Growth quantities

The structure and functioning of the assimilation apparatus of tapioca has been studied in three varieties of tapioca by Williams and Ghazali (1969) and Williams (1971, 1972 and 1974). The crop supports a leaf-area index of about 4 and belongs to that group of plants in which assimilation into non-photosynthetic parts continues for an extended period after the formation of the assimilation apparatus (the canopy) is complete. Differences in the form and structure of the canopy, as illustrated in Fig. 10.1c, have been associated with yield, in so far as low yield has been associated with broad and horizontally disposed foliage, and high yield with more vertically orientated leaves with narrow leaflets. Leaf form and orientation also affect the response

TABLE 10.2   Yields of cereals and root crops in terms of kg wheat equivalents per hectare

Crop	General Estimate for all Africa	
	Crop Kg/ha	Kg wheat equivalent/ha
Maize	1,790	1,340
Eleusine millet	953	650
Cassava (tapioca)	14,280	3,285
Sweet Potato	6,080	1,824
Yams	9,790	
Rice (rough)	795	646

Source.   Clark and Haswell, 1964.

to spacing (see p. 153), as might be expected. However, potential assimilation rates as indicated by stomatal conductivity are very low in tapioca. Table 10.1 shows, for comparison, the stomatal conductivities (as measured with a falling-pressure porometer) of tapioca (peak rates) and various other crop species, and it is apparent that conductivity is low compared to most of the other species. Improvement of assimilation potential could be the most important genetic aim in increasing productivity. Fig. 10.5 shows assimilation rates and relative growth rates of three varieties of tapioca, together with that of wheat, as an example of a crop with relatively high assimilation rate (from Williams and Williams, 1968).

Even so, tapioca is often considered to be a high-

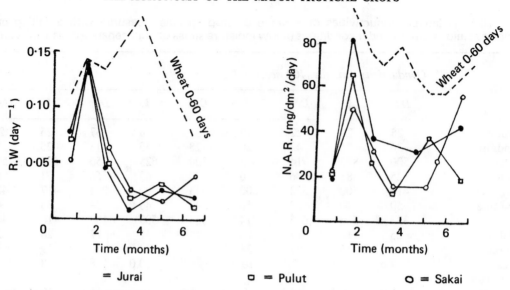

= Jurai   □ = Pulut   ○ = Sakai

Fig. 10.5 Net assimilation rates and relative growth rates (Rw) of three tapioca varieties and of wheat
Source. From Williams, 1972.

yielding crop in terms of starch, and this must be attributed to the relatively long growing period (average 12 months compared with 3 months for annual crops), and to a high crop index. Table 10.2 shows yields of tapioca (from Africa) and, for comparison, a number of other starch crops, and Fig. 10.6 indicates the distribution of assimilates into the various plant parts of three tapioca varieties which are high, medium and low-yielding respectively, indicating the relatively high crop-index or ratio of harvested material to total plant material, particularly in the high-yielding variety.

Differences in assimilation rate between varieties (as indicated in Fig. 10.5) have been attributed to a difference in the sink strength of the tubers for assimilates. Major yield differences between the three varieties of Fig. 10.1c and Figs. 10.5 and 10.6 can be associated with tuber size and the degree of expansion of the storage tissue cells as mentioned above. The major yield component appears to be tuber size (particularly thickness) rather than tuber number, and in these three varieties, quantitative difference in cell expansion as indicated by Table 10.3 is closely related to tuber yield which, in these varieties, average around 5, 10 and 15 tons per acre respectively.

Large cell size of the harvested organs and leaves is also associated with yield in bananas and pineapples (see Chapters 2 and 3), where it is associated with ploidy.

Fig. 10.6 Changing weights and distribution of plant parts in three varieties of tapioca which are high (Jurai), medium (Pulut) and low-yielding (Sakai) respectively

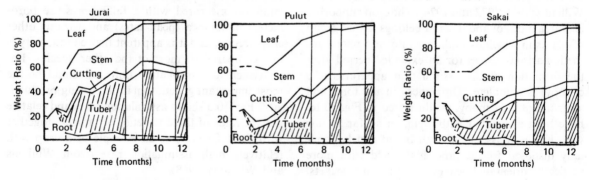

Source. From Williams, 1972.

TABLE 10.3 Radial expansion of storage tissue cells of the tubers, and of lignified root xylem cells of high (Jurai), medium (Pulut) and low-yielding (Sakai) tapioca varieties

	Radial diameter of cells (mm)		
	Jurai	Pulut	Sakai
Storage Cells	0.108	0.080	0.068
Lignified Cells	0.024	0.023	0.025

S.D. for storage cells, P, 0.05=0.016

CLIMATE AND SOILS
Temperature and rainfall

There is very little information on the comparative performance of tapioca varieties in different climatic situations but it is known that in the genus *Manihot* there are species adapted to a wide range of habitats, from the swamp forest environment to semi-desert regions. The cultivated tapiocas however (in spite of their probable origin in the dry tropics and their drought resistance characteristics —see above) are mostly grown in relatively humid regions of the tropic compared, for example, to crops such as maize, sorghum and millet. In India, culture of tapioca is confined to the relatively humid states of Kerala and Tamil Nadu, and Abraham (1970) states that tapioca is adapted to tropical and sub-tropical countries with well-distributed rainfall in the range 70"–100" per annum. In Africa, it is grown in rather lower rainfall regions on the eastern side of the continent and Jennings (1970) considers that 30" of rain per annum is satisfactory for tapioca production, although it is often grown in regions of lower rainfall than this. Yields, however, are rather low in Africa in general. In West Africa tapioca is again grown in higher rainfall areas and is considered unsuitable for semi-arid regions. In Ghana it is mainly grown in the more humid south (Doku, 1969) and in Nigeria it forms the main subsistence crop also in the more humid southern parts. Fig. 10.7a and b indicate respectively the extent of cultivation of various food crops in Nigeria, and the occurrence of wet and dry seasons in those parts. In West Africa the yields of tapioca are generally higher than in East Africa, and average yields in northern Nigeria (dry) and eastern Nigeria (humid) are stated by Chadah (1962) as being 2–3 tons and 3–5 tons/acre respectively. In Tanzania (Scaife, 1966) reports declining yields from 20 to 5 tons/ha with delay of planting into the wet season, which dramatically illustrates the effect of moisture on productivity.

Probably the most useful analysis of the effects of climate on tapioca yield comes from the Malagasy Republic (Arraudeau, 1967) where tapioca is extensively grown in a range of climatic locations from sea level to high elevations. Average yields of local varieties and of improved hybrid clones in different climatic regions is indicated in Table 10.4 (after Arraudeau, 1967).

From these results there is a clear indication that tapioca is best adapted to humid regions when high productivity is the objective and there is also the

Fig. 10.7a Crop cultivation zones of Nigeria

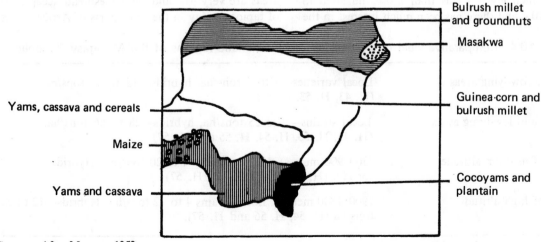

Bulrush millet and groundnuts

Masakwa

Guinea-corn and bulrush millet

Yams, cassava and cereals

Maize

Yams and cassava

Cocoyams and plantain

Source. After Morgan, 1959.

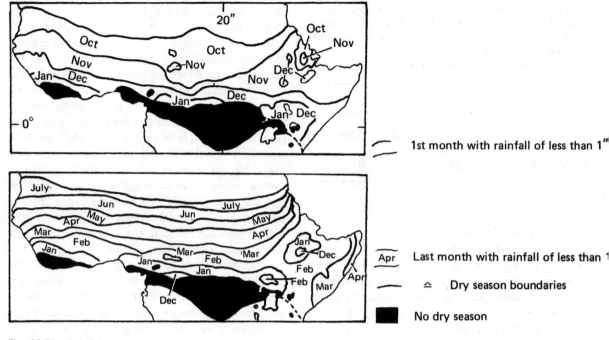

1st month with rainfall of less than 1"

Apr   Last month with rainfall of less than

≃   Dry season boundaries

◼   No dry season

Fig. 10.7b   Rainfall zones
Source.   After Bowden, 1964.

suggestion that regions of medium altitude and temperatures slightly lower than those of the humid lowland tropical zones might also be beneficial to yield, possibly because of effects of temperature on the distribution of material to the tubers and tops in the crop. Yields of up to 78 tons/ha have been recorded in the Cauca Valley of Colombia which is climatically equitable with moderate temperatures.

Tapioca grown as a famine crop in the drier regions of Africa and other tropical countries produces much lower yields than those indicated in Table 10.4 and there has been much interest in the

breeding of more drought-resistant varieties or varieties which are more highly productive under dry or marginal conditions. Nichols (1947) in Africa, intercrossed *M. utilissima* with the *M. glaziovii* (Ceara rubber) and *M. melanobasis* and eventually obtained selections which were both drought resistant and resistant to mosaic virus in Tanzania (see Childs, 1961). These crosses have also proved useful in Ghana (Doku, 1969). *M. glaziovii* has also been used in India for producing strains with resistance to drought (Abraham, 1970). Even so, yields in dry areas are very low and the widespread acceptance of tapioca today in the drier parts of Africa seems

TABLE 10.4   Comparative yield of tapioca in different climatic regions of the Malagasy Republic

Dry and low-lying areas	: Local varieties—7 to 8 tons/ha, hybrids—12 to 35 tons/ha (H. 43, H. 53, H. 54)
Humid and low-lying areas	: Local strains—9 to 25 tons/ha, hybrids—28 to 66 tons/ha (H. 45, H. 53, H. 54, H. 56 and H. 57).
Areas of medium altitude	: (300–900 metres). Local strains—4 to 20 tons/ha. Hybrids 30 to 80 tons/ha (H. 35, H. 49, H. 54, H. 57)
Areas of high altitude	: (900–1300 metres). Local strains 4 to 12 tons/ha. Hybrids—12 to 25 tons/ha (H. 54, H. 56 and H. 57).

Source.   After Arraudeau, 1967.

partly related to the resistance of the crop to locust attack and the fact that tapioca serves as a perennial reserve crop in case of failure of annual species.

There is apparently no information on the effect of frost on tapioca but it seems likely that it would be sensitive to severe cold stress in common with most other tropical species.

Photoperiod

Tapioca is not generally recognized as being sensitive to photoperiod but Bolhuis (1967) working in controlled environment conditions with six varieties of tapioca has demonstrated a short-day photoperiodic requirement for tuber initiation. In the varieties studied, photoperiods between 10 and 12 hours showed little difference, but from 14 hours and greater, very few tubers were formed.

Soils

Tapioca is grown on a wide range of soil types, but being a root crop, does best on soils of a friable nature which permit expansion of the tubers. Sandy soils are suitable with fertilization and adequate rainfall. In South-East Asia, tapioca is sometimes grown on organic soils (lowland oligotropic peats) where it grows well with adequate drainage and fertilization (see Chew, 1970). The loose texture of these soils greatly facilitates harvesting which is one of the main cost inputs in production.

PLANTING
Land preparation, spacing and yield

Tapioca is usually planted from cuttings on mounds of soil or on broad ridges, and this heaping up of the soil is often reported to increase yields and reduce the task of harvesting, presumably because of improvement of the soil for the expansion of tubers.

The usual spacing for tapioca is about 3 ft and for a crop harvested at 12 months or more, this seems adequate. For harvesting after a shorter period, a denser spacing will probably give higher yields, and furthermore, it seems likely that there is also a genotype interaction in regard to spacing with thin-leaved and vertical-foliage types having a closer optimal spacing. The effects of spacing on the yields of three varieties of tapioca over 9 and 12 month growing periods are shown in Figs. 10.8 and 10.9 respectively.

Yields of tubers have already been indicated. The highest recorded yields are of the order of 80 tons/

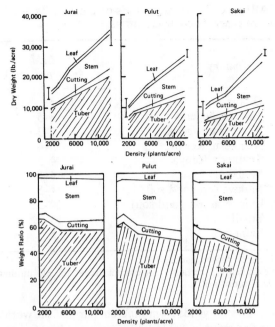

Fig. 10.8   Response of three varieties of tapioca to planting density after a nine-month growing period

Source.   After Williams, 1972.
Note.   Significant differences P=0.05 shown.

Fig. 10.9   Response of three varieties of tapioca to planting density after a twelve-month growing period

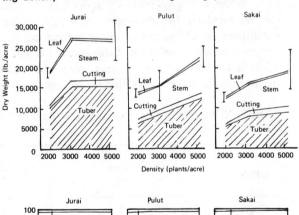

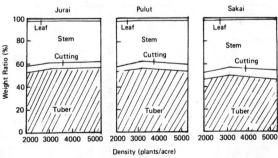

Source.   After Williams, 1972.
Note.   The variety pulut is a vertical-leaved type and may respond to closer spacing.

TABLE 10.5   Average yields of cassava (tapioca) in selected tropical countries

Country	Yield (tons per acre)
Ghana	5
Eastern Nigeria	3–5
Northern Nigeria	2–3
Northern Kenya	3–4
Southern Kenya	4–5
Mauritius	4–5
*Malaya (on good soil)	4–5
Indonesia (on good soil)	10–25
Brazil	15–19
India	5

Source. After Chadha, 1962.
*This figure is certainly not correct except for old mining land (poor soils).

ha. Average yields from different parts of the world are indicated in Table 10.5. Great yield variations are related to climate, fertilization and general standards of husbandry.

Crude protein levels of tapioca foliage have been reported as about 15 per cent and this material is widely used as feed for animals in tropical countries. When it is grown closely spaced (1 ft) and clipped regularly (approximately 4 weeks) under conditions of adequate rainfall or irrigation, the canopy assumes a hedge-like habit and yields up to 9,000 lb dry leaf per acre per year. With heavy nitrogen fertilization (c. 2,000 lb urea/acre/year) and clipping of the young leaf material, protein* yields above 30 per cent have been obtained.

Planting material and planting

Planting is generally carried out by hand although much attention is being directed to the use of machinery at the present time. Beeny (1970) has reported that a modified cane-planter can be used for planting tapioca cuttings.

Cuttings of length of about 20–25 cm bearing several nodes are planted at about 3 ft spacing. There is however a genotype/spacing/time-to-harvest interaction for optimal productivity in tapioca which was discussed before.

The size and type of cuttings (whether from the base of the stem or the tip) sometimes affects establishment and yield. Enyi (1970) reports that cuttings from the lower part of the stem give the highest yield and this was related to the dry-matter content of the pieces and probably therefore with reserve materials. But Chan (1970) could find no difference in the performance of basal middle and top cuttings and concluded that cuttings of about 23 cm length gave the best yield in all clones irrespective of their origin (at 3 ft spacing) and were superior to those of lengths 15 and 8 cm. More often, however, the age of cutting seems to affect yield and Table 10.6 (after Huerto, 1940) indicates the effect of cutting origin on yields and germination. This seems to suggest that the yield effect is related to bud germination and probably there-

*Crude protein = N × 6.25%.

TABLE 10.6   Percentage germination, yield of roots and starch in tons as affected by age of cassava cuttings

Group	Cutting No.	Germination %	Yield of roots tons/ha	Starch yield tons/ha
Top	1	15.1	4.4	0.57
	2	45.2	15.6	1.81
	3	63.9	19.6	2.25
Middle	4	74.1	18.2	2.21
	5	77.0	19.7	2.12
	6	82.4	19.2	2.07
Base	7	82.1	19.4	2.02
	8	82.5	18.5	2.01
	9	82.7	21.4	2.31

Source. After Huertao, 1940.

fore to stand density. In later investigations Chan (personal communication) has found that yield increases with cutting size up to about 2 ft at conventional 3 ft spacing.

Since cuttings bear several nodes and produce more than one stem it would seem likely that compensatory growth could occur to affect differences due to spacing, and that the factor of cutting size could be added to the genotype/spacing/time interaction mentioned above.

A considerable number of investigations have been carried out on the method of planting cuttings, usually comparing the following three methods:

1. Placing cuttings horizontally an inch or two below the soil surface.
2. Planting them vertically with a short section of the stem protruding above the soil surface, and
3. Planting them slanting, again with a short section of the stem protruding above the soil.

There seems to be a general conclusion that horizontal placement is the best, but this would seems to be conditioned to some extent by the heaviness and drainage of the soil and rainfall. Chan (1970) could find no difference in the method of planting the cuttings in the relatively rainfall-sufficient regions of West Malaysia.

Manuring and irrigation

Fertilizer response in tapioca is extremely variable and it appears to a great extent to be determined by the natural fertility of the soil. It also seems likely that genotype/fertilizer interactions occur in different tapioca clones. Silva and Freire (1968) report that, on poor sandy soil, nitrogen and phosphorus effects were small but potassium increased yields. Scaife (1966) reports that nitrogen, phosphorus, potassium and calcium as well as farmyard manure did not increase yields in western Tanzania, but the lack of fertilizers response here could be conditioned by a relatively dry climate which may impose limitations on the degree of nitrogen response. Chan (1970) reports a significant response in Malaya which is relatively humid, but not for the other major elements phosphorus and potassium. Amon (1967), in western Nigeria, also reports a high nitrogen response on granitic soils but a variable potassium and phosphorus response. Obi (1967) reports considerable reduction in yields on successive cropping on leached soils in eastern

Nigeria, which could presumably only be maintained by fertilization. On peat soils in Malaya, which are very low in cationic nutrients and available nitrogen, Chew (1970) reports that liming was necessary, and with 14 clones studied the optimal fertilizer levels were of the order of 200 kg N, 55 to 65 kg P_2O_5 and 120 to 135 kg K_2O per hectare. High nitrogen and potassium requirements are reported by Silverstre (1967) in Madagascar, and analysis of plant parts indicates that a crop of 40 tons per hectare removes approximately 285 kg of nitrogen 132 kg of P_2O_5, 460 kg of K_2O and 224 kg of CaO in the plant parts (Arraudeau, 1967). These figures included stem material and clearly appropriate quantities of nutrients must be made good in continued cultivation to replace those removed by the crop. Green manuring with legume crops is quite commonly practised in rotational cultivation of tapioca.

However, high levels of nitrogen fertilizer have been reported to promote excessive leaf growth and to reduce tuber yields in pot cultures (Krochmal and Samuels, 1967). These workers demonstrated the importance of high phosphate in a 2^3 factorial experiment in which a high phosphate ratio in the nutrient mixture greatly increased tuber yields, while high nitrogen increased the yield of leaves and tops. They attribute the high phosphate requirement to the role of this element in the phosphorylation process in the enzymatic synthesis of starch (Malavolta et al. 1955). Normanha and Pereira (1949) also report significant phosphate responses in Brazil.

From the point of view of the timing of fertilizer application, the nature of canopy development, which occurs most rapidly in the first few months of growth and then declines or even becomes negative towards the latter stages, and the effect of nitrogen on the top/tuber ratio (Krochmal and Samuel, 1967) suggest that nitrogen applications should be given at the time of planting or in split applications during the early phases of growth (e.g. Normanha et al. 1968). Phosphorus should be applied in split applications throughout development, particularly on soils that inactivate phosphate. Potassium, because of its relative resistance to leaching in soils of moderate CEC, could be applied at planting or in split dressings at planting and during tuber growth.

In regard to fertilizer placement, the general finding is that broadcast or lateral placement gives the

TABLE 10.7 The effect of virus infection on cassava yield

Region		Estimated yield loss due to virus infection[1]	Authority
Congo		20%	Muller, 1931
Nigeria	(a)	33%	Golding, 1936
	(b)	43%	
Nigeria		34%	Beck and Chant, 1958
Nigeria	(a)	33%	Ekandem, 1964
	(b)	14.5%	
Madagascar	(a)	83%	Cours, 1951
	(b)	nil	
Zanzibar		76%	Tidbury, 1937
Zanzibar		95%	Briant and Johns, 1940
Tanzania	(a)	65%	Jennings, 1960

Source. After Jennings, 1970.
Note. Results listed (a) are for plants or cultivars which had more severe symptoms than (b).

best results. The practice of putting fertilizer in the planting holes or a furrow is not to be recommended since it causes damage to the cuttings (Normanha *et al.* 1968).

In Africa the practice of fallowing between periods of cropping is widely undertaken. Scaife (1966 and 1968) reports a beneficial effect of tapioca on subsequent yields of cotton when tapioca is used as a 'fallow crop', this being equivalent to natural bush fallow in its regenerating effect on the soil. The experience in Malaysia has been very different and under the high rainfall condition of this region, leaching and erosional losses occur under tapioca. Cultivation on hill-sides should be avoided on unstable soils.

Irrigation is hardly ever practised in tapioca growing but it is clear that high yield can only be obtained in rainfall-sufficient areas. If tapioca is grown as an industrial crop, as distinct from a famine or reserve crop, consideration could be given to irrigation in areas of marginal rainfall.

Harvesting

Harvesting of tapioca is at present carried out almost entirely by hand and its commercial production is therefore dependent on an abundant supply of relatively cheap labour. The often elongated form of the tubers and the fact that they have tendency to break have rendered the production of machines for harvesting largely unsuccessful up to the present time (e.g. Beeny, 1970). However, mechanical harvesting by means of heavy tractors and lifting implements is carried out on commercial estates in the Malagasy Republic, but with significant losses of crop (Arraudeau, 1967). The mechanical harvesting of tapioca in normal soils is necessarily linked with the development of plants with short compact tubers located near the base of the stem, and clones with superficially located tubers (see Martin, 1970). The peat soils of the humid tropical regions mentioned earlier represent a special case in which it will be possible to develop mechanical harvesting techniques with present-day varieties because of the light texture of these soils.

MAJOR DISEASES

The major diseases of tapioca are mosaic virus and vascular wilt (caused by the bacteria *Xanthomonas manihotis*). Of the two, mosaic is much more serious and appears to be endemic in many parts of the world, particularly Africa.

Resistant varieties provide the means of control and there is much interest in the breeding of resistance to this disease. Table 10.7 (after Jennings, 1970) indicates the effect of virus infections on yield in various parts of Africa.

Xanthomonas vascular wilt occurs under conditions of excessive humidity and is quite serious in some parts of the world. It is also controlled by means of resistant varieties, and by improvement of the soil's physical condition. A virulent strain which attacked a supposedly resistant variety and also spread to the stem and leaves of the plant was recently reported by Pereira and Zagatto (1967).

Tapioca is relatively free of insect pest problems, (apart from the spread of mosaic virus by insect vectors) probably because of the content of hydrocyanic acid glycoside in the sap.

11 COCONUTS [Cocos nucifera]

INTRODUCTION, VARIETIES

THE coconut (*Cocos nucifera*) and the oil palm (see following chapter) belong to the palm family and together they possess the unique feature of having only a single growing point on each tree. This affects the growth of the community in that the tree cannot adjust to variable spacings to the same extent as branching species so that spacing is more critical than in freely-branching species.

The coconut is undoubtedly one of the most important plant species in tropical countries but the research and improvement of this crop by organizations devoted to its study have been disappointing. Consequently, the coconut is today experiencing difficulty in maintaining itself in estate-organized enterprises for copra production. This state of affairs has been accentuated by periodic poor copra prices but is primarily due to the fact that there has been virtually no genetic improvement of yield in plantation agriculture, and the palms grown today, with the exception of certain hybrids (see further on) are virtually the same as the open-pollinated unselected palms used in ancient times. (The history of improvement of the oil palm has been quite different where a several-fold increase in yield has been achieved by the simple process of selection, and yet greater yield improvement is anticipated from controlled breeding programmes in the future.)

The coconut, however, is widely used in tropical countries for culinary purposes and in a host of other ways (as distinct from copra production) and in many of the distant island groups of the world, particularly in the Pacific, it remains the most important commodity of trade. For the plantation industry, however, the future seems to depend on the improvement of yield, the mechanization of copra extraction, the manufacture of by-products from the husk and the intercropping of estates with other crops such as cocoa (see Chapter 7). The unbranched habit of the coconut makes it impossible to improve the yield of the crop by vegetative multiplication procedures at the present time, as is possible with most of the other major tropical crop species discussed in other chapters. Future possibilities for vegetative propagation are considered further on. The basis for initial improvement is probably simple and does not require the long periods of time envisaged by some coconut breeders. Substantial improvement could undoubtedly be achieved by planting seed from controlled pollinations of selected high-yielding mother trees with pollen from high-yielding pollen donors, and it is gratifying to see that some progress in this direction is being made at the present time. Zuniga (1969), for example, reports that only three years yield recording is sufficient for testing specific progeny crosses provided only the top 10 per cent yielding palms are used as parents.

All of the main commercial varieties of coconut are now distributed on a pan-tropical basis and the origin of the palm is much disputed. This was at one time considered to be the Americas, on the argument that other closely related species are found only there (Cook, 1901, 1910). However, the discovery of a fossil pliocine *Cocos* species in New Zealand, and the fact that a wider range of types exist in the East (see Child, 1964) now lends support to the notion that the coconut is of eastern origin. Purseglove (1968) supports the view that the coconut might be Melanesian in origin.

The concept of varieties in a highly heterozygous species such as the coconut with a high degree of out pollination is difficult to maintain but there are nevertheless a number of distinct natural varieties, the most important of which are the dwarfs and the talls. Dwarf coconuts are distinct with a degree of self-pollination. Tall varieties are mostly cross pollinated and exhibit a gradation of characters. Reyne (1948) considers that there are three basic nut-colour varieties, pale yellow, green and orange, all of which hybridize to produce intergrading colours. Among dwarfs types three colour varieties are also recognized; namely green, yellow and red-fruited types (see .Whitehead, 1966a). The 'Malayan dwarf' has now become recognized as a distinct varietal type (Whitehead et al. 1966; Parham, 1968); and colour differences in the dwarf coconuts have further been related to vigour since Satyabalan et al. (1968a) have reported that tall x dwarf orange hybrids are superior to tall x dwarf green hybrids. The 'Malayan dwarf' is resistant to lethal yellowing disease—see further on.

Whitehead et al. (1966) have shown that in Malayan dwarf × tall crosses, the red and yellow petiole colour of the dwarfs (which parallels nut colour) is recessive to the colour of the talls (useful in hybrid seed production).

A distinct type of dwarf from Fiji is recognized on the basis of its hybridization with the Malayan dwarf (Parham, 1968). This dwarf has the desirable characteristic of large nut size, closer to that of talls, unlike the Malayan dwarf which has small fruits (Whitehead, 1966b).

Numerous 'strains' of tall coconuts have evolved from geographical isolation and numerous cases of hybridization between them are recorded (e.g. Satyabalan et al. 1968; Zuniga, 1965; Duff, 1968; Ratnam, 1968 etc.).

The normal coconut possesses a three-carpelled fruit, but Smit (1967) has reported the occurrence of a high frequency of a two-carpelled type at a location in Peru, and this feature would merit varietal status. The size of nut is another recognized varietal feature since large-fruited types occur which have been given the varietal name macrocarpa (see Child, 1964).

Nut shape has also been used for varietal classification, and other features include the thickness of husk and shell, pink-coloured husks and edible husks (see Mao, 1959). A thick-shelled small-nut type and a dwarf yielding numerous small nuts are also recorded (see Child, 1964; Whitehead 1966b), and a semi-tall coconut referred to as the 'king coconut' has also been assigned varietal status.

Child (1964) considers that the dwarf coconut is of little commercial interest, but recent experience in Malaya has shown that early yielding and very high production are possible from dwarfs grown at much closer than conventional spacing. This is supported by observations in Jamaica (Whitehead, 1966a; Shaw, 1971; Smith, 1970). Hybrids between Malayan dwarfs and talls are of great commercial interest because of the disease-resistant properties of the dwarf mentioned before, and the hybrids are vigorous with broad stems and a high rate of leaf production. They come into bearing early, producing large nuts characteristic of or better than the tall parent (e.g. Satyabalan, et al. 1968b).

BOTANY

The vegetative plant

The coconut is one of the most stately trees of the tropics and has come to be symbolic in the western mind of the supposed easy life of those parts. The cover of this book illustrates a fringe of coconut palms near the sea in the Seychelle Islands.

The crown of the coconut palm is its most characteristic feature. When mature it consists of up to about 50 large radiating fronds bearing rows of some 100 leaflets (in the tall form) on each side of a tough rachis (Fig. 11.1). The leaflets are all orientated in one plane, unlike the oil palm (see Chapter 12) and range up to $4\frac{1}{2}$ ft in length. The leaves of the crown are arranged in a roughly $\frac{2}{5}$ phyllotoxis with consecutive leaves about 140° apart, forming a spiral relative to the trunk (Patel, 1938). Satyabalan et al. (1968b) have reported that left and right-hand spirals occur in approximately equal numbers but that this factor has no bearing on yield. However, Davis (1963, J. Genet., 58, p. 42) reports a considerable yield difference in favour of the left-hand spirals. In oil palms there does not appear to be any yield difference (Chapter 12) under Malayan conditions. However, according to Davis, the direction of the spiral may be determined by geophysical forces; at higher latitudes in the northern hemisphere left-hand palms predominated and were higher yielding.

All the leaves and stem tissues arise from the single apex located within the crown, and in addition to the emerged leaves, there is a continuous series of unformed leaves down to the primordial leaves around the apex, much the same as in the oil palm (Chapter 12). The petioles or leaf stalk are expanded at the base round the trunk and in the young palm form the major supporting tissue of the aerial parts of the palm, being several feet in height. Old scenescent leaves become detached at the .base of the crown and fall off. The rate of leaf production (see below) is considered to be an indication of vigour and yielding ability. In young palms up to about two years, the leaves tend to remain entire (without splitting into leaflets).

The trunk of the coconut does not become ap-

Fig. 11.1a A stand of tall coconuts (*c.* 12 years)

Fig. 11.1b The Malayan dwarf coconut

Fig. 11.1c Inflorescence in leaf axil

Fig. 11.1d Inflorescence in male phase at anthesis (left)
and with young developing nuts (tall variety)

parent for about five years in the tall form (less in dwarfs) because of the overlapping leaf bases. After this the trunk can be seen at the base of the crown, and thereafter does not increase much in thickness (Fig. 11.1). Small variations in thickness can be correlated with climatic and nutritional conditions of the palm at the time of formation in the crown. Being monocotyledonous, the trunk when formed cannot expand and an interesting feature, not at all understood, is that the living tissues of the trunk must survive the whole life of the tree if these remain functional in the movement of assimilates from one part to another. In the movement of water from the roots to the top, root-pumping pressure is apparently of central importance and in one investigation, Davis (1961) showed that the maximum heights to which water could be moved by suction from the crown (through transpiration) and by root-pumping pressure were 20.1 cm and 1250 cm respectively. Davis considers this to be an adaptation to the tree habit in the large palms in which the formation of conducting tissue is severely restricted.

The outer layers of the trunk consist of a type of cork tissue.

The root system

The root system is entirely adventitious as is characteristic of monocots. The basal part of the trunk becomes swollen into a bole by the formation of numerous root bases, particularly in the tall type of coconut. The main roots are uniform in thickness (about 1 cm) and bear numerous fine laterals. The root system is largely superficial and may extend laterally for up to 30 ft, according to Child (1964). Penetration in depth varies with soil conditions and is often limited by the water-table. Being unable to expand laterally, coconut roots do not break up hard-pan to the same extent as in dicotyledonous species. The root tip bears a root cap and the actively growing area behind it is an absorbing region. There are no root hairs, absorption taking place at the root surfaces. Behind this the root surface becomes hardened and thickened. Roots bear white soft protrusions thought to be breathing organs and indicate a degree of adaptation to flooding (Davis, 1968).

Inflorescence

Inflorescences are produced in the axils of the leaves and these will first appear in about the sixth year in tall palms; sooner in dwarfs. The inflorescences of some leaf axils may abort when young but generally each will develop to maturity.

Each inflorescence is enclosed in a sheath. After emergence from the subtending leaf axil the sheath or spathe splits and the inflorescence becomes exposed. Each inflorescence is branched and the branch bears male flowers towards the apex and a number of female flowers towards the base (Fig. 11.1). Male flowers open first over a period of a few weeks in tall varieties, followed by the ripening of the female flowers, largely preventing self-pollination. In the case of Malayan dwarf varieties, within-spadix overlapping of male and female phases results in the high degree of self-pollination mentioned earlier (see Whitehead, 1966a). (The Fiji dwarf exhibits out-pollination. For controlled pollination as in genetic work, or in the production of specific crosses for field planting, collection of pollen is best done 3–8 days after opening of the spathe and this can be stored after freeze-drying for periods of at least 6 months (Whitehead, 1965a).

Pollination occurs both by agency of wind and insects (e.g. Reyne, 1948). A high degree of abortion occurs following pollination and remaining fruits mature after about a year (see below).

Growth quantities

There have been few analytical studies on the growth of coconuts reported in the literature. The commencement of fruit production and yield seems to be closely related to the development of the photosynthetic apparatus of the crown. Romney (1968) reports that seedling growth during any period tended to depend on the size of the plant at the beginning of that period. He found that the relationship between length (L) weight (W) of the Malayan dwarf palm fronds, changed from $W(g)=3.6L$ (in)-52 for seedlings to $W=5.9L-100$ at 14 months age. Observations on semi-tall coconuts in the Ivory Coast (Fremond and Brunin, 1966) indicate that production of nuts and early bearing were associated with the number of leaves produced in the first two years. The rates of leaf production are rather low in the coconut (compare oil palms, Chapter 12), consequently the size of the assimilation apparatus would be closely related to leaf production rate. Fremond and Brunin report that 6–8 leaves per annum were produced in early years, rising to 10–12 per annum from about five years after planting; and Jaunet (1968), reported similar rates for tall palms reaching 12.5 per annum in fifteen-year-old palms. Silva and Abeywardena (1970) further report that bearing of fruit caused a reduction in the size of leaves in palms from about 10 years in age. The rate of leaf production of the Malayan dwarf is reported (in Jamaica) to be higher than any tall varieties (Harries *et al.* 1969) and the Fiji dwarf is known to produce a dense leaf canopy (e.g. Whitehead, 1966b) It is difficult to obtain comparative information on the rate of leaf production from different environmental locations, particularly in relation to rainfall distribution which affects canopy development strongly in oil palms (see Chapter 12). Satyabalan *et al.* (1968c) have shown that high-yielding palms are also more regular bearers and irregular bearing is associated with low yield, which suggests that ability to withstand adverse climatic fluctuations may have an important bearing on yield in some locations.

Stomatal measurements (Boyer, 1965) show that coconuts possess drought avoidance characteristics indicated by early stomatal closure under increasing moisture stress. This may indicate a relatively low potential for assimilation.

There has been much interest in coconuts in relating seedling growth quantities to subsequent yield. Liyanage (1953) and Liyanage and Abeywardena (1957) have reported that faster seedling growth is associated with higher copra yields in mature palms, as indicated before. Ninan *et al.* (1966) and Foale (1968a) have demonstrated between-progeny differences in seedling growth, and Foale has shown that such differences are not because of nut size since growth differences due to

this factor do not persist. He states that there is an active compensatory mechanism in that the assimilation apparatus of seedlings supplies less when more food materials are available from the nut. This suggests that in the seedling, sink demands by the growing apex may be limiting growth.

There is abundant evidence for hybrid vigour in coconuts both in relation to seedling growth and yield, in crosses between tall populations and tall × dwarf (e.g. Satyabalan *et al.* 1968 a and b, and 1970; Fremond and Nuce, 1971; Liyanage, 1969).

Development of the fruit. Development of the fruit following pollination to the stage of ripeness to harvest is indicated below for tall coconuts. The time for Malayan dwarf coconuts is about 360 days (Whitehead, 1966b).

For maximum oil yield (which is 60 per cent or more by weight of the dry kernel), fully-ripe nuts must be harvested. Copra yield from dwarf nuts require some 5–9 thousand per ton as opposed to 4–6 thousand of tall palm nuts (Harries, 1970). Nambiar *et al.* (1969) report that growth of the nut after pollination is at first slow but accelerates to a maximum rate from about the third to seventh month and then declines. The final copra yield was related to the growth during the fast growth phase. Following germination, seedlings show a high initial relative growth rate (R_w) for about 4 months, corresponding to the consumption of endosperm and thereafter R_w declines to a constant level (Foale, 1968c). This is similar to oil palm—see Chapters 1 and 12.

In a study of seasonal production over a 10-year period, Abeywardena and Fernando (1963) showed

Fig. 11.2 Development of nut from flowering to harvest

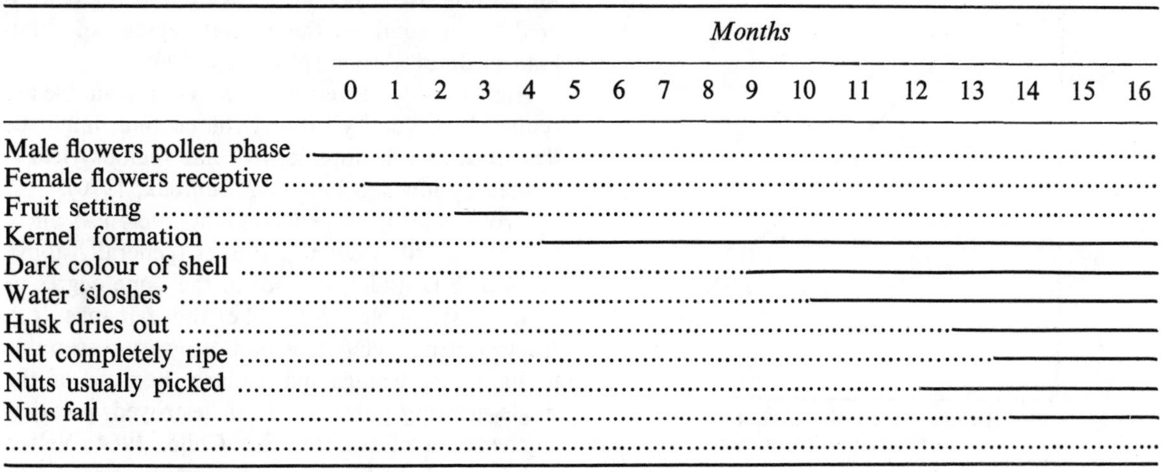

		Months															
	0	1	2	3	4	5	6	7	8	9	10	11	12	13	14	15	16
Male flowers pollen phase																	
Female flowers receptive																	
Fruit setting																	
Kernel formation																	
Dark colour of shell																	
Water 'sloshes'																	
Husk dries out																	
Nut completely ripe																	
Nuts usually picked																	
Nuts fall																	

Source. After Child, 1964.

that the yield differences brought about by seasonal variation in rainfall was mainly due to variation in fruit setting percentage, followed by the number of female flowers per inflorescence, nut weight and number of inflorescences. Park (1934) and Patel and Anandan (1936) report that the growing points of the inflorescence are affected by drought leading to a higher male flower component and to whole inflorescence abortion under extreme conditions (see 'Climate' below). Yield-wise, the effect was seen up to two or more years later (compare also oil palms, Chapter 12).

CLIMATE AND SOILS

Climate

The economic production of coconuts is confined to the humid tropics. Child (1964) states that a well-distributed rainfall in the range 50″–150″ is required for optimal yield, but points out that in some coastal locations with nearby hills, rainfall lower than 50″ is satisfactory because seepage water from inland is present. As explained, seasonal fluctuations in rainfall affect production some time later. Figs. 11.3 and 11.4 illustrate the effects of earlier moisture supply on yield, and the yield components nut number and copra content. Based on the expected evaporation in humid-tropical regions (*c.* 5″–6″ per month), 50″ rainfall would be marginal for economic production. Marginal rainfall would

Fig. 11.3   Relationship between antecedent soil water status and copra yield

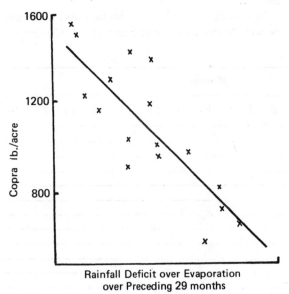

Source.   After Smith, 1966.

Fig. 11.4   Fall in production of nuts and their copra content following a severe drought period, Celebes

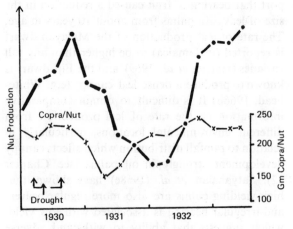

Source.   After Child, 1964.

be particularly important on poor, shallow-profiled soils (e.g. Maas and Wondenberg, 1970). There are many reports of rainfall limitations to coconut production (e.g. Savin and Christoi, 1965).

Because of the general low yield and economic problems of the coconut industry, production for trade is not extended far outside the humid tropical latitudes at the present time. Most coconut production is, furthermore confined to lowland coastal regions. The highest elevation at which coconuts are commercially grown for the copra trade is about 3000 ft in South India (13°N). In cold locations coconuts take longer to reach the bearing stage, they are shorter with poor foliage and the nuts are smaller.

The use of coconuts for culinary purposes and sale of fresh nuts is likely to allow production on a smaller scale at higher latitudes and it is interesting to note that coconut growing is being considered as far south as the coastal regions of Zululand in South Africa (Nyenhuis, 1965) .

High sunshine levels when water is available are required for good yield and the coconut might be described as a heliophile. In some regions yield is limited by low sunshine (e.g. Sumbak, 1970).

Strong winds, as generated by cyclones, affect coconuts in the cyclone latitudes (generally above 10°N or S latitude), but not to the same degree as highly susceptible crops like the bananas (e.g. Abeywardena, 1969). Coconuts are very susceptible to lightning damage, particularly because of the single growing point which, if destroyed, prevents regeneration of the tree (see Child, 1964; Velsen and Edward, 1970).

There is some evidence that photoperiod affects coconuts in that long days are reported to increase the number of spadix primordia (Wickremasuriya, 1968).

Soils and drainage

Because of the fact that the majority of coconut industries are established in coastal habitats and small islands, certain types of soils predominate. Throughout the Pacific, coral sands are often used for coconuts. Moisture supply from rainfall or seepage water is essential to high yields. These soils require K for good yields (e.g. Foale, 1965; Manciot, 1968), and Fe and Mn (Pomier, 1968). On silica sands potassium is often also required and in fact K is the most important nutrient in coconut growing. Volcanic ash soils are also used for coconuts and are considered good soils. However, K deficiency is also encountered (e.g. Foale, 1967) and sulphur deficiency is widespread.

Alluvial river and estuarine soils, when rich in nutrients and freely-drained, are stated by Child (1964) to be the best coconut soils. When heavy clay soils are used extensive drainage is required (e.g. Briolle, 1969). Delorme (1966) reports that N and K deficiencies are most frequent on hydromorphic soils in Dahomey. Acid sulphate soils and peat soils formed under certain conditions are generally not suitable for coconuts. The optimal soil reaction for coconuts ranges between about pH 5 and pH 8 and acid sulphate and peat soils are often much more acidic. Overdrainage should not be practised in acid sulphate and peat soils and in the former sea water may have an ameliorating effect. Foale (1969) showed that at a salinity of 1000 ppm coconuts were extremely productive. The coconut possesses some tolerance to salinity but is killed when salt levels are high. Cassidy (1968) reports death of coconuts in a situation where chloride fluctuated up to 2380 ppm.

Inland tropical soils are used only to a lesser extent for coconuts; on the iron-rich latosols of the humid tropics, cations, nitrogen and boron are likely to be deficient (e.g. Santhirasegaram, 1967) and P nutrition will be problematical.

PLANTING
Land preparation, spacing and yield

Because coconut acreage on a world-wide basis is not increasing to the same extent as with many other tropical crops, clearing usually involves the removal of old coconut stands for replanting. Poisoning of old groves with sodium arsenite or 2:4-D injected into the trunk is sometimes carried out prior to clearing (e.g. Migvar, 1965). Stump removal by burning is desirable where rhino-beetle infestation occurs since the stumps constitute a breeding ground for the larvae.

Conventional spacing varies for dwarf and tall coconuts. For talls, spacings of 8–9 m (c. 140–160/ha) have been used (e.g. Liyanage, 1963; Douglas, 1965, etc.). Cumulative yields are not generally found to increase at closer spacings because of the requirement for high insolation by the coconut. This also leaves a rather open canopy and offers opportunities for intercropping (see p. 166). In a spacing trial with talls ranging from 6.7 to 10.7 m in Jamaica (Whitehead and Smith 1968), nut yield (the number of nuts per acre) increased at the denser spacings but copra yield was not significantly increased at spacings closer than the conventional 9 m. However Foale (1968b) in a spacing trial using 171, 218 and 267 palms/ha showed higher yield (no decline in nuts per palm) at the intermediate density up to the sixth year.

For dwarf varieties denser plantings ranging from 240 (7 m) to 280 per acre are required for optimal yields and the canopy is much denser and less suited for intercropping (e.g. Smith and Romney, 1969; Smith, 1970). When intercropping is contemplated in dwarfs a wider spacing is needed (e.g. 21 × 28 ft with a banana inter-crop—Smith, 1968).

Square planting is traditionally used for coconuts (unlike oil palms, Chapter 12) but in low sunlight locations triangular (equilateral) spacing with the rows orientated in a north-south direction is said to increase yields (R. Manicot–cited by Whitehead, 1966b).

In many parts of Asia, coconuts are grown on rice bunds at a spacing of 6 to 8 m.

In most coconut handbooks, planting holes of about 1 m cube are recommended for coconuts, these being filled with top-soil, wood ashes and various other substances. This highly costly practice has been questioned by Kroon (1970) since in Sumatra and elsewhere, good coconuts are established with very small holes on light soils and holes of 0.6 m on heavy soils.

The yields of copra from present day coconut estates are not high (mainly for the reason discussed before, that out-pollinated, virtually unselected nuts

Fig. 11.5   Coconut nursery in Malaya

are generally used for planting). One ton of copra per acre (c. 4000 nuts of the tall variety) would be a good average yield. With dwarf and hybrid plantings yields are higher. When controlled pollinations of superior trees of the tall variety are used for planting material yields of up to 12,000 nuts (about 3 tons of copra) have been achieved in Ceylon (Mr. Ray Wijiwardene—personal communication), and even higher yields seem possible if the procedure of planting superior trees is followed in the future.

Nursery and planting out

Coconuts exhibit a short dormancy of a few weeks after the nuts are mature. Genetic variations in rate of germination also occur (Whitehead, 1965b). The usual method of raising nursery plants is in beds of sandy soil as shown in Fig. 11.5, but the use of polythene bags seems to be penetrating the coconut world (Rognon, 1971). The advantage of these with the large-seeded coconut which, at the conventional transplanting time, has only about seven stout roots and is easily handled, however, seems dubious. Seedbeds should have loose sandy soil to facilitate the lifting of the nuts, and unshaded nurseries are generally used except in extreme conditions.

Only vigorous seedlings are recommended for planting. The principles involved in the selection of seedlings are discussed in Chapters 1 and 12 in more detail. In coconuts 'weak' seedlings are rejected and this improves the field performance somewhat. Seedlings are transplanted from the 'crowsbeak' stage (c. 18 weeks) up to about the 4-leaf stage, and earlier planting seems better generally (e.g. Sumbak, 1968; Charles, 1969), but this would

depend on environmental conditions of drought, etc.

Horizontal planting is superior to vertical (e.g. Varisai Muhammad et al. 1971) and fertilization, if practised, should be carried out after 4-6 weeks when the seedlings have been established.

Controlled pollination

As stated before, controlled pollination of high-yielding mother palms (e.g. those producing an average of more than 150 nuts/annum) with pollen from similar high-yielding trees of other genetic origin is highly desirable. A pamphlet on how to do this is produced by the *Coconut Research Institute*, Ceylon (Leaflet No. 47, 1966). Normally male-flower removal is carried out after which the inflorescence, pruned to about 30 female flowers, is bagged with a polythene sleeve and pollinated 21–24 days after spathe opening (see also Liyanage, 1966).

Vegetative propagation

Vegetative propagation is of great commercial interest in both coconuts and oil palms but seems impossible at the present time. The subject has been discussed by Williams and Hsu (1970). Research on this topic has been cloaked in some secrecy but the possibilities seem to be in the following areas. A small proportion of both oil palms and coconuts naturally produce bulbils or suckers, either from the base (Child, 1964), the roots, or from the inflorescence buds (Davis, 1969a). There is a possibility that suckers can be induced in normal palms by treatment with various growth regulators. A second possibility is the dividing of the apex, which sometimes produces a branched shoot (e.g. Davis, 1969b), and a third is by tissue culture. Tissues from the growing point of coconuts have been reported to produce roots after a 4-week dark period followed by 'suitable light' for 1–2 months (*CELOS Bull.* 13 (1971), p. 10). Hampden (1973) recently reports successful culture of coconut tissue at Wye College, England, in which shoot-like organs have been produced. If it turns out to be possible to produce whole plants in this way the prospects of the commercial coconut industry will be completely changed.

Manuring and irrigation

Fertilizers and cover crops are widely reported to increase yields in coconuts. Responses, however, take from one to three years to occur (e.g. Charles,

1965) probably because of the long period between primordial formation and nut yield in the coconut. Potassium deficiency is most commonly encountered in coconut areas, and K application should commence in the immature stages for full effectiveness (Fremond and Ouvrier, 1971). Potassium increases both the number of nuts and their copra content. Annual application rates recommended range in the order of 0.7 to 1 kg of KCl per tree for bearing palms (e.g. Charles and Douglas, 1965; Nethsinghe, 1966a; Nathanael, 1969).

Nitrogen is required particularly in the young stages (e.g. Smith, 1969a) but is often adequately supplied by legume cover crops (e.g. Fremond and Goncalves, 1967). It increases the number of female flowers and nut yield, but excessive N reduces copra content of the nuts (Kunhi Muliyar and Nelliat, 1971; Smith, 1964/5, 1965/6).

Phosphate responses are less frequent in coconuts possibly because of the coastal hydromorphic and sandy soils commonly used do not inactivate P as do inland ferraltic soils. However, responses do occur (e.g. Brunin, 1968), and frequently NPK mixtures are recommended for coconut (e.g. Silva, 1967; Fenwick and Maharaj, 1967).

Magnesium applications have been recommended to offset the antagonistic effects of K fertilization (Brunin, 1970); and sulphur deficiency has been reported in Papua New Guinea, this reducing copra quality and oil content (Southern, 1967). On coral soils Fe and Mn deficiencies occur. Pomier (1967, 1969) recommends 10 g of ferrous sulphate and 5 g of Mn SO_4 per year, application being by trunk injection (or husk injection in seedlings), or spot placement in the root zone; broadcasting is not effective.

Boron deficiency has been associated with disease incidence in coconuts (Chen, 1966; Davis and Pillai, 1966); and there is some evidence that sodium may be a nutrient or osmotic regulatory element in coconuts (Smith, 1969b; Roperos and Bangoy, 1967).

Apart from Fe and Mn applications, fertilizer response is generally found to be better when broadcast in the root zone or weeding circle of palms rather than in band placement (Charles and, Douglas, 1965; Nethsinghe, 1966b; Silva, 1968), and biennial application has been reported as more advantageous than annual (e.g. Charles and Douglas, 1965; Nathanael, 1969).

Salgado and Abeywardena (1964) report that K

and P levels of coconut water reflect soil levels and adequacy of these elements, but generally the levels in the 14th leaf are used for nutrient deficiency diagnosis. Critical values of nutrients in per cent dry weight are of the order N:1.8–2.0, P:0.12–0.13, K:0.8–1.0, Ca:0.5, Mg:0.3 per cent (Fremond and Nuce de Lamothe, 1968), which are IRHO recommendations. These levels have been questioned on the basis of trials in the West Indies (e.g. Fenwick, 1968). The critical level of Fe is considered to be about 50 ppm (*Oléagineux* 24, p. 273).

Because of the relatively low profitability of copra growing, irrigation is not generally practised with coconuts. However, irrigation can undoubtedly increase yields in marginal environments where prolonged dry seasons occur (e.g. Shanmugham, 1966) and is particularly useful in the establishment phase (Nelliat, 1968).

Weed control and covers

In flat land mechanical weed control and hand circle weeding are most effective in newly-planted coconuts before the root system spreads to the inter-rows, particularly in the control of *Imperata cylindrica*. However it is necessary to use herbicides early if damage to the roots is to be avoided. The herbicides paraquat, diuron, atrazine and dalapon have all been found to be safe for coconuts (Romney, 1968). Picloram, 2:4-D, 2:4:5-T and bromacil are unsafe for coconuts (Kasasian and Smith, 1968); although Hoyle (1968) states that 2:4-D amine (as opposed to the ester form) is safe for mature tall coconuts (presumably when the spray does not reach the foliage or living parts of the tree). This, however, is unsafe for the dwarf variety.

Leguminous covers produce variable results in coconuts depending on moisture conditions. In West Africa, with a marked dry season *Centrosema* delayed growth of coconuts in the immature stages, particularly in dry periods (Fremond and Brunin, 1966) but the advantage in terms of costs of upkeep was considered to be favourable. In Jamaica, Smith (1964/5–1965/6) reports yield increases due to *Pueraria triloba* cover. In areas with a marked dry season, as in parts of Ceylon, disc-harrowing the covers in the dry season is necessary to prevent yield losses as a result of moisture competition.

Interplanting and grazing

Because of low profitability of coconuts interplanting and intergrazing are often practised. The case of cocoa has already been mentioned (Chapter 7).* In tall coconuts, the planting density is such that a high proportion of sunlight penetrates the coconut canopy which makes them ideal for intercropping. The effect of intercropping and grazing on coconuts is related to moisture supply and nutrition. Interplanting reduces yield if moisture supply is limiting and when adequate fertilization is not carried out. In some circumstances lowered coconut yield is tolerated because intercropping or grazing is more profitable (e.g. Christoi and Souza, 1966). In humid areas, interplanting pasture grasses is possible without markedly reducing coconut yields when appropriate extra fertilizer applications are made (e.g. Santhirasegaram, 1967b).

Mechanization of copra extraction and coconut by-products

In the growing of coconuts in plantations, as opposed to small acreages, the de-husking task is one of the most labour intensive and absorbs meagre profits from copra. Consequently, there is a present need for mechanization of this process but unfortunately no such mechanization of copra extraction seems in the offering at the present time. The need to increase earnings from coconuts has, however, resulted in the investigation of a whole range of actual and possible by-products including, coir and coir yarn (e.g. Khosla, 1965) as well as coir dust, rubberized coir, active carbon (from the shell); and by distillation, acetic acid, phenol, creosote and pitch. From the coconut water, alcohol, sugar and acetic acid are obtainable (see review by Nathanael, 1966).

Coconut press-cake is, of course, used in animal feeds, and more recent investigations have shown that coconut can produce a high quality protein concentrate for human food as well as a cheese-like product (Prasanna *et al.* 1969; Somaatmadja and Ali, 1969).

MAJOR DISEASES AND PESTS

With coconuts (and oil palms, Chapter 12) it is necessary to mention one pest which is of wide occurrence—the rhinoceros beetle (*Oryctes rhinoceros*). This beetle causes major damage in many areas, and may also be a vector in the spread of

*See also *Cocoa and Coconuts in Malaya*, Inc. Soc. of Planters.

red-ring disease (see below). The beetle feeds on young leaf and heart tissue and may cause death of the tree through destruction of the growing point and greatly reduce yield through defoliation. Various types of control have been tried, including biological methods and attractants, but at the present time, practical control centres on field hygiene and the removal of stumps and other field debris in which the larvae may develop, or the treatment of these breeding grounds with insecticides.

As far as diseases are concerned, several are of great economic importance. In the West Indies (notably Jamaica) a virus or mycoplasma disease called lethal-yellowing causes widespread losses (e.g. Price *et al.* 1967). In this disease, light-brown, irregular water-soaked areas appear at leaflet tips of bud leaves and thence spread to the heart; at a later stage yellowing of older fronds occurs (Martinez, 1965). Carter (1966) has shown that diseased palms exhibit phloem necrosis. The Malayan dwarf coconut possesses resistance to lethal-yellowing, and is therefore of great interest for future planting in affected regions.

In India (and probably Ceylon) a serious 'root-wilt' disease occurs which also appears to be virus-caused (Shanta *et al.* 1964). Diseased palms deteriorate and become susceptible to leaf and other diseases (e.g. *Helminthosporium*—Rodha and Lal, 1969).

A disease of unknown cause occurs in the Philippines and elsewhere, known as 'cadang-cadang' or 'yellow-mottle-decline'. Yellow water-soaked depressions occur on the undersides of leaflets, leaves become smaller and finally the crown is only a short tuft. Cadang-cadang has been attributed to many causes including soil condition and nutrient status and a virus. Breeding of resistant coconuts seems to be the main control approach (e.g. Bigornia and Infante, 1965).

Red-ring disease is caused by the nematode *Radinaphelenchus cocophilus* which is transmitted to the palms by various insect pests. Yellowing and shedding of leaves, starting with the oldest, are the external visual symptoms. Within the lower trunk a band of reddish tissue, in which the nematodes are concentrated, is formed. Phytosanitary procedures prevent spread. These include the poisoning of infected trees with sodium arsenite followed by burning and vector control. Certain nematocides have shown promise in treatment of trees (see Hoyle, 1971).

12 OIL PALMS [Elaeis spp.]

INTRODUCTION, SPECIES AND VARIETIES
THE oil palm industry is the newest of the plantation industries of tropical countries and one of the
best organized, due largely to the activities of the
West African Institute for Oil Palm Research in
Nigeria and a number of large government and
commercial organizations. However, long before
the development of plantation industries which
dates only from the early part of the present
century, trade in palm oil obtained from wild and
semi-cultivated groves in West Africa was carried
out between West African countries and Europe.
The earliest trade records date from the sixteenth
century (see Hartley, 1967) and by 1911 export of
palm oil from West Africa amounted to some
87,000 tons, and palm kernels (see Fig. 15.5d) some
232,000 tons. Thus, there was an established demand
for oil palm products in Europe, and the growth
of the soap and margarine industries in the second
decade of the present century led to even greater
demand and the initiation of the first plantation
industries. Oddly, the plantation cultivation of oil
palms began in the Far East and only later developed in Africa. The first commercial planting is
attributed to M. Adrian Hallet, a Belgian, who in
1911 established a small plantation in Sumatra.
From this the industry has grown steadily in South-
East Asia. The original planting in Asia was established from four palms introduced *via* Holland to
the Botanical Gardens at Buitenzorg, Indonesia.
The origin of these palms is uncertain, but they are
most likely West African palms. Two facts became
apparent at an early stage; firstly the fruits of the
palms, now referred to as Deli palms or Deli Dura's
from the location in Sumatra where seeds from the
Buitenzorg palms were planted, were found to be
relatively thin-shelled compared to African palms,
and secondly, the yield of the palms in the more
equitable climates of Sumatra and Malaya (without
marked dry season) was found to be higher than
that obtained in most African localities.

The plantation industry in Africa initially developed in the Belgian Congo in the 1920s and was
based on the *tenera* oil palm (see p. 168). Africa has
subsequently declined in the production of oil palm
products for export and has lagged behind South-
East Asia in development of the plantation industry
as distinct from the exploitation of wild palm
groves. The world export figures for palm oil and
palm kernels for 1965 are given in Table 12.1 (from
Hartley, 1967).

TABLE 12.1 Oil palm products exported in 1965
(in 1000 ton units)

Origin	Palm oil	Palm kernels
Africa:		
Nigeria	159	415
Ivory Coast	1	15
Dahomey & Togo	13	31
Other West African sources	13	106
Congo (Kinshasa)	81	–
Angola	14	14
Other Central African sources	6	13
Asia:		
Indonesia	115	33
Malaysia	139	12
TOTAL	541	639

This table emphasizes the importance of Malaysia and Indonesia in the oil palm trade. However,
the high ratio of palm kernels to palm oil exported
from African countries indicates the greater extent
to which palm oil is used for local consumption
in Africa.

In recent years oil palm cultivation on a plantation basis has been introduced to many South and
Central American countries (see Savin, 1966; Ollagnier and Martin, 1967), African countries, Madagascar and New Britain.

There are actually two definite species of oil palm: *Elaeis guineensis* Jacq. from Africa, and *E. melanococca* Gaertner (=*Corozo oleifera* (HBK) Bailey) from South America. A third species *E. madagascariensis* Becc. from Madagascar is now thought to be a variant of *E. guineensis*. Commercial plantations are, at the present time, almost entirely based on the various varieties of *E. guineensis*, which is the West African natural-grove oil palm, but there is interest in *E. melanococca* for breeding purposes, since it hybridizes readily with *guineensis* (e.g. Hurtado and Nunez, 1970). *Elaeis melanococca* is dominant for certain characteristics; particularly large leaves with the leaflets all set in one plane, unlike *E. guineensis* (see below), a higher oil content of the pericarp and a shorter habit.

The occurrence of *E. guineensis* in natural groves, and its introduction to South-East Asia has already been mentioned. Fig. 12.1 illustrates the distribution of these groves in the so-called palm belt of West Africa, which corresponds closely to the humid tropical zone as delineated by Blummenstock and Thornthwaite (1941), to which the oil palm is most adapted (see 'Climate', p. 175 and Chapter I).

Like the tall coconuts, the oil palm is highly heterozygous. At the same time, however, many of the commercial palms, particularly in Asia, possess a degree of consanguinity, due to the common origin of all Asian Deli palms used for producing seeds. Two distinct colour varieties of *E. guineensis* occur; one has green fruits when immature, ripening to bright orange (*var. virescens*), and the other more common form has black fruits ripening to deep red.* Within each colour form three distinct types can be recognized on the basis of shell thickness; there is the *Dura* which is thick-shelled, the *pisifera* which is a shell-less form, and the *tenera* which is thin-shelled and is a hybrid between the other two. The *tenera* is now commonly used for commercial planting, and the 'families' of Duras derived from the original Deli stock (the Deli Duras) which are relatively thin-shelled among Duras are much favoured as parental material, both in South-East Asia, as well as in parts of Africa (see Ollagnier and Gascon, 1965; Hardon, 1969). *Pisiferas*, used as the male parent in seed production, are mostly from Africa. Unlike the position with coconuts (see previous chapter) breeding work based on the production of *teneras* from specific crosses of high-yielding *Duras*, and the shell-less *pisifera* has led to the production of the superior commercial material which is now widely planted.

A dwarf form of *E. guineensis*, derived from selection in Malaya (Jagoe, 1952) and known as the 'dumpy' has a highly heritable short trunk characteristic which is of interest in the genetic improvement of the oil palm, but is not used commercially at the present time.

BOTANY

The vegetative plant

The gross morphology of the oil palm is rather similar to that of the coconut. There is a single growing point which produces the crown of foliage leaves or fronds. The growing point is located deep

*Various other fruit colours are also recorded.

Fig. 12.1 An impression of the distribution of natural *E. guineensis* groves, (hatched area) and various isolated points of colonization

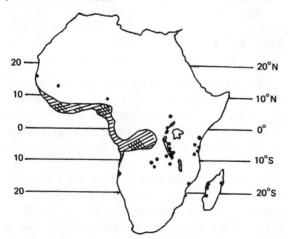

Source. From Hartley, 1967, after Zeven, 1967.

Fig. 12.2 Leaf primordia of the oil palm

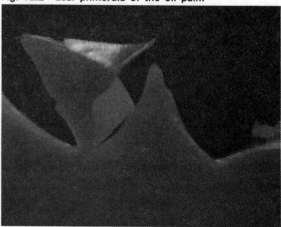

Source. From Williams and Hsu, 1970.

Fig. 12.3a Young seedling at early nursery stage producing entire leaves

Fig. 12.3b Older seedling (*c.* 6 months) beginning to produce divided leaves

Fig. 12.3c A stand of mature oil palms, harvesting 'in progress

within the crown tissue and produces a continuous series of foliage organs from the primordia (Fig. 12.2), which develop into the mature leaves. The apex gives rise to very little stem tissue, which is formed at a later stage from a meristem that is continuous with the bases of the fronds (Strasburger, 1926). In the first few years of growth, the main supporting tissues of the crown consist of the expanded leaf bases (as with the coconut palm, Chapter 11). Fig. 12.3 illustrates stages in the development of the palm from germination to maturity.

The mature frond of the oil palm (*E. guineensis*) bears two ranks of leaflets on either side of the stout rachis, unlike the coconut and the South American oil palm (*E. melanococca*) in which all the leaflets are set in the same plane. The degrees of displacement of the leaflets has a bearing on the efficiency of the canopy as discussed later. The leaflets bear stomata predominantly on the lower surface.

As the palm grows it becomes apparent that the leaves are arranged in a series of eight spirals originating from the apex. The frond bases remain attached to the trunk for a much longer period than with the coconut palm, and as ageing fronds are pruned off in the course of harvesting the bunches, the spiral pattern can be seen from the arrangement of pruned bases, as illustrated in Fig. 12.4.

Fig. 12.4 Oil palm trunks showing spiral whorls of frond bases, alternately shaded

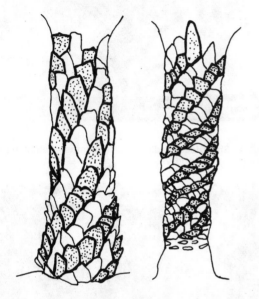

Notes. Left: palm *c.* 12 years age; right: palm, about 20 years, is beginning to show the naked trunk from which the frond bases have fallen.

Left-hand and right-hand spirals occur with approximately equal frequency in Malaya (*c.* 52–53 per cent lefts) and this factor does not appear to be related to yield there (Arasu, 1970—see Coconuts, Chapter 11).

The rate of leaf production in the oil palm is higher than in coconuts, but is also affected by climate. In Malaya, with an adequate and even rainfall distribution, frond production varies between 25 and 35 per year. In Nigeria, with a marked dry season, rates of 20.5 to 23.1 per year have been recorded (Broekmans, 1957). In dry conditions the rate of expansion of the central youngest emerging frond (the spear leaf) is also lower and these spear leaves delay unfolding of the leaflets until wet conditions prevail.

The root system

The root system of the oil palm is relatively superficial, few roots occurring below 18″ (Purvis, 1956;

Fremond and Orgias, 1952). In addition, in permeable soils and absence of a high permanent watertable, a number of descending roots may reach depths of 10 ft or more and are mainly for anchorage (Ruer, 1969). The main surface roots radiate to considerable distances (in isolated palms roots have been found 62 ft from the tree) and produce ascending and descending secondary, tertiary and quarternary roots, the latter being the main absorptive regions. Concentric zonation of root concentration occurs about the trunk which may be related to the history of fertilizer placement (e.g. Purvis, 1956a), which initially causes inhibition of root growth, followed by proliferation of a mass of feeding roots.

Hydathodes, or breathing roots, also occur in oil palms (as with coconut) and indicate an adaptation to flooded conditions.

Root pressure has not been investigated in oil palms, but by analogy with other palms (Davis, 1961), it might be expected that root pressure is important in the moisture economy of the tree.

The inflorescences

Male and female inflorescences are formed in separate leaf axils in the oil palm, usually in cycles of several months duration, depending much on climate and conditions of the physical environment. Fig. 12.5 illustrates the inflorescences and fruit of the *tenera* oil palm which, as explained before, is a hybrid between a shell-less form (the *pisifera*) and a thick-shelled type (the *dura*). The *pisifera* type of palm generally produces a high proportion of infertile and poorly-developed fruit because a growth factor important in development of

Fig. 12.5a Male inflorescence at stage of pollen release

Fig. 12.5b Female inflorescence at maturity for fertilization

Fig. 12.5c A fully-developed oil palm bunch

Fig. 12.5d Fruit of *pisifera, tenera* and *dura* palms

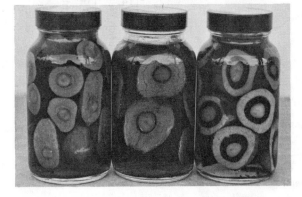

the fruit appears to be associated with the presence of shell. Some *pisiferas* however do produce fertile bunches with fully-developed fruit and Tang (1971) has recently suggested the exploitation of *pisifera* × *pisifera* material from these exceptional trees.

Apart from the thickness of shell, the *dura* (which is no longer planted commercially) closely resembles the *tenera*.

Pollination occurs both by the agency of wind and insects. Pollen has a strong aniseed odour. Sometimes assisted pollination is carried out to increase fruit-set (see p. 182). Following pollination, 5 to 6 months elapse before the inflorescence develops into the bunch and is ripe for harvesting. Oil formation takes place with ripening and only reaches maximum levels when the fruits become bright red or orange and are easily detached from the bunch (Rajaratnam and Williams, 1970).

Growth quantities and yield components

In the oil palm, as also the coconut palm, the leaf area index of the canopy can be manipulated much more precisely by varying planting density, and through plant genotype than is possible with freely-branching species. Certain leaf-area characteristics have been found to be closely related to growth and productivity. Hardon *et al.* (1969) showed that as the palm develops, successive leaf areas increase to a maximum at about 8 to 9 years (Fig. 12.6) after which a threshold is reached that correlates closely with a yield plateau achieved at about the same time (Fig. 12.7). Similarly, within a number of genotype populations, yield was correlated with individual frond area. On the basis of LAI, assuming an average frond number of 40 per palm, optimal LAI for the palm under the equitable growing conditions of Malaya appeared to be about 7, rather higher than is generally achieved in plantation practice. Increasing LAI by increasing planting density has its limitations because of the parallel increase in trunk and vegetative material apart from the assimilation apparatus, and from the etiolation effect of dense spacing which increases trunk height. Probably increasing LAI by the selection and breeding of palms with larger individual fronds represents the best way of achieving higher LAI values. Hardon *et al.* (1972) indicate that LAI has a reasonable level of genetic control. Furthermore, increase in the angular displacement of the two ranks of leaflets (mentioned above) to the maximum of 45° should increase light penetration and make for both

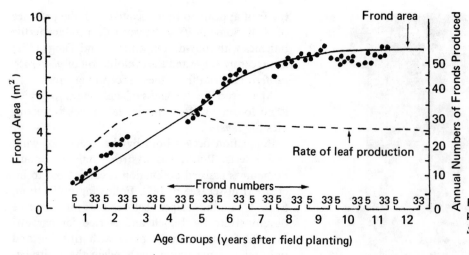

Fig. 12.6 Frond areas of palms of different ages

Source. After Hardon *et al.* 1969.

a higher optimal LAI and a more productive canopy.

The yield plateau (Fig. 12.7) may be maintained for a number of years, but eventually a slow decline occurs. This may be attributed to increased respiration of the trunk which gains in height by one or more feet per year and to increasing moisture stress of the crown as it becomes more elevated.

The number of leaves produced by the oil palm (20 to 35 per year, in contrast to the coconut) indicates that this is not a limiting factor in canopy development. The number of bunches produced is generally well below the number of fronds (Fig. 12.7). Under plantation conditions in which fronds are pruned during harvesting, a palm usually has about 40 fronds present at any time, which appears

Fig. 12.7 Variation of yield and its components, average bunch weight and annual number of bunches with the age of an oil palm planting (var. Deli *dura*) in South Malaya

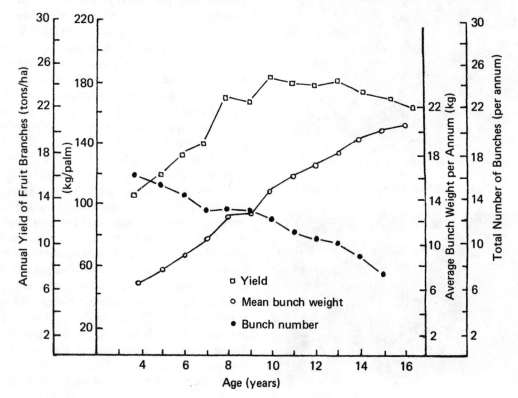

Age (years)

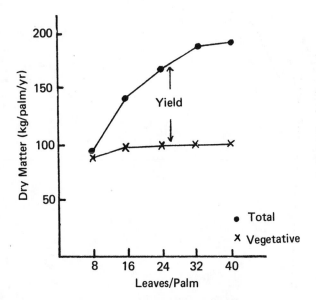

Fig. 12.8 Effect of defoliation on total dry matter production, vegetative dry matter and bunch yield

Source. From *Progress Report* 1970, Oil Palm Genetics Laboratory, Malaysia.

to be near optimal. Experimental pruning of fronds reduces yield in the manner illustrated in Fig. 12.8.

The major yield bottle-neck in oil palms (in common with most perennial crop species) appears to be a relatively low assimilation rate (Rees, 1958), and stomatal closure under conditions of moisture stress (see 'Climate' below). Under equitable conditions there appears to be potential for the genetic improvement of this character (*Progress Report*, 1971, Oil Palm Genetic Laboratory, Malaysia). Growth data from various sources are shown in Table 12.2. Corley *et al.* (1973) have recently found

evidence for genetically-controlled variation in photosynthetic rate in oil palms using $C^{14}O_2$.

The data of Table 12.2 indicate the extreme sensitivity of bunch yield to moisture stress (see 'Climate' below). Bunch yield is a function of bunch weight and bunch number; the latter being particularly sensitive to environmental conditions. Bunch number is determined by the relative production of male and female inflorescences (see 'Sex determination' below), and by abortion of the female inflorescences. Pronounced peaks in abortion rate can be seen from the occurrence of empty leaf axils. Throughout development of the bunch inflorescence from the primordial stage, there is a slow increase in size for an extended period followed by a rapid increase when the inflorescence reaches a certain stage. Abortion occurs most frequently after this change in growth rate as illustrated in Fig. 12.9.

There has been interest in the selection and rejection of seedlings at the time of planting out from the nursery, and subsequent performance. Selection based on a single scoring of 'vigour', height, length of the newest fully-emerged leaf etc. has produced variable results, sometimes with the selected seedlings producing significantly higher-yielding plantings (e.g. *Annual Report* West African Inst. for Oil Palm Research, 1957–8). Williams and Hsu (1970) point to the need for relative measure of seedling performance based on two successive scorings because of the exponential form of the growth curve in nursery material (see also 'Nursery' below and Chapter 1).

The success of the *deli* × *pisifera* palm mentioned earlier has detracted from the question of inbreeding depression on growth and yield, arising from

TABLE 12.2 Growth quantities for oil palms from various sources

	CROP GROWTH RATE tons/ha/yr.	LAI	Assim. g/dm²/wk.	Bunch Index*	Dry matter (vegetative)
Corley *et al.* (1971) Malaya—adequate moisture	29.83	3.61	0.160	0.423	16.5
Ng *et al.* (1968) Malaya—adequate moisture	28.2	3.72	0.130	0.447	15.2
Rees and Tinker (1963) Africa—moisture stress	19.5	4.93	0.078	0.267	14.3

Source. Adapted from Corley *et al.* 1971.
*Bunch index=harvest index (ratio of bunch yield to total yield).

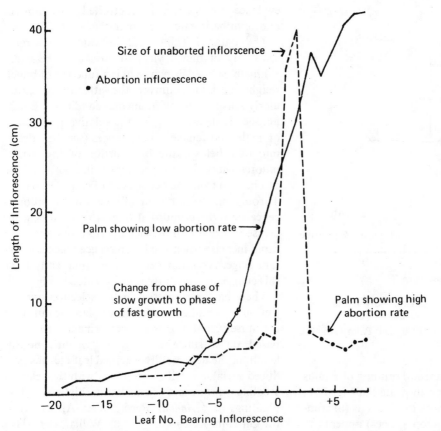

Fig. 12.9 Length and abortion of inflorescences in two palms in relation to leaf

Source. After Broekmans, 1957.
Note. O=spear leaf, positive and negative values for emerged and primordial leaves respectively. Abortion occurs about one year before bunch maturity.

the uniformity of all *deli* material. In Malaya, Hardon (1970) showed that inbreeding has a depressing effect on yield and bunch weight.

Sex determination

The inflorescence primordia are formed in the axils of the primordial leaves and sex determination takes place about two or more years before emergence and maturity of the bunch. A high sex-ratio or proportion of female inflorescences (and hence yield) occurs under good physical conditions particularly of high sunshine and absence of moisture stress. Experimental pruning of the fronds, which reduces assimilation, also causes a marked depression in sex-ratio some two years later as illustrated in Fig. 12.10.

Cycles of male and female inflorescence production usually last several months. In a uniform equitable climate, female inflorescence production usually predominates and cycles are often unrelated to seasonal factors. Under sub-humid conditions,

male inflorescence production predominates and cycles are often linked to seasons. At the time of change from one type of inflorescence to another, hermaphrodite inflorescences are formed. These differ according to whether the change is from male to female or vice versa (Williams and Thomas, 1970). In the former, relatively massive bunch-like inflorescences are formed with a few male inflores-

Fig. 12.10 Changes in sex-ratio of pruned and unpruned palms

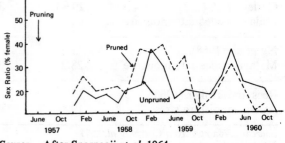

Source. After Sparnaaij *et al.* 1964.

cence 'fingers' replacing some of the spiklets. In the latter, the inflorescence resembles a normal male, but the fingers bear either a few or many female flowers or truncated spiklets. It is thought that the carbohydrate or reserve status of the palm effects differentiation of the primordia.

CLIMATE AND SOILS
Climate

For optimal yield, absence of moisture stress and high sunshine are essential. In monsoonal regions where a pronounced dry season occurs and the wet season is heavily overcast, yield may be related to the amount of dry-season rainfall. Sparnaaij et al. (1964) related yield to hours of 'effective sunshine' which was that received when adequate water was available assuming a consumption of 4″ per month (based on lysimeter studies on seedlings and probably an under estimation). The relationship is shown in Fig. 12.11. In West Africa the climate is generally sub-optimal for oil palms and Vossen (1969), for example, considers that areas with less than 400 mm rainfall deficit over evaporation are 'moderate' for cultivation. Under Asian conditions, bunch yields generally reach 10 tons (c. 400 lb/palm) and regions with a rainfall deficit would not generally be recommended for oil palms. In the South and Central American region variable climates occur. Fig. 12.12 illustrates the moisture budget of a number of major oil palm districts and yields.

In addition to bunch yield and sex-ratio, oil production is reduced by moisture stress (Ng and Goh, 1970).

The effects of temperature on the oil palm are not well understood because its cultivation has not been extensively attempted outside the equatorial latitudes or at high elevations. With seedlings, growth is arrested at 15°C (Hartley, 1967). At Tela in Honduras (15° North) the cold months are thought to delay maturing of the bunches (min. temperature below 20°C in cold season). Oil palm cultivation, however, is envisaged at a number of high latitude locations including Madagascar (Bouchard and Damour, 1971), and it seems likely that lowered yields or delayed maturity can be expected. In Sumatra it has been reported that palms above 500 m (c. 1,500 ft) come into bearing at least a year later than at lowland elevations (see Hartley, 1967).

Hurricanes. Oil palms should be relatively resistant to strong winds. However, Tailliez and Valverde (1971) observed differences between oil palms from different origin in susceptibility to hurricanes in Colombia.

Soils and drainage

Oil palms are grown on a wide range of soil types, and performance is often related to soil type. Even in rainfall-sufficient regions, soil type and water-table affect productivity and response to and requirements for fertilizers are much affected by soil.

In Asia, those commonly regarded as the best oil palm soils are coastal clays derived under a marine environment. They are relatively nutrient rich and have a relatively high water-table. A typical analysis is shown in Table 12.3 (from Hartley, 1967). Yields on such soils may reach 12 tons/acre of fruit bunches, and response to fertilizers may not be great. On such coastal clay soils access to the water-table may also be important in relation to high yield. Acid sulphate soils occurring in coastal regions and deep peat soils are unsuitable for oil palms (see Williams and Hsu, 1970) .

Inland soils which are deep profiled, permitting root exploration, are also high-yielding under the generally adequate rainfall regime of the South-East Asian regions where oil palms are grown, but yields generally are not as high as 10 tons/acre. Deep profiled soils derived from basalt are probably considered best among inland soils,* along with the less fertile granitic soils which also have excellent structure. Typical analysis for these soils are shown in Table 12.4.

*Recent experience shows that some of these soils, when excessively drained, produce lower yields than originally expected.

Fig. 12.11 Regression of bunch yield per palm on hours of 'effective' sunshine

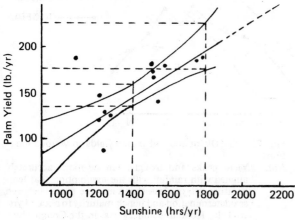

Source. Sparnaaij *et al.* 1964.

Fig. 12.12a Moisture budget of major oil palm areas and yield distribution

Telok Anson, Malaysia 4° 02′N

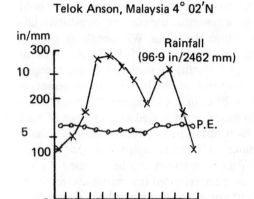

Waifor Main Station, Nigeria 6° 30′N

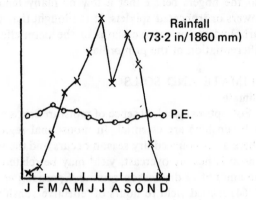

Yangambi, Congo 0° 48′N

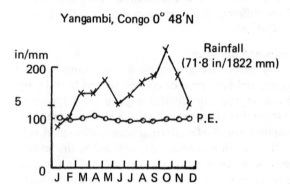

Barrancabermeja, Columbia, 7° 04′N

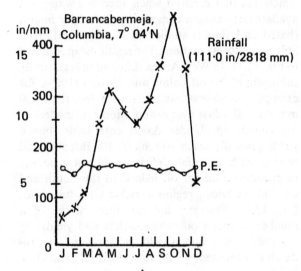

Tela, Honduras 15° 43′N

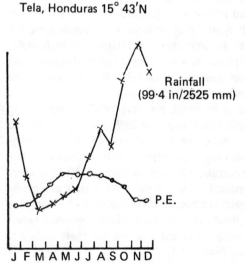

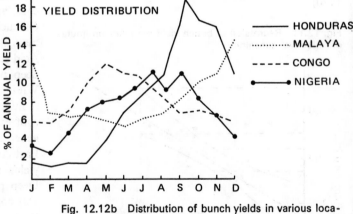

Fig. 12.12b Distribution of bunch yields in various locations

Note. Yield peaks and troughs can be more accurately aligned with rainfall with the assumption that temperature delays plastochron and the period between sex determination and bunch maturity to about 3 yrs. in Tela, Honduras and 2½ yrs. in the Congo which also experiences low temperatures.

Source. After Hartley, 1967.
Note. P.E.=pan evaporation.

TABLE 12.3 Analysis of marine coastal clay, a superior oil palm soil

Depth, inches	0–6	6–12	12–24	24–41	41–53
Clay %	80	79	81	70	68
Silt %	18	17	17	18	20
Sand %	2	4	2	12	12
C %	1.31	1.04	0.82	1.01	1.44
N %	0.20	0.18	0.11	0.11	0.09
pH	4.7	4.3	4.3	5.4	7.7
Exch. K m eq/100 g	1.57	0.93	0.78	0.80	2.34
Exch. Ca m eq/100 g	4.8	3.9	5.1	8.1	10.7
Exch. Mg m eq/100 g	14.2	10.0	10.7	13.8	17.2
Cation exch. cap. (C.E.C.) m eq/100 g	32.5	32.5	30.0	30.9	52.0
P, easily sol. in NaOH ppm	65	84	125	128	42
P, conc. HCl sol. ppm	226	207	244	322	335

In spite of a relatively high clay content, the water-holding capacity of these soils is not high and moisture stress may occur, particularly with seedlings and young palms in which root exploration is more limited.

Soils derived from shales and sandstones are also suitable if deep profiled and if rainfall is adequate. All inland soils in the Asian region require fertilizers. Soils with massive lateritic concretions and with truncated profile are not suitable for oil palm athough if gravel formation is not massive root exploration can take place (e.g. Bouchard and Damour, 1971).

In some parts of Sumatra and in New Britain recent nutrient-rich volcanic soils are planted with oil palms and support excellent growth.

The soils used for oil palm cultivation in Africa have been well reviewed by Hartley (1967). Generally they are more sandy than in Asia. In the Congo large expanses of country have parent material consisting of coverings of wind-borne sand and in West Africa, parent material of coarse sandstone is common.

The soils used for oil palms in the South and Central American region are variable and the crop has not been established long enough to permit generalizations.

Generally oil palms will tolerate temporary flooding for several weeks and the hydrophytic nature of the root system has already been mentioned. Further, the natural habitat of the palm is often swampy. However, long standing water affects the palm and water movement is necessary in heavy soils subject to frequent inundation. Generally field drains of 2 to 3 ft in depth are considered adequate.

PLANTING
Land preparation, spacing and yield

In many oil palm regions new plantings are being undertaken on previously uncultivated land, and since the climatic regions suitable for oil palm are humid, the natural cover is normally jungle. Clearing of jungle is carried out by the felling and burning technique to achieve a degree of clearing for access to the field by foot for the initial planting operations. Complete clearing is not generally attempted, since rotting of unburned stumps usually occurs naturally by the time access for harvesting is necessary (about 3 years). When old palm stands are cleared, however, (both oil palms and coconuts) stump removal is necessary to prevent build-up of disease organisms particularly *Ganoderma* (e.g. Menuel and Gerard, 1967). Usually stump burning is carried out, but there is evidence that pathogens arising from old stumps do not infect new plantings if the stumps are removed from beneath the soil. The technique of clearing using explosives is discussed by Williams and Hsu (1970). Layout of access roads for transport of the heavy fruit bunches at harvesting is important in oil palms. Current estate road spacings are about 16 chains. (c. 350m). This has been discussed by Bevan et al. (1966), Pralain (1969), Williams and Hsu (1970), Hartley (1967) and others. There is also current interest in vehicular collection of bunches at the palm (e.g. Beeny, et al. 1967), which requires a higher standard of clearing.

In parts of Africa, rehabilitation and improvement of natural groves are carried out by replacing palms from the old stand with superior palms

TABLE 12.4   Characteristics of basalt and granite-derived inland soils

Basalt derived

Depth, inches	2–9	9–16	16–26	27–37	37–48
Clay %	90	93	95	95	95
Silt %	7	4	2	2	2
Sand %	3	3	3	3	3
C %	1.76	0.88	0.66	—	—
N %	0.22	0.10	0.09	—	—
pH	5.1	4.9	5.0	5.0	5.0
Exch. K m eq/100 g	0.17	0.08	0.05		
Exch. Ca m eq/100 g	0.25	0.08	0.08	<0.05	<0.05
Exch. Mg m eq/100 g	0.93	0.59	0.50	0.42	0.40
Cation exch. cap (C.E.C.) m eq/100 g	14.9	11.1	10.6	8.6	6.8
P, easily sol. in NaOH	89	82	78	55	35

Granite derived

Depth, inches	0–3	3–12	12–24	24–36	36–48
Clay %	43	46	58	65	66
Silt %	6	6	2	4	4
Sand %	51	48	40	31	30
C %	1.49	0.71	0.49	0.40	0.46
N %	0.16	0.11	0.07	0.07	0.06
pH	4.6	4.2	4.2	4.5	4.5
Exch. K m eq/100 g	0.41	0.17	0.12	0.09	0.09
Exch. Ca m eq/100 g	0.08	<0.05			
Exch. Mg m eq/100 g	0.33	<0.05			
Cation exch. cap. (C.E.C.) m eq/100 g	7.3	5.0	5.3	6.0	6.3
P, easily sol. in NaOH	39	22	19	19	8
p, conc. HCl sol. ppm	161	113	121	114	124

Source. Malaysia—after Hartley, 1967.

while retaining part of the old stand for a number of years (Sheldrick, 1965).

Optimal spacing for oil palms has been investigated in Africa and Asia and has been reviewed by Hartley (1967), Williams and Hsu (1970) and recently by Corley *et al.* (1972). When considering the total growth of dry matter produced by the palm, it will be apparent that this species will be sensitive to planting density because of its inability to branch. With regard to bunch or oil yield, however, within a rather broad range, cumulative yields tend to be fairly constant since denser spacing reduces individual palm yield linearly, and differences in individual palm development, particularly in relation to leaf area, with age, allows wider spacings to gain in later years. In addition, crown develop-

ment will vary with physical conditions, thus affecting optimal planting density. Production at any density (Px) can be obtained if the reduction in cumulative yield of individual palms with increasing density (competition factor—Prevot and Duchesne, 1955) is known as follows:

$$Px = [P_1 \pm (Dx - D_1)C]Dx$$

Where P_1 is the production per palm at the lowest density (minimal interpalm competition), $Dx - D_1$ is the difference in density between the lowest and that under consideration and C is the reduction in individual palm yield for unit increase in density. Using data from Africa, Prevot and Duchesne (1955) calculated the effect of density on cumulative yield (Fig. 12.13).

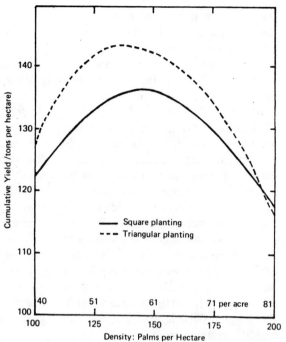

Fig. 12.13 Cumulative yield per hectare in relation to planting density

Source. From Prevot and Duchesne, 1955.

Fig. 12.14 Cumulative yields after 25 and 30 years in three environments in Malaya

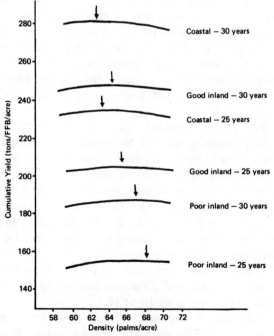

Source. After Corley *et al.* 1972.
Note. The arrows indicate the optimal.

Triangular (equilateral) spacing has always been found to give the highest yields in oil palm.

Using data from various sources, Corley *et al.* (1972) estimated optimal densities for a range of conditions and for plantings of 25 and 30 years duration in Malaya as shown in Fig. 12.14. Within the range 58 to 72 palms acre the optimum is seen to be fairly broad. A density of about 61 palms/acre appears to be best under good physical conditions ranging to 67 on poorer inland soils. Under African conditions (Fig. 12.13) competition for moisture may favour slightly lower densities.

Varying density, by thinning, has not proved to be economically viable in oil palms.

In general, yields from good *tenera* palms under favourable physical conditions (as in Asia) are of the order 10–12 tons/acre (2–2.4 tons palm oil, plus kernels). In Africa, yields are generally much lower than these, although high yields have been reported with newer planting material (e.g. Carriere de Belgarric, 1967; Bachy and Regaud, 1968).

Seed production and nursery

The success of the oil palm industry is intimately linked with the production of high-yield *tenera* seeds which are produced from controlled pollinations of high-yielding *duras* with appropriate *pisiferas*. Yield recording of trees for the purpose of selecting suitable parents requires an extended period. Progeny testing is also necessary for new combinations although at the present time most seeds produced by the organizations engaged in this activity have been widely planted on a commercial scale and are known to be superior. Since the *pisifera* parent often produces abortive fruit, the potential value of these may be judged from the performance of the siblings (the *duras* and *teneras* of a *tenera* × *tenera* cross).*

The crossing of oil palms for seed production is carried out under strictly controlled conditions (see Purvis, 1956b). Both male and female inflorescences are surface sterilized with formalin solution before maturity and bagged for collection of pollen and prevention of stray pollination. Pollen storage for later use requires drying or freeze drying and desiccator or refrigerated storage (e.g. Bernard and Noiret, 1970). A few days after pollination the bag is removed and the inflorescence is allowed to mature before cleaning of the seeds (see Bevan *et al.* 1966) and germination.

*In fact very little selection has been done with *pisiferas*.

Seeds require heat treatment before they will germinate (Rees, 1959a & b, 1962). This may be carried out in a sand bed exposed to the sun, or more recently, in controlled temperature cabinets at 38–40°C (see Bevan *et al.* 1966). Surface sterilization of dried seeds at *c.* 10 per cent moisture with methyl bromide is desirable (Mok, 1970), after which moisture is adjusted to about 18 per cent prior to bagging and heat treating for 40–60 days. Tailliez (1971) however, points out that excessive heat treatment retards initial growth and that this should be reduced to the minimum. After heat treatment moisture level is raised to *c.* 22 per cent by soaking and then the bagged seeds are allowed to germinate at ambient temperature. Seeds are sometimes treated with a mixture of 'thiram' and streptomycin to reduce microorganisms during germination (see Turner and Gillbanks, 1974).

Various nursery techniques are used for oil palms (see Sparnaaij and Gunn, 1959; Bevan *et al.* 1966; Hartley, 1967; Williams and Hsu, 1970). The most popular method in use is the polythene bag nursery (Fig. 12.15), using bags of size 12″–15″ × 18″–24″ laid flat, filled with a suitable soil (e.g. Delvaux, 1967). Jungle topsoil is a favourite. Seedlings are usually planted out from about 12 months from germination in a suitable season. Oil palm seedlings require high fertilization with complete fertilizer at rates from about ¼ oz to 1½ oz per month with advancing age (Bull, 1966a). A polythene bag nursery requires an adequate watering system capable of delivering about 5 mm/day under dry conditions (e.g. Coomans, 1971).

Field nurseries are also used for oil palms. Generally seedlings take longer to reach planting out

Fig. 12.15   A polythene-bag oil palm nursery in Malaya

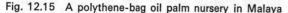

stage and must be root pruned, prior to raising them with a block of soil (e.g. Ruer and Rosselli, 1967; see also Chapter 1).

Seedling selection in oil palms has already been mentioned. The 'roguing out' of obviously abnormal seedlings is commonly practised (e.g. Gascon *et al.* 1965; Courtois, 1968; Turner and Gillbanks, 1974). A real advance in oil palm growing, however, will be achieved when a reliable index for high yield potential is discovered for the nursery. Unlike vegetatively propagated crops, oil palm seedlings show a considerable range of field performance and individual palm yields in the field may range from very low or zero to as much as 800 lb fruit per annum. Consecutive estimates of leaf area for relative measurement, from certain linear dimensions (since leaf area is related to yield under equitable conditions), have been suggested by Williams and Hsu, 1970—see also Chapter 1. Bachy and Martin (1967) note that the length of the longest leaf and collar girth are highly correlated measures.

Manuring and irrigation

The practice of manuring in oil palms is advanced compared to many other crops. Manuring is normally practised and good yield responses obtained. Oil palm manuring has been recently reviewed by Hartley (1967), Williams and Hsu (1970), Ng (1970), Ollagnier *et al.* (1970), Turner and Gillbanks (1974). In determining nutrient requirements, Ng (1970) has stressed the importance of taking into account all available evidence including climate, genotype, soil type and nutrient status and nutrient composition of the palm, particularly the leaf which is the most important diagnostic technique used today.

Phosphorus nutrition, using rock phosphate, is widely practised and field deficiencies are no longer encountered; similarly yield responses are uncommon because of the generally adequate use of this fertilizer, although in the early oil palm days frequent P responses occurred (May, 1956). The yield of fruit from oil palms, as already explained, rises to a maximum at *c.* nine years, and thereafter progressively declines. According to Bull (1966b), this decline can be arrested by heavy use of fertilizers. An example of nutrient removal by the oil palm crop is given in Table 12.5.

Examples of commercial recommendations for fertilizers on two soils are shown in Table 12.6 (cf. Tables 12.3 and 12.4).

TABLE 12.5   Total macronutrients used per acre in assumed yields of fruit

Assumed yield (tons per acre)	Nutrient (lb per acre)				
	N	P	K	Mg	Ca
1	6.5	1.0	8.2	1.8	1.7
2	13.0	1.9	16.4	3.6	3.4
3	19.5	2.9	24.6	5.4	5.1
4	20.0	3.9	32.7	7.1	6.8
5	32.5	4.8	40.9	8.9	8.5
6	38.9	5.8	49.1	10.7	10.2
7	45.4	6.8	57.3	12.5	11.8
8	51.9	7.8	65.5	14.3	13.5
9	58.4	8.7	73.7	16.1	15.2
10	64.9	9.7	81.8	17.9	16.9
Standard error for 10 ton level	±1.0	±0.3	±1.6	±0.09	±0.80

Source.   After Ng and Thamboo, 1967.

Rather different fertilizer responses have been noted between Africa and Asia which might be related to climatic factors. Potassium response is much more common on sandy soils in Africa, dependent on a satisfactory P level, Ochs (1965). Bachy (1969) also reports the benefits of K, and Holland (1967) in a component analysis of a factorial N, P, K, Ca, Mg trial draws attention to K/Ca Mg antagonism. Nitrogen response under sub-optimal conditions is limited (e.g. Walker and Melsted, 1971).

In Malaysia, N response, particularly in young palms is common. Mollegaard (1971) also reports a P dependent K response from Malaysia. Marti-neaux et al. (1969) points to the importance of N/K, N/P and P/K ratios in yield. Boron deficiency often occurs in oil palm, particularly in the first few years of bearing.

Foliage analysis is the common diagnostic tool in determining nutrient needs, usually using leaf 17, a mid-crown leaf for sampling. IRHO recommendations (see Oléagineux, 23, pp. 11, 713; 24 pp. 1) and others are as follow for major nutrients.

N=2.5 per cent (up to 2.7 per cent in Malaysia);
P=0.15 per cent;
K=1.00 per cent; Ca=0.60 per cent; Mg=0.24 per cent.

The critical boron level is considered to be about

TABLE 12.6   Examples of fertilizer recommendations for oil palm on two soils in Malaya

Stage	Coastal clay and muck soils, Nutrient rich				Granite derived inland soil			
Prior to planting	2 cwt per acre CIP				4 cwt per acre CIP			
Planting hole	8 oz CIP per palm				1 lb CIP per palm			
	lb S/A	lb CIP	lb MP	lb Kies	lb S/A	lb CIP	lb MP	lb Kies
1st year in field	1.1	0.6	0.6	0.3	1.7	1.0	0.9	0.4
2nd year in field	1.5	0.7	1.4	0.4	2.2	1.1	2.1	0.6
3rd year in field	3.0	1.4	2.8	0.8	3.3	1.6	4.2	0.9
4th year in field	4.5	2.1	4.2	1.2	4.8	2.2	5.2	1.2
5th year in field	5.2	2.5	4.9	1.4	5.5	2.7	6.3	1.5
Total for five years	15.3	11.8	13.9	4.1	17.5	17.6	18.7	4.6

Notes.   CIP=rock phosphate; S/A=sulphate of ammonia; MP=muriate of potash.

18 ppm (Rajaratnam, 1971). Ng *et al.* (1969) record Mn levels from 105–455 ppm Cu from 6.1–14.6, Zn from 9.5–39.2 and Fe from 60–109 ppm.

Recently Ollagnier and Ochs (1971) have suggested that Cl may be an important nutrient for oil palms, with an optimal level of 0.5–0.6 per cent dry weight! There is only one case of S deficiency recorded, which might have been complicated by excess boron (Brosehart *et al.* 1957). However, an increasing importance of this element may become apparent with increasing use of urea as the main N source in oil palms, replacing ammonium sulphate.

In sampling, Ng and Walters (1969) recommend the random selecting of nine palms, but uniformity studies (Poon, 1970) suggest higher numbers are necessary.

The nature of the relationship between nutrient levels and yield, and the concept of 'critical levels' above which yield responses do not occur are indicated diagrammatically in Fig. 12.16. Initially a decrease in level with increased growth may occur due to dilution (A); subsequently a curve obeying the law of diminishing returns (C) can be recognized and a threshold (D) followed by toxicity (E).

Generally application of fertilizer is by spreading in circles around the palms (Ruer, 1967): annual or bi-annual applications are commonly used. Irrigation is not practised in oil palms cultivation at the present time but there is currently great interest

in this, even in the relatively rainfall-sufficient regions of Asia. This has also been the subject of discussion at IRHO in Africa (*Oléagineux* 24, p. 389).

Pollination

Under certain conditions, artificial pollination may increase fruit set in oil palms; particularly so with high-yielding families, in which female inflorescence production predominates, and particularly in young palms which tend to produce too few male flowers and in which the canopy is not elevated to receive adequate wind circulation. In favourable environmental conditions, as occurring in Asia, male inflorescence production is further reduced. Under sub-humid conditions generally adequate male flower production occurs. The need for artificial pollination has been reviewed by Grey (1965) and Williams and Rajaratnam (1971). The effect of long-term pollination may best be considered in relation to the theoretical optimal yield of an area, the optimum being a function of climatic factors, soil and nutrition and the genotype of the palm (Fig. 12.17). At one extreme if the level of natural pollination is zero, then the gain to be expected by achieving full pollination is equal to the theoretical maximum yield. At the other extreme, with optimal natural pollination, a gain cannot be expected under any circumstances. Under favourable environmental conditions it is quite likely that natural polli-

Fig. 12.16   Relations of nutrients and yield in oil palms

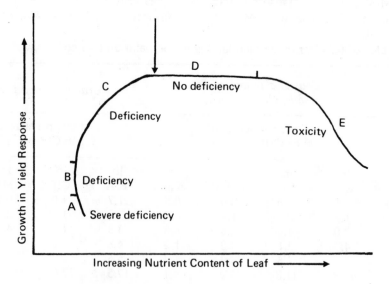

Source. From Hartley, 1967.

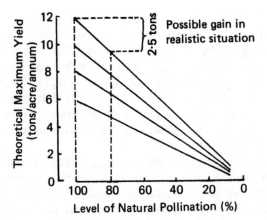

Fig. 12.17 Relationship between level of pollination and yield for fields of different potential maximum

Source. After Williams and Rajaratnam, 1971.
Note. 100 per cent represents, in each case, the level of pollination necessary to achieve maximum yield.

nation will be below optimal. However, continued high level assisted pollination has led to an increase in male inflorescence production and declining yield over a period of seven years (Grey, 1965). The actual level which will maintain continued optimal production must be determined by experience and should be that required to bring yield up to the expected maximum obtainable from the environment. Full pollination is considered to be achieved when pollination is carried out 3-daily (the life of the female flower); optimal yield may be achieved at a frequency of pollination lower than this.

The normal method of assisted pollination is by use of a hand puffer bulb filled with dried pollen which is puffed onto the mature female flowers. Williams and Rajaratnam (1971) investigated the application by a hand duster machine using a pollen and talcum mixture which is less labour intensive and possibly more satisfactory for long-time use. Pollen may be stored, after drying, at + 5 to − 5° C for periods of 12 months and is collected during male inflorescence cycles (e.g. Williams and Rajaratnam, 1971; Bernard and Noiret, 1970). Assisted pollination also reduces the incidence of bunch-rot under humid conditions.

Hardon and Turner (1967) showed that wind pollination occurs to a considerable extent, although insect dispersal also undoubtedly occurs and may be more important than generally realized.

Weed control, covers and mulch

The oil palm, like the coconut, is sensitive to translocated hormonal herbicides of the type 2:4-D and these must be used with great caution. They should not be used in young material, but only in mature palms with the crown elevated from the ground. In addition, drift to the living parts must be avoided. Mixed with sodium chlorate, the 2:4-D amine can be used in mature palms with the precautions mentioned above to control particular weeds such as *Mikania* (½ pt 2:4-D amine+8 lb sodium chlorate/acre—e.g. Ramachandran, 1969). Dalapon and 2:4-D may be used for pre-clearing nursery areas, but not when seedlings are present; for young seedlings ametyrene has been recommended (e.g. Tailliez, 1971), and manual weeding (Ramachandran, 1969).

'Triazines' have been reported to be the best herbicides under certain conditions in Indonesia, being non-phytotoxic to the palm (Mangoensoekardjo and Nurdin, 1970).

Traditionally, sodium arsenite was used in oil palm and gave excellent low-cost weed control; but this has now been replaced by the less toxic arsenical MSMA (Barnes and Tan, 1969). Various herbicide cocktails have found to give effective and long-term control: a mixture based on paraquat (0.38 kg/acre) and diuron (0.38 kg/ha) and MSMA (1.79 kg/ha) is recommended by Seth and Abu Bakar (1970), and Sheldrick (1969) recommends 2.2 kg/ha paraquat with 3.4 kg atrazine, monuron or diuron.

Mulching with bunch refuse (from the oil extraction plant) has been recommended in oil palm (e.g. Bachy and Regaud, 1968), both for weed control and as a source of K; but this is seldom practised. Cover legumes are usually planted with oil palms immediately after clearing. Commonly used legumes are *Puerana phoseoloides*, *Centrosema pubescens* and *Calopogonium mucunoides*, usually as a mixture (Turner and Gillbanks, 1974).

MAJOR DISEASES AND PESTS

The rhinoceros beetle (*Oryctes rhinoceros*) has already been mentioned in relation to coconut (Chapter 11). Rat problems are sometimes serious in oil palms because rat population build-up is encouraged by the crop (also with coconuts). Tree banding with a 18″ band of aluminium to prevent rodents entering the crown is practised in coconuts but is less satisfactory in oil palms because of the rough nature of the trunk when the frond bases are not dislodged (see 'Botany' above). Satisfactory rat control, however, is obtained by sequential baiting with anticoagulant poisons mixed with various foods (e.g. Wood, 1969).

Various fungal and bacterial diseases affect oil palms, sometimes seriously. Seedling blast disease (a condition following invasion of the root system by species of *Pythium* and *Rhizoctonia*, Robertson, 1959), can cause widespread nursery losses with inadequate watering. It is evidenced by wilting and death of the seedlings but is satisfactorily controlled by adequate watering. Blaak (1969) reports success in breeding for resistance to blast.

In Africa, *Fusarium oxysporium* (*var. elaeidis*) causes widespread losses in some areas in mature palms. Rapid wilting of the fronds and death of the trees occur. The disease is endemic, even in virgin soils. Invasion occurs through wounded tissues (Renard, 1970).

In both Africa and Asia a *Ganoderma* rot can be serious, particularly on replanting of oil palms and new planting on old coconut land. Removal of old trunks from the soil (generally with burning) is used to control the disease. Akbar *et al.* (1971) reports differences in susceptibility to the disease in palms of different genetic composition. Various other root-rot diseases also occur but none are as serious as the above.

Bunch-rot (generally following inadequate pollination and humid conditions) is sometimes a problem in oil palms, particularly in Asia. Various fungi are responsible including *Marasmius palmivorus*. Turner (1967) reports success in treatment with phenyl-mercury acetate, but adequate pollination is the best insurance against bunch-rot.

Red ring disease (see Coconuts, Chapter 11), has also been reported in oil palms in South America.

13 CITRUS [Citrus spp.]

INTRODUCTION, SPECIES AND HYBRIDS

THE geographical origin of citrus is thought to be the old-world tropics, principally the Indo-Pacific region. Most citrus species are adapted for growth in full sunlight and possess some xerophytic properties which have led crop botanists to think that the genus may have been evolved in sub-humid regions although a number of species occur as rainforest understory trees (e.g. *Citrus macroptera* and a recently described species *C. halimii*—Stone *et al.* 1973). The major citrus-producing areas at the present time are the sub-tropical and frost-free temperate regions of the world; particularly of the United States, Argentina and Brazil and other South and Central American countries, the Mediterranean, India and Pakistan, Japan, South Africa and Australia. A wider range of species, however, can be grown in the tropics because of the absence of temperature stress, and there is emphasis on the development of citrus industries in many tropical countries. In the tropics the main species grown at the present time are mandarins, limes, pomelos (pummelos) and grapefruits but there is also much interest in the development of sweet oranges for the warmer regions.

Although considerable confusion exists in the classification of *Citrus* the main citrus species commonly recognized are *C. sinensis* (sweet oranges), *C. paradisi* (grapefruits), *C. limon* (lemons), *C. aurantifolia* (limes), *C. medica* (citrons), *C. reticulata* (mandarins and tangerines), *C. junos* (the Satsuma orange), *C. grandis* (pomelos). The clementine mandarin (*C. clementina*) and Tahiti lime (*C. latifolia*) are also regarded as a separate species. There are many varieties of the major species which are propagated clonally and by the use of nucellar seedlings (see p. 187–8) and each citrus area has developed its own commercial planting material, the chief source of new varieties being the production of chance gametic seedlings by bud mutations which are common in citrus and by controlled breeding programmes.

In addition to the above species, various hybrids are also used commercially for fruit production. The most important are the tangelo (*C. reticulata* × *C. paradisi*), the citrange (*C. sinensis* × *Poncirus trifoliata* or trifoliate orange) and the limequat (*C. aurantifolia* × *Fortunella marginata*). The ortanique, which appears to have arisen as a chance cross between *C. sinensis* and *C. reticulata*, is now an important citrus in Jamaica (e.g. Nugent *et al.* 1967). The 'mandarin' Ellendale is also a *C. sinensis* × *C. reticulata* cross. A new type of citrus with characteristics of both the orange and the grapefruit is the Chironja (see Moscoso, 1970).

Citrus is propagated mostly as grafted scion material onto stock seedlings (see p. 187), and a greater range of species are used as root stocks. In addition to all the above species, *C. aurantium* (sour orange), *C. karna* (see Srivastava and Arya, 1969), *C. macrophylla* (e.g. Burns and Coggins, 1969), *C. volkamariana* (Le Bourdelles, 1968) and others are used to produce stock seedlings.

Various closely related citrus genera also used as rootstocks are the trifoliate orange (*Poncirus trifoliata*), *Fortunella marginata*, and species of *Microcitrus*, *Eremocitrus*, *Citropsis*, *Atalantia*, *Oxanthera* and *Clymenia polyandra*, a true citrus native to the sub-arid regions of Australia. Stock material based on *Eremocitrus* has properties of drought and salt resistance and the trifoliate orange has properties of cold resistance. Various other characteristics of common stock species and hybrids are considered further on. Some are used predominately in the countries in which they are native while others have been used more widely. In Japan the chief stock material is based on *C. junos* × *P. trifoliata* (Inoue, 1971).

BOTANY

The vegetative plant, root system and polyembryony

Citrus is among the most well-studied crops of tropical regions, though most of this work has been carried out in the developed regions of the sub-tropical and temperate zones, and the development of sound citrus industries can be directly related to the degree of research and development effort in each case. There are a number of excellent books dealing with citrus of which one of the most comprehensive is an old text 'The Citrus Industry' by Batchelor and Webber (1948) in two volumes.

Some citrus varieties illustrate well the importance of sink activity in crop yield, and development of the fruit is probably a major limiting factor to yield under favourable agronomic conditions in the parthenocarpic types. Also, citrus which is principally grown as grafted material demonstrates the importance of stock/scion relations in this type of culture, unlike rubber (Chapter 9) in which stock/scion relations are of minor importance in the growth and productivity of the crop.

Morphologically, commercial citrus species vary from much-branched shrubs or small trees up to 15′ in height such as most lime varieties, to trees of 30 ft or more in height such as the grape fruits and pomelos. Limes characteristically have a small crooked trunk and a dense branch canopy. Pomelos and grapefruits are usually open-crowned with drooping branches. Citrons and mandarins are shorter trees (12–20 ft) with irregular or open crowns. Lemons and oranges are intermediate in height (c. 20 ft) and generally have a fairly dense rounded crown. Some wild species of citrus e.g. *C. halimii*, which is a rain-forest tree, reach a height of 60 ft or more. Leaves are generally ovate and vary in length from 3 to about 20 cm; many varieties having a glabrous surface. The branches are conspicuously spiny, although there are obviously great genetic differences between different species and clones in relation to moisture requirements, irrigation studies on some clones indicate that citrus yield may not be affected until about 60 per cent of available water is used. Maitin (1967) and Inoue *et al.* (1971) report that full vegetative and reproductive growth of young Satsuma oranges was maintained at about 70 per cent of field capacity. Consumptive water use under optimal conditions has been estimated at about 4 mm/day under irrigation (in Lebanon, Vink *et al.* 1971) and about 0.8E* (Maitin, 1971).

The root system of cultivated citrus varies considerably with the rootstock variety used and is also influenced by the scion. Under soil conditions allowing unrestricted development of the root sys-

*E = open water (pan) evaporation.

tem, different stocks may produce roots predominately in the surface horizons or with conspicuous tap roots. For example, of the important rootstock species, sour orange (*C. aurantium*) generally produces a deeper root system with a strong tap root while rough lemon (*C. limon*) tends to produce abundant lateral roots in the top foot or two of soil. Limes also have a superficial root system (Brown, 1920). Under conditions of strong winds sour orange rootstocks are often more resistant to uprooting. Generally the size of the root system limits the crown size and *vice versa*, so that good stock/scion combinations which allow vigorous crown growth will generally have a more extensive root system. The root systems of citrus trees grown from cuttings and marcotts (in common with most other plant species) are much more superficial and will support only small tree crowns. Furthermore, they are more susceptible to wind damage. Sour orange, however, may develop marked tap roots even from cuttings (Halma, 1934).

Flowers are borne seasonally or continuously depending on climatic conditions and are solitary or in small clusters.

Many citrus varieties are polyembryonic, producing as many as 30 nucellar embryos as well as the normal gametic ones (see Ochse *et al.* 1961). This is an advantage in the production of uniform stock seedlings (see below) but is a barrier to breeding work. According to Veno and Nishiura (1969) the production of nucellar embryos is markedly temperature sensitive, being less when temperature is low because of reduced division and development of nucellous cells. Parthenocarpic development of the fruit is also common in some cultivars.

Stock/scion relations, common stocks

In general stock/scion compatibility is expressed in morphological changes. A successful union involves the fusing of new tissues produced by the cambia of the stock and scion and the inter-connecting of conducting elements to permit the movement of elaborated products of photosynthesis and of water and nutrients and possibly hormonal substances, but there is apparently little or no passage of organized protoplasmic bodies or nucleii. There is considerable variation however in the movement of nutritive substances due to stionic (stock/scion) factors. The effect of stock and scion on the root system and general scion growth has already been mentioned. In a highly compatible union the region of the graft is characterized by an even or tapering union without marked stock or scion overgrowths. Such morphologically smooth unions usually produce long-lived healthy trees. When scion overgrowths occur (as with some scions on sour orange—see below) trees may be dwarfed and short-lived. Scion overgrowths have been attributed to a reduced ability of photosynthates to move across the graft union, to poor growth of the scion following accumulation of photosynthates, and to a reduced supply of nutrients and water from the poorly-developed stock root system (see Jensen *et al.* 1927). In the case of stock overgrowths (as with some clones on grapefruits—see below) difference in size may be attributed to natural differences in growth rate, but more often to an enhanced growth of the stock with particular scions. However, the stock overgrowth does not usually affect scion growth adversely on grapefruit. On the other hand, overgrowth of the stock of trifoliate orange, which is normally a slower-growing species, with some orange, grapefruit and lemon varieties is so marked that dwarfing of the scion occurs, presumably as a result of a higher photosynthate supply by the scion to the stock. The tissues of the typical stock overgrowths of trifoliate orange are softer and possess larger cells (Batchelor and Webber, 1948).

Apart from the stionic reactions of the graft union, general growth of both stock and scion are effected by different combination. This is well illustrated by the classical experiment of Brown (1920) on the growth of Malta orange on four different rootstocks, as shown in Table 13.1. In the same experiment Santara mandarin made good growth on sweet lime and sour orange but was markedly dwarfed on rough lemon and citron.

In addition to affecting growth, stionic reactions also affect precocity and fruit quality. Generally a vigorous growing rootstock delays maturity and stocks that have a dwarfing effect induce precocity. Further characteristics and effects of different rootstocks follow:

Rough lemon (C. limon). This is probably the most widely-used rootstock on a world basis. As described above, it produces a superficial root system and is therefore useful on truncated-profile soils and is also recommended for light sandy soils. It is less cold resistant and used in many tropical countries, generally showing good scion compatibility and good yield, e.g. with oranges and mandarins in Colombia (Rios-Castano *et al.* 1968 and Camacho *et al.* 1968), Troyer citrange and orange in Egypt (El Wakeel and Soliman, 1969). Rough lemon is also one of the principal rootstocks of the West Indies (e.g. Hosein, 1969) and tropical Central and South America. Samson (1966) reports that it is recommended for inland soils in Surinam. Although yield on rough lemon is usually good, fruit quality as expressed by sugar content of the juice is generally low on this rootstock although this can be overcome to some extent by good fertilizer practice. The fruits of rough lemon are highly polyembryonic producing about 90 per cent nucellar seedlings which make it a genetically very uniform rootstock. Rough lemon is susceptible to gummosis disease, and Eid (1972) reports a disease complex of Fusarium and a nematode with rough lemon rootstocks in Tanzania.

Sour Orange (C. aurantium). A widely-used rootstock in tropical countries (e.g. Said and Inayatullah, 1965; Hosein, 1969) but susceptibility to tristeza virus has promoted investigation of alternative stocks. For example, Blondel (1967) reports that in Corsica, trifoliate orange, Troyer citrange and Cleopatra mandarin (see further on) are now recommended as substitutes for sour orange. Sour orange however shows resistance to gummosis and *Phytopthora* foot-rot and with its strong root system (see above) is suitable for heavier soils. Nucellar seedling production is about 80 per cent, slightly lower than rough lemon. There is a tendency to produce marked scion overgrowths with Navel

TABLE 13.1 Growth of Malta sweet orange on four different rootstocks

Rootstock	Height of tree	Breadth of crown	Fruits per tree
Rough lemon	11 ft 10 in	9 ft 6 in	200
Sweet lime	5 ft 9 in	5 ft 4 in	25
Sour orange	5 ft 9 in	5 ft 0 in	10
Citron	3 ft 6 in	3 ft 0 in	16

and Valencia oranges and also lemons as discussed before (Batchelor and Webber, 1948).

Sweet Orange (C. sinensis). This rootstock is used to some extent in tropical countries (e.g. W. Indies—Hosein, 1969; Colombia —Rios-Castano *et al.* 1968). It is adapted to rich, well-drained soils and is susceptible to root diseases. It generally has good compatibility and produces fruits of superior quality. Rios-Castano *et al.* (1968) for example, report that the best quality fruit of Indian River and Washington oranges was obtained on 'lerma' sweet orange stocks in Colombia, but yield was lower than on rough lemon. Different varieties show nucellar seedling production between 40 and 95 per cent.

Mandarins or tangerines (C. reticulata). Mandarins have been traditionally used as rootstocks in China and other Asian countires but are now much more widespread, chiefly because they show some resistance to tristeza virus. In Corsica for example, Cleopatra mandarin is recommended to replace sour orange (Blondel, 1967). The mandarins are now important rootstocks in many tropical countries (e.g. Hosein, 1969) but are considered to be not highly drought resistant and therefore unsuitable for sub-humid and strongly monsoonal climates without irrigation. Samson (1966) reports that Cleopatra mandarin is now recommended in Surinam for planting on lowland coastal clays. Ranjpur lime (a mandarin variety) is recommended for inland sandy soils in Surinam. Mandarins are generally resistant to gummosis and foot-rot but susceptible to scab. Salibe (1966) reports that the calamandarin (or calamondin, *C. reticulata* var. *austera* —see Ochse *et al.* 1961; sometimes considered to be a separate species, *C. madurensis*—see Veno and Nishiura, 1969) is the major rootstock of the Philippines and is highly resistant to foot-rot and also has some resistance to scab and canker. The calamondin is also an important rootstock in Indonesia.

Mandarins are reputedly slow to develop as seedlings. Mature trees on mandarins come into bearing later, but catch up with trees on other rootstocks after a few years (Ochse *et al.* 1961). Rios-Castano *et al.* (1968) reports that Cleopatra mandarin offered intermediate conditions of growth, production and quality between rough lemon and sweet orange (see above) in Colombia. Ranjpur lime has been found to be incompatible with some orange varieties and is reputedly short-lived (Ochse *et al.* 1961). Nucellar seedling reproduction varies between 10 and 100 per cent.

Grapefruits and Pomelos (C. paradisi and *C. grandis).* These two citrus species have similar characteristics as rootstocks, being adapted to rich heavier soils and performing poorly on sandy soils. Grapefruit stocks tend to produce stock overgrowth particularly with certain lemon Varieties (see Batchelor and Webber, 1948) and Valencia and Navel oranges. Scion growth may be good but in general fruit yields and quality are poor. Camacho *et al.* (1968) however report good production of Oneco mandarins on grapefruit rootstock. Grapefruits produce 60–95 per cent nucellar seedlings but pomelos are generally monoembryonic and therefore unsuitable for the production of genetically uniform stock. Both are susceptible to tristeza virus.

Citrange (C. sinensis×P. trifoliata) and *P. trifoliata.* These rootstocks are known best for their cold hardiness and are therefore mostly used in sub-tropical and temperate regions. They have some tolerance to tristeza and foot-rot but are susceptible to canker and scab. They have been used in the West Indies and elsewhere (e.g. Hosein, 1969; *Citrus Res.* (Trinidad), p. 27, 1969). In Trinidad *P. trifoliata* has been reported to lead to higher N levels in the foliage of Valencia orange. Marked stock overgrowths occur with some scions, particularly some orange, grapefruit and lemon varieties (see Batchelor and Webber, 1948), but fruit quality is generally satisfactory. Some cases of complete incompatibility are reported (Ochse *et al.* 1961).

Sweet lime (C. aurantifolia). The sweet limes are little used as rootstocks except in Egypt and Israel. They are strong growers and precocious but reputedly short-lived and sensitive to foot-rot. El Wakeel and Soliman (1969) report that fruit size of Egyptian oranges was consistently smaller on sweet lime rootstock. Limes have also been tried as rootstock material in Brazil (Manica and Anderson, 1969). They are predominantly used as rootstocks for Jaffa oranges in Israel where they are adapted to light sandy soils.

Others. *Citrus karna* is used as a rootstock in India (Srivastava and Arya, 1969). *C. macrophylla* and *C. volkamariana* have been used in Guadeloupe in a humid region at 250 m elevation and the former has been tried in California. Species of *Citropsis* which are highly resistant to foot-rot are used as rootstocks in the Congo, and various other closely related citrus genera are used in countries where they are native to the land. There may be considerable potential for the exploitation of new species and genera for suitable stock material in tropical countries.

Fruit development, flowering and photosynthesis

In the past and at present citrus culture has usually been at wide spacing with wide rows between the trees, so that the question of canopy efficiency (which is so important in the cultivation of many crops, including the tropical crops of rice, maize, tea, tapioca and others) has not arisen. In recent years, however, concepts of spacing have been changing in citriculture, as evidenced by much closer spacing used in a number of cases (see p. 192). If this turns out to be a general trend then canopy efficiency may become very important in the breeding and agronomy of citrus. At the present time, however, one of the main factors in citrus yield is development of the fruit. Self-incompatibility in some cultivars and inadequate fruit set and fruit development, limit yield. For example, in Early Imperial mandarin and in Ellendale (*C. reticulata* × *C. sinensis*), fruit yield may be more than doubled by cross pollination due both to percentage set and fruit size, the latter being correlated with seed number (Stafford, 1970). Delange and Vincent (1970) also report a positive and significant correlation between fruit size and seediness in Washington Navel when out-pollinated with *C. grandis*. The seeds of citrus, as with most fruits, are important centres for the production of hormones concerned with development of the fruit. Fruit development hormones can also be produced from other tissues within the fruit (e.g. see Papayas, Chapter 14) but in many seedless and few-seeded varieties of citrus, the absence of seeds appears to be a growth-limiting factor for the fruit. Krezdorn (1966) has reviewed the factors contributing to the fruit setting problem in citrus. The fruit set problem derives from both incompatibility (the failure to pollinate and produce seed) and the inability to produce sufficient fruit by parthenocarpic development alone. Iwamasa (1967) reports that seedlessness in Valencia was due to the production of a high proportion of non-viable gametes through segmental interchange in the chromosomes. In Satsuma orange strains (*C. junos*) suppression of anthers led to sterility, and in 'Orlando tangelo' self incompatiblity was largely due to slow pollen tube growth in the style (Carlos and Krezdorn, 1968).

Various techniques for overcoming the compatibility/parthenocarpy problem include mixed planting (and here the presence of insect pollinators is important e.g. Barbier, 1967), girding of branches, an old technique which leads to higher photosynthate availability to the fruit (e.g. Church, 1933), and the use of growth regulatory substances.

Moss (1970a) showed that gibberellic acid applied early reduced flowering and the number of flowering inflorescences (at 25 ppm) and increased fruit set (presumably through reduced competition). Applied at petal-fall stage and later at 300 ppm, gibberellic acid increased fruit set, presumably through the promotion of parthenocarpy, and fruit volume and reduced mean fruit size. Promotion of parthenocarpy by gibberellic sprayed at about 100 ppm at the small-fruit stage has been reported in 'Clementines' (Bertin, 1968), limes (Roy *et al.* 1970), Navel oranges (Franciosi and Ponce, 1970) and many others. However, Moss (1970) does not consider the use of gibberellic acid to be practical to improve sweet orange yields. Applied late and at low concentrations, gibberellic acid delays maturity and colour development and retains fruit longer on the tree. This aspect could be of interest in tropical countries when rapid maturation causes low quality in citrus (Torres and Rios-Castano, 1968). However at 500 ppm at the mature fruit stage, gibberellic acid promoted mature fruit drop in Australia (El-Zeftawi, 1970).

Other hormonal chemicals 2:4-D and napthalene acetic acid (NAA) when used at low concentrations (*c.* 20 ppm) in pre-harvest spraying reduced fruit drop and delayed maturity in many citrus varieties (e.g. Bravo, 1969; Dantur and Aso, 1969). At higher levels (e.g. 200–500 ppm) with NAA, when applied early, increased fruit drop has been reported (e.g. Cassin, 1967; Hirose, 1970; El-Zeftawi, 1971). 2:4:5-T is also effective in causing fruit thinning but at much lower concentrations (about 10 ppm) —e.g. Cassin (1967). These substances also have residual parthenocarpic effects when used for fruit thinning since remaining fruit may be larger than when thinning is carried out by hand (Hirose, 1970). At lower concentrations when applied early, 2:4-D and 2:4:5-T may produce greater fruit set (e.g. Roy *et al.* 1970). Combinations of 2:4-D and gibberellic acid also delay maturity and increase fruit drop and (by analogy with other crops) there is a possibility of synergism between these two types of growth regulators.

In sub-tropical and temperate regions where sea-

sonal variations in climate result in seasonal cropping, and also in the tropics where rainfall variation
may induce seasonality, the problem of biennial
bearing arises (e.g. Moss, 1972a; Cassin, 1967;
Marchal and Lacoeuilhe, 1969). A pattern of biennial bearing may be initiated by a particularly good
or bad season and may then be continued by internal competition factors (e.g. Moss, 1970b). This
problem can be considered from two points of view;
firstly, increasing production in the off year, and
here the fruit promoting properties of growth regulators might be utilized (e.g. Moss, 1972b), and
secondly, reducing overbearing in the heavy-bear-
ing season by use of growth regulators for fruit
thinning as discussed above. Heavy bearing also
affects fruit size and marketability. Since the cause
of biennial bearing seems to involve the supply of
assimilates and depletion of reserves in the heavy-
bearing season, fruit-thinning procedures seem to
offer the best possibility of overcoming this problem. Photosynthesis of the inflorescence leaves is
apparently important in supplying the photosynthetic needs of those inflorescences, since Moss
(1970c) reported that fruit set of leafy inflorescences
was about three times that of leafless ones. In addition to reducing flowering (see above) gibberellic
acid also increases the proportion of leafy inflorescences.

In addition to the growth regulators mentioned,
ethylene ('Ethrel') and ascorbic acid are also effective in fruit thinning (e.g. El-Zeftawi, 1970). Biennial bearing is also affected by mineral nutrient exhaustion (see p. 195).

The effect of temperature on flowering is important in citrus and many varieties have critical temperature requirements for flowering and fruiting
and are adapted to specific conditions. In oranges
(*C. sinensis*) low temperature is often more important. With oranges in Australia for example,
Moss (1971) reports that marginal flowering conditions are obtained at 24/19°C day/night temperatures and good flowering conditions are obtained
at 18/13°C day/night temperatures. In Malaysia,
the Fukien and Teo Chew varieties of mandarin
orange will flower and fruit profusely at 3,500 ft
elevation but not at sea level, the difference being
largely attributed to temperature. Torres and Rios-
Castano (1968) consider that continuous bearing
of locally-adapted orange, lemon and mandarin
varieties is due to the more or less uniform temperatures throughout the year in Colombia and the

same situation undoubtedly applies to other humid
tropical locations.

CLIMATE AND SOIL
Climate

Since we are dealing with such a great range of
genotypes in citrus it is difficult to generalize in
regard to climatic requirements. As explained before, the major citrus industries are established in
the sub-tropics. In general oranges and lemons
thrive best at temperatures lower than those obtained in the equatorial tropics and the question of
temperature for flowering of oranges has been
mentioned above. Garcia Benavides (1971) has reviewed the literature on the requirements of oranges
based on climatic data from orange-growing regions
extending between 37°S and 44°N latitude. Tangerines (or mandarins) are grown in both sub-tropical
and tropical latitudes, while grapefruits, pomelos
and limes appear in general to be better adapted to
the warmer conditions of the equatorial lowlands.
Krezdorn (1970) has reviewed the suitability of
different *citrus* species and varieties for growing
in the tropics. Praloran (1968) considers that about
12°C is the minimum temperature for vegetative
growth of any citrus on the basis of data from fifteen different citrus-growing regions of the world.

In tropical regions, as already mentioned, fruiting is more continuous but rapid fruit maturation
due to the absence of cold temperatures during
ripening leads to poorer quality and poor development of colour. Reuther and Rios-Castano (1970),
on the basis of a comparison of citrus varieties in
California and tropical Colombia, conclude that
fruits developed in the tropics are larger and have
a higher juice content but the skin is greener and
there is a lower dry extract ratio and acid content.
They also state that quality of fruit left on the tree
deteriorates more rapidly under tropical conditions
and consider that in Colombia, elevations between
1100 and 2200 m are required for Navel oranges.
Late Valencia however can be grown from sea level
to 1750 m and pomelos from sea level to 1300 m.
Rojas and Zambrano (1969) compared the quality
of Valencia and 'Nativas' oranges grown at a range
of elevations between 300 and 1800 m in Colombia
also, and found increasing quality with elevation.
However, most citrus varieties also require high
sunlight and this is often not obtained at high elevations in the tropics. In Malaya, mandarin oranges
grown at elevations of about 4,000 ft mature well

at one location with about 1800 sunshine hours per year but poorly at another location which is frequently overcast. Under very dry and hot conditions citrus may suffer loss of quality from sunburn (e.g. Minessy *et al.* 1970).

At the other extreme, citrus is not well adapted to survive frost and in this regard the trifoliate orange has proved to be the most cold resistant genotype. The citrus industry of Florida is periodically affected by frost, for example in1 962/3 when severe losses were suffered (*Citrus* Fruits: *Monthly Bull. Agr. Econ. Stats.* FAO 15(9), p. 16). However, citrus cultivation is undertaken at some relatively high latitude sites. Petrov (1967), for example, reports that citrus cultivation is possible in Czechoslovakia provided plants are grown in deep ditches protected from frost. Salazar (1966) in a study of four *Citrus* species showed that cold resistance was related to the numbers of calcium oxalate cells in the leaves and also to the ratio of central vein width to lamina width.

The moisture requirements of citrus have been considered before. Most citrus cultivars would do well under annual rainfall regimes of about 50", fairly well distributed for non-irrigated cultivation: but in many citrus-growing areas with lower rainfall regimes, irrigation is extensively used and this is considered in more detail further on. Under humid tropical conditions, the occurrence of short dry seasons and high light levels and the consequent development of slight moisture stress might aid the maturation and colour development of the fruit and lead to a higher sugar content.

In an interesting study of the effects of rainfall in a marginal area of Israel for non-irrigated citrus growing, Kalma (1968) showed that of an annual rainfall of 650 mm, 7–11 per cent was intercepted by the foliage in fully-developed citrus with a closed canopy and loss to the soil/plant system was estimated as 6–11 per cent of total precipitation.

Citrus is very sensitive to salt spray (e.g. Lomas and Gat, 1967), but can be grown close to the sea when protective wind-breaks of salt-tolerant trees are first established.

Soils and drainage

Citrus is grown on a very wide range of soil types throughout the world and some indication of the adaptability of the various rootstock species to soil types has already been given. By and large, however, the main citrus industries are established on relatively light friable soils and not on heavy clays and generally good drainage is required in the root zone. Kawamura *et al.* (1970) state that optimal bulk density for citrus soils in Japan is in the range 1.1–1.3 with a solid ratio of 40–50 per cent. Many citrus stock species have superficial roots but it has been found that there is a high correlation between tree volume and the depth of soil available to the roots. Minessy *et al.* (1971) also showed a relationship between growth and soil depth as affected by the water-table. Inland tropical latosols are suitable for citrus growing (Mello *et al.* 1969). Citrus, however, is not well adapted to very acid soils. Kawamura *et al.* consider pH 4 to be the lower limit for oranges in Japan and Reyes *et al.* in the Philippines showed that acidification of the soil to pH 3.8 by continued use of ammonium fertilizers produced poor citrus growth. The optimal pH is around 6.5 or near neutrality. Citrus will also grow in slightly alkaline soils but may exhibit micronutrient deficiencies.

Citrus is sensitive to high salinity and high chlorine levels although irrigation with sea water in Israel at levels of 1200–1400 ppm Cl on sandy soils did not affect yield (Stylianou and Orphanos, 1970). Bhambota and Kanwar (1970), however, report that sweet oranges were sensitive to NaCl, Na_2SO_4, $NaHCO_3$ and borax levels higher than 550 ppm on heavier soils. They considered that a total salt concentration of 550 ppm was safe provided the NaCl content did not exceed 250 ppm.

PLANTING

Land preparation, spacing and yield

In general land preparation for citrus involves clearing to a high standard to allow frequent entry to the field for pruning, weeding and other field operations. In addition, under the sub-humid conditions of the major sub-tropical citrus-producing areas, the field is also prepared for irrigation and a variety of mechanical operations involving the entry of tractors to the field. Under tropical conditions much of this work may be done by hand labour and the standard of clearing may not be so high. Furthermore, in the humid tropics, irrigation would not be practised except on very sandy soils.

In citrus cultivation spacing has traditionally been relatively wide with spaces between the trees, largely, in the more developed countries of the subtropics for the entry of tractors etc. for field operations and harvesting. Table 13.2 (from Ochse *et*

TABLE 13.2 Minimum and maximum planting distances and populations in citrus trees

	Distance (m)	Trees per ha
Orange	6.7 × 6.7	222
	7.2 × 7.2	192
	10.0 × 10.0	100
Grapefruit	9.0 × 9.0	120
	12.0 × 12.0	70
Mandarins	6.0 × 6.0	270
(tangerines)	9.0 × 9.0	120
Pomelos	10.5 × 10.5	90
	12.0 × 12.0	70
Lemons	6.7 × 6.7	222
Limes	4.5 × 4.5	496
	6.7 × 6.7	222
Citrons	6.0 × 6.0	270

al. 1961) gives an indication of minimum and maximum planting distances used in various *Citrus* species, based primarily on tree size.

Cody (1969) reports the use of 6.6 m square spacing for oranges in Honduras on fertile soils and 7 m square on poorer soils; a situation opposite to that pertaining in some crops (e.g. oil palms) where closer spacing is adopted on poorer soils (on the reasoning that poorer soils will support smaller trees which need to be more closely spaced).

In recent years there has been a trend to adopt closer spacings for increased early yield, followed by appropriate thinning at a later stage of development. Stannard (1970) reports that a double-spacing trial for Valencia oranges (to 375 trees/ha) has shown over 70 per cent increase in production over the first five years. Similarly Burger (1967) reports that higher yields are obtained in South Africa over the first fifteen years of plantation life from close spacing, after which time thinning is necessary. In Florida, Campbell (1968) reports that very high yields of over 50 tons/ha/year of Tahiti lime (*C. latifolia*) are obtained at high tree populations of 500 to 750 trees/ha. It is likely that the trend towards closer spacing in citrus will be accentuated in the future, and this may lead to consideration of canopy efficiency as mentioned before.

Yield of citrus varies greatly, due more to cultural factors than to species or variety. The national average production of oranges and mandarins in Pakistan (former West Pakistan) is reported by Khan (1967) to be about 10 tons/ha annually. At the other extreme yields of 50 tons/ha or more may be obtained (e.g. Campbell, 1968). Potentially, the tropical regions, with the correct varieties and cultural conditions, should produce higher citrus yields than sub-tropical and temperate regions where low temperatures limit production for part of the year. An estimate of yield from Honduras, for example under good cultural conditions indicates levels in excess of 50 tons/ha (Cody, 1969). Under tropical conditions, bud-grafted citrus would come into economic bearing in about the fourth year with good husbandry, earlier in precocious species such as limes propagated from cutting or marcotts.

Propagation and nursery

The two main methods of propagation used in citrus are budding or grafting stock seedlings, and by cuttings or marcotts. The use of cuttings has been predominately confined to Asia in the past.

The characteristics of the main stock varieties have already been considered. In propagation by graftage bud grafting is mostly used and so is cleft grafting. In bud grafting the inverted T method is used as illustrated in Fig. 13.1 (from Batchelor and Webber, 1948). This differs slightly with the age of bud wood used as indicated in the figure. Under tropical conditions, bud wood should be obtained fresh if possible but can be stored for short periods (about 10 days) without significant loss of viability (e.g. Teixeira, 1971). In the selection of bud wood several precautions must be taken. Firstly, bud mutants (which occur quite often in citrus) should be avoided for routine seedling production where they can be recognized (by morphological differences). For experimental work however, the propagation of bud mutants is practised. Secondly, freedom from disease should be ensured by obtaining bud wood only from healthy trees. And thirdly, continued vegetative production of successive generations of bud-wood clones may lead to a decline in vigour in many citrus varieties so that it is necessary to return to the original trees or to undertake re-establishment of propagation material through a seed generation. The reason for vegetative decline is not clearly understood but may be due to progressive virus infection (see Ochse *et al.* 1961).

Nursery techniques. For propagation by budding, stock seedlings must first be grown. Storage of seed for a suitable planting time can be effected for about 6 months or more at low temperature (about 5°C) in a humid medium, e.g. Vermiculite

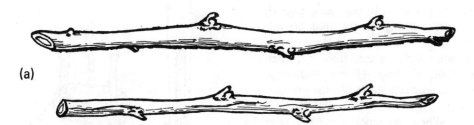

Fig. 13.1a Bud wood at round mature stage and angular immature stage

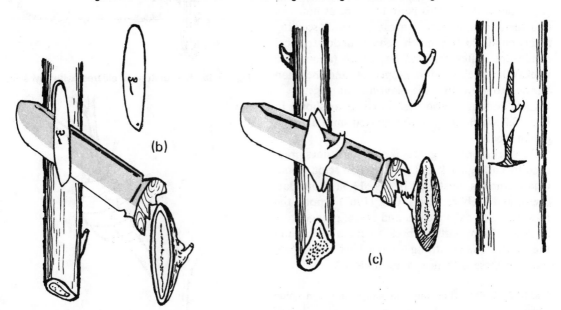

(b)

(c)

Fig. 13.1b Cutting of bud from round bud wood

Fig. 13.1c Cutting of angular bud and insertion into stock (side slipping)

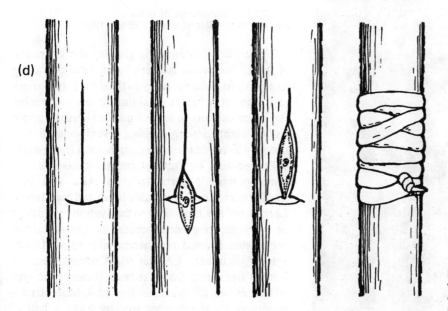

(d)

Fig. 13.1d Insertion of round wood bud in normal budding of citrus by the inverted-T method, and taping

Source. From Batchelor and Webber, 1948.

and sand—Ferreira (1970) or in polythene bags to prevent loss of water—Mungomery (1966). Germination of seeds is usually carried out in closely sown beds of potting soil which are then lined out when they reach a height of about 15 cm. Formerly, the nursery was usually established in the ground, but this has given way to the use of polythene bags of suitable soil. When the stock seedlings have reached 10–12 mm diameter they may be budded (e.g. Cohen, 1967). This period may be as much as two years in high latitude areas but stock seedling reach buddable size much sooner under tropical conditions. After the bud has taken (about two weeks) the stock shoot is pruned off and the scion allowed to grow until the planting out stage.

Cleft grafting is also used in citrus as shown in Fig. 13.2 which is mandarin grafted onto a rough lemon stock.

Re-working of old groves is also sometimes practised by cutting back and grafting new scion wood. One method is illustrated in Fig. 13.3a (from Batchelor and Webber, 1948). Cox (1967) reports the re-working of old Valencia and Navel orange trees in Australia by cutting back to main limbs and then, after twelve strong shoots have developed, budding these with new scion material.

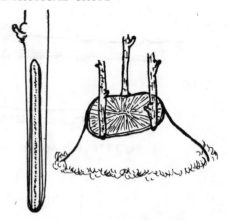

Fig. 13.3a   Bud grafting of old trees with new scion wood

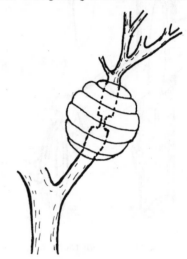

Fig. 13.3b   Marcottage of citrus
Source.   After Batchelor and Webber, 1948.

Fig. 13.2   Cleft graft of mandarin orange on rough lemon stock

While the scion shoot is growing the operation of frame formation is carried out, firstly by tying to a stake to obtain erect growth of the future trunk and then, at a height of about 70 cm by pruning the branch shoots to leave a number (usually about six or eight) strong-growing, evenly-spaced side shoots which will form the main branch framework.

Propagation by cuttings and marcotts is sometimes practised, particularly in Asia and with smaller citrus species; notably limes and mandarins. Large trees are not suited to cultivation by cuttings or marcotts in general because of the superficial root system formed and susceptibility to wind damage (cf. Rubber, Chapter 9). Cuttings tend to branch frequently from the base, forming the type of framework illustrated in Fig. 13.4. Marcotts are produced on the mother tree by tying a ball of earth around a girded side shoot (Fig. 13.3b) and

Fig. 13.4 Mandarin orange grown from a cutting, showing branching of the main shoot and framework

after the roots have been formed they are transplanted to the field. This is rather more time consuming than the making of cuttings. With limes, Gangwar and Singh (1965) showed that semi-hardwood cuttings gave the best take and that treatment with Indolyl butyric acid at 100 ppm stimulated root development. Another method of propagation in citrus which is seldom used commercially is layering (e.g. see Verma et al. 1970).

In recent years, propagation of citrus in polythene bags has greatly facilitated transplanting operations, but the technique of balling of ground-planted seedlings is still used to some extent. Recently mechanization of this procedure has been introduced (e.g. Corns, 1966) similar to that used in rubber (Chapter 9).

Manuring

Citrus generally requires relatively heavy fertilizer dressings for high yields. Nitrogen is one of the most important nutrients required under tropical conditions and, unlike with many other crop species, is sometimes increased with advancing matu-

rity. For example Ochse et al. recommend general fertilizer dressings of $N:P_2O_5:K_2O$ ratio of 6:4:8 for immature trees and 8:0–6:6–10 for mature-bearing trees, the variation in P_2O_5 and K_2O levels depending on soil conditions. Rates applied increased with age up to about 500 kg/ha/annum. Cody (1969), in Honduras recommends a 20:10:8 (N:P:K) or 14:14:14 mixture, depending on the soil, increasing to 800 kg/ha at the bearing stage, and Hosein (1967) a 20:10:10 (N:P:K) mixture rising from 70 kg/ha in the first year to about 400 kg/ha in the fifth. Under conditions where phosphate inactivation occurs, relatively high P levels may also be required. For example, Villezar and Oyardo (1968) recommend a 12:24:12 mixture in the Philippines, rates up to 1000 kg/ha/year still producing a profitable yield response in 12-year-old mandarins. Schatz and Drescher (1971) also report that N response depended on a higher P level in the fertilizer. High P nutrition is associated with the fruit qualities of low acidity and high juice content (Moss, 1972b).

In the Caribbean area, Wier (1969a) reports that N generally is the only element appreciably affecting fruit production, generally increasing fruit number (e.g. Schatz and Drescher, 1971). Nitrogen may increase or decrease fruit size, depending on bearing and competition factors (e.g. Aso and Stein, 1967; Aso and Dantur, 1970) and generally increases soluble solids and acid levels of fruit but may also delay maturity.

The build-up of soil acidity by the use of nitrogen fertilizers, particularly ammonium sulphate (e.g. see Reys et al.) can be controlled by liming. Adding lime along with the nitrogenous fertilizers produced the best pH control in the root zone (e.g. Wier, 1969 and Citrus Res. (Trinidad), 1969, p. 13). Urea has a lower acidifying effect on the soil than ammonium sulphate. The adverse effect of N on fruit acidity is considered to be less marked under humid tropical conditions than under arid conditions (Wier, 1969a).

Under conditions where seasonal bearing occurs, heavy yields cause strong mineral depletion, particularly of N and P, and this is associated with biennial bearing (see p. 190) (e.g. Marchal and Lacoeuilhe, 1969).

High levels of K have been reported to reduce fruit yields under tropical conditions (Wier, 1968), and to affect uptake of Mg (e.g. Mello et al. 1966; Wier, 1969b) through antagonism. Potassium appears to be the strongest antagonist in the three-way competition between K, Mg and Ca (Wier, 1969b). Of the rates of 0, 0.9 and 1.8 kg/tree of kieserite and potassium chloride, the medium level gave the best yields (Wier, 1968) in a factorial trial on Valencia oranges in Trinidad. Potassium at moderate levels increases fruit size (e.g. Wier, 1970), but generally it also increases acidity and reduces sugar and juice content (Wier, 1970, Moss, 1972b). Magnesium, however, has no appreciable effect on quality.

Magnesium, which is a very important nutrient for citrus, has been supplied as dolomitic limestone, but this form is slower in correcting deficiencies than magnesium sulphate except on acidic soils (e.g. Mello *et al.* 1969; Anderson, 1970).

The effect of calcium on citrus is complicated by the pH changes associated with the use of limestone. Aso and Dantur (1971) report that liming increased fruit number but reduced size and increased peel thickness in Argentina. Gallo *et al.* report that high calcium leaf level is associated with high yield.

Micronutrient deficiencies are commonly reported in citrus. Manganese, zinc and copper are most commonly reported (e.g. Feucht and Arancibia, 1967; Smith, 1968; Wier, 1966; *Agr. Gaz. NSW.*, 82, p. 36, etc.). Iron deficiency is also common (e.g. Comas, 1966). Copper deficiency symptoms (multiple sprouting) have also been confused with virus symptoms (Schwarz, 1970). Copper toxicity, however, has been reported in Florida through excessive application of this element together with its use in disease control (Spencer, 1966; Feucht and Arancibia, 1967). Copper toxicity symptoms are moderated by high P fertilization. Smith (1968) reports that Cu, Zn and Mn deficiencies are adequately controlled by applications of 11, 34 and 67 kg/ha respectively, and Cu and Zn are often supplied by foliar spray. Deficiency symptoms obtained in sand culture of a number of minor elements are described by Bar-Akiva and Lavon (1967). Srivaman-Nair and Mukherjee (1970a and b) found that in spray application of Zn, each flush required treatment because of poor mobility or rapid utilization. Khanna *et al.* report that Zn spray affects the uptake of Cu. Molybdium deficiency in citrus has been reported on serpentine soils due to high levels of nickel antagonizing molybdium uptake which was overcome by foliar sprays of ammonium molybdate (Sato, 1969).

More frequently major nutrients are supplied as surface soil dressing around the trees (in Angola, injection of fertilizers as 20 per cent solutions into the soil has been recommended—*Gaz. Agr.* (Angola) 14, p. 247). Micronutrients are frequently applied as foliar spray as indicated above, at levels ranging from 0.02 per cent for molybdium, 0.12–0.9 per cent for zinc and copper salts; manganese has also been used at 2.7 kg $MnSO_4$ to 375 litre water along with 1.13 kg lime (*Agr. Gaz. NSW.* 82, p. 36). Application of major elements by foliar spray is also sometimes undertaken. Bar-Akiva and Kaplan (1967) report that foliar application of urea, usually when new leaves are forming at 1.6 per cent, is widely carried out in Israel. However, Inoue *et al.* (1971) report that when young trees of Satsuma orange were adequately supplied with N *via* the roots, urea did not increase growth and at high N levels to the roots, growth was depressed. Cutuli (1968) reports that K deficiency could be corrected with 3–5 per cent KNO_3 sprays with a wetting agent during leaf flushing and Bar-Akiva *et al.* report that magnesium nitrate (about 1.2 per cent) corrected Mg deficiency in Israel.

Foliage application may be particularly useful in correcting deficiencies in instances where stock/scion compatibility affects nutrient movement. For example, it has been reported that Navel oranges on rough lemon rootstock show Mg deficiency symptoms in Australia, but not on trifoliate orange (*Agr. Gaz. NSW.* 78(8), 1967, p. 484). Similarly Inoue (1971) reports that N levels in Satsuma orange were significantly higher on trifoliate rootstock than on *C. junos* (Satsuma orange). A study carried out in Trinidad (*Citrus Res.* (Trinidad), 1969, p. 27), showed complicated stock/scion effects with certain varieties on sour orange, rough lemon, Rangpur lime, Cleopatra, Troyer citrange and trifoliate rootstocks. In one trial the first four above induced higher K and lower Ca and Mg levels in the scions; trifoliate induced higher N levels and Troyer citrange low Zn and Mn.

Foliar analysis is widely used to assess fertilizer needs in citrus, but because of the great variations in optimal levels for different species, varieties and for different climatic situations, and in relation to the stage of tree and fruit development, it is not practicable to give any actual levels here. Readers are referred to papers by Nadir (1965), Mello *et al.* (1966), Allbrook (1967), Drescher (1970), Kiely *et al.* (1972) and to the many books dealing with citrus cultivation. Martin-Prevel *et al.* (1966) report that leaves from fruiting branches had a lower N, P and K contents and higher Ca and Mg contents than on non-fruiting branches. Further, shaded leaves had reduced P levels and higher K levels. Nadir (1965) reports that additional sampling of leaves at the time of fruiting is necessary because of rapid nutrient removal, especially of K. In Argentina, Aso and Dantur (1969) report that seasonal variations in leaf nutrient levels were greatest at the times of new leaf flushing and decreased thereafter, and leaf N was abruptly lowered by moisture stress. Calcium level increased with leaf age, as would be expected with its relatively immobile nutrient; Mg also increased.

Irrigation

The moisture requirements of citrus for optimal growth and yield have been mentioned before. Generally citrus appears to be able to continue optimal growth and yield down to about 70 per cent field capacity, though with some varieties particularly those grown from marcotts or cuttings in which the root system is more superficial, it is possible that higher soil-moisture levels may be required for optimal performance.

Under sub-humid conditions, as prevailing in many of the important citrus-growing regions, irrigation is essential for successful production and is widely practised. Under such conditions efficient water use is also essential. On flat lands and on soils which are not too porous to permit movement of water along furrows, surface or basin irrigation is used. Sub-soil trickle irrigation is also sometimes practised, and on steep land and porous soils, overhead irrigation is principally used.

Studies on consumptive moisture use, based on measured evaporation and transpiration and measurements of moisture depletion of the soil, have been used to estimate irrigation needs (Maitin, 1967; Vink *et al.* 1971 etc.). Generally irrigation is most efficiently carried out on the basis of computed water requirements, and the time to irrigate on the basis of soil-tension measurements or on plant condition (wilting etc.). Beilorai and Levy (1971), for example, showed that yield response to irrigation of grapefruits was associated with the number of days in which soil moisture tension in the top 90 cm of soil exceeded 1 atm, and irrigation every 18 days gave higher yields than irrigation every 40 days. Similarly, El-Nokrashy *et al.* (1966)

in Egypt, comparing irrigation treatments based on consumptive water use, plant condition (the onset of wilting) and the 'conventional' method of watering when the soil cracked, showed much more efficient water use and higher yields when irrigation was determined by the two former methods.

Pruning

Pruning in the nursery stage to establish a good tree framework has already been mentioned. During nursery and establishment phases the removal of water shoots from the stock in grafted material is also carried out. Pruning of citrus in the field is primarily aimed at the maintenance of optimal balance between vegetative and reproductive growth and to allow access to the field. In developed countries pruning costs have promoted the use of cheaper mechanical and other methods of pruning citrus in the field. For example Cary (1971a) reports the use of a tractor-drawn topping and hedging machine in Australia and Burns *et al.* (1970) report the possibility of using growth inhibitors such as maleic hydrazide and dimethylhydrazide to reduce the rate of top and shoot growth and consequently reduce the frequency of pruning required.

Weed control and cover crops

Traditionally cover crops have been used in citrus growing between trees for all or part of the year, in the latter case being worked in by tillage at appropriate times (see Batchelor and Webber, 1948). At the present time, however, inter-tree soil cover and weed growth have been largely replaced by minimal cultivation techniques involving herbicide use. For example, Cary (1971b) reports that highest yields of citrus have been obtained from non-tillage bare-surface weed control using herbicides. Furthermore, in frost-prone areas, frost damage is reduced by the insulating effect of the moist compact surface of killed weeds obtained in inter-rows by herbicides use (Kaye, 1970; Coomber and Kaye, 1971). However, under erosion-prone conditions complete weed control may not be desired in inter-rows, and a system that gives periodic control using contact herbicides may be best (e.g. Hasida, 1969). A wide range of herbicides has been found safe for citrus. Kasasian (1967), for example, reports that paraquat (on clayey soils), dalapon, 2:4-D amine and bromacil are safe for citrus, and for the pre-clearing of nurseries an atrazine or diuron and TCA mixture is recommended. However, Americanos (1971) reports that bromacil caused temporary damage in Valencia oranges in Cyprus, but simazine, diuron and monuron and terbacil were recommended. Other reports of the use of the above herbicides as well as triflurain, dichlorbenzil, linuron, fluometuron and chloroxuron are given by Davis (1966), Donadio and Moreira (1969) Forsyth (1969), Moreira *et al.* (1969), Turpin *et al.* (1970) and many others.

Harvesting and post-harvest operations

Manual harvesting is used in citrus. The problem of harvesting cost in tropical areas is different from that in developed countries. There is great interest in the sub-tropical and temperate citrus-producing countries to develop satisfactory abcission-inducing techniques to facilitate mechanical harvesting, for example by shaking the tree (e.g. Burns, 1971).

Improvement of fruit colour is important in citrus, particularly in the tropics. Ethylene is predominately used to induce colour development (e.g. Hirose, 1970; Teisson, 1970; Daito and Hiros, 1970 etc.). To prolong the storage life of fruit (also important in the tropics), low temperatures, controlled atmosphere (CO_2, biphenyl) waxing, fungicides and treatment with growth regulators have been effective (e.g. Ben-Yehoshua, 1967; Pantastico *et al.* 1968; Soule, 1969; Vakis *et al.* 1970; Seberny and Hall, 1970; Shaukat Ali and Ehsan, 1971; *Agr. Gaz. NSW.* 79(2), p. 112; Norman *et al.* 1971; Laville 1971, etc.).

MAJOR DISEASES AND PESTS

Numerous virus diseases occur in citrus which have caused major set-backs to the industry. The relationships of some of these diseases are not clearly understood. Tristeza virus (orange tree quick decline) is one of the most important virus disease and has been mentioned before when considering rootstocks. It causes phloem necrosis and swellings on the stock. A range of virulence and tristeza resistance in various species and varieties of citrus occurs, with some virulent strains of the virus attacking 'resistant' stocks (e.g. Namekata and Rossetti, 1969). The disease is insect-transmitted and widespread in occurrence. The use of resistant stocks and clean bud-wood along with vector control are important in control of the disease.

Psorosis is considered to be a complex of viruses causing chlorosis of leaves, scaly bark and other symptoms (e.g. Broadbent, 1970). It occurs widely in South American countries and elsewhere (e.g. Weathers *et al.* 1969; Moreira, 1969; Nour-El-Din and El-Bauna, 1967). Psorosis is bud-transmitted and controlled by rigorous quarantine measures. For example, the citrus bud-wood scheme in Queensland has virtually eliminated this disease (*Queensland Agr. J.* 98, p. 58).

Xyloporosis (cochexia) virus which causes honeycombing of the bark and tree leaning (also referred to as cachexia disease) occurs, particularly in sweet lime rootstocks. Rough lemon is also apparently susceptible. Exocortis is another virus which causes scaling and cracking of the lower bark. Trifoliate orange rootstock is particularly susceptible and possibly Ranjpur lime. Weathers *et al.* (1969) report its synergistic reaction in regard to virulence in trees co-infected with tristeza. Resistant rootstocks and clean bud-wood are the control measures for both xyloporosis and exocortis.

Stubborn disease is a virus which is widely distributed causing a brush-like appearance and stunting of trees, which is apparently bud-wood transmitted. It is commonly thought to be caused by a virus but Ghosh *et al.* (1971) report the occurrence of mycoplasma-like bodies associated with stubborn disease. Calavan and Christiansen (1966) report that sour orange and sweet lime show some resistance.

Greening disease is yet another virus complex which causes yellow-spotting and sometimes yellow-veining of leaves and stunting. It is insect-transmitted (e.g. Catling, 1970; Capoor *et al.* 1967). Greening disease has been thought to be the same as yellow-leaf-mottle (Salibe and Cortez, 1966; Martinez and Wallace, 1968).

There are also a host of other virus diseases in citrus (e.g. see Weathers *et al.* 1969; Azeri and Herper, 1972; Rondon *et al.* 1970; Zelcer *et al.* 1971; Schwarz, 1970; etc.). Much work needs to be done on identification procedures in citrus virus diseases particularly in the tropics (e.g. see *Agr. Handbook* USDA, No. 333, 1968).

Vector control is important in insect-transmitted diseases, principally by use of insecticides, but there is much interest in the use of biological control measures based on studies on population dynamics (e.g. Hanna, 1967; Norman *et al.* 1968; Bramley, 1968). Infra-red photography has been tried for early disease detection in citrus in Florida (Norman, 1966). Apart from insects affecting virus transmission in citrus, insects affecting the crop, sometimes seriously, are fruit flies, various mites and scale insects and others, and control methods range from insecticide use to biological methods. In the case of fruit-spoiling insects, fruits such as pomelos are frequently wrapped in paper in tropical countries. Spoiled fruit should be buried. Various trunk borers also affect citrus.

Citrus suffers from many fungal and bacterial diseases. The most important of these have been recently described by Hanna (1969). In tropical countries the most serious diseases are bacterial canker (caused by *Xanthomonas citri*—see Fouget, 1967; Brun, 1971), *Phytopthora* root-rot (e.g. Wong and Varghese, 1966; Granada and Sanchez, 1969) which is difficult to control except by use of resistant material, and citrus scab caused by the fungus *Elsinoe fawcetti* which is controlled by copper and other sprays (e.g. Phelps, 1969; Maliphant, 1970). *Phytopthora* also attacks the lower trunk causing death of the bark and frequently gum exudation, when it is referred to as gummosis. In British Honduras, a premature fruit drop caused by *Colletotrichum gleosporoides* is serious but can be controlled by fungicide treatments (Fagan, 1971).

A number of nematodes affect citrus, of which the most serious is *Radopholus similis* (see DuCharme, 1968).

14 PAPAYAS [Carica spp.]

INTRODUCTION, SPECIES

THE papaya or paw-paw is a widely grown pan-tropical fruit used for local consumption and also assumes some importance in trade in Hawaii, Australia, South America and elsewhere where it is exported from the growing regions to large consumer centres. It is also widely used in the preparation of canned tropical fruit mixtures and recently as a puree in the frozen foods trade. There is obviously much scope for the expansion of export trade of papaya products. The commercial papaya is based on *Carica papaya* which is native to tropical America. The main problems of papaya growing are associated with a host of diseases (see later). For this reason investigations have been initiated on the use of other species for resistance, and also for general genetic improvement (e.g. Torres and Rios-Castano, 1968a; Adsuar, 1971). A mountain papaya species (*C. candamarcensis*) has been found resistant to strains of papaya mosaic virus. Three other species, *C. goudotiana*, *C. monoica* and *C. cauliflora* are also of interest and some inter-specific crosses have been found to produce viable seeds. Micheletti (1967) reports that *C. cauliflora* is resistant to distortion ringspot virus disease.

For the American market, and also potentially for Europe and elsewhere, there is interest in uniform small-fruited types and for this reason hermaphrodites (see below) have been utilized. A papaya bank of germaplasm has been established in Peru. (Calzada-Benza, *et al.* 1967).

BOTANY

The papaya is a fast-growing species usually producing a single-stemmed short tree with a crown of large simple lobed leaves on long petioles (Fig. 14.1). The trunk is softly fibrous and the roots are also soft and without protective layers. The fruits develop successively in the leaf axils (Fig. 14.1a) and under uniform tropical conditions ripen successfully, producing about one mature fruit per week.

Trees may live for many years but in commercial production, they are usually replaced after two or three years, often because of build-up of disease and also because of declining yield with increased height of the trunk.

The fruits vary considerably in size, shape and colour. The two basic flesh colours are yellow and reddish and there are many intergrades. Fruits are thin-skinned and easily bruised and the edible portion is a thick melon-like layer of pericarp. Within the hollow centre is a variable number of mucilaginous round black seeds. Fruits vary from small melon-shaped varieties of about 1 lb weight to large pendulous pyriform types weighing 10 lb or more.

The papaya is dioecious, producing male, female and hermaphrodite trees. From the agricultural point of view it is obviously important to secure plantings of high female ratio. Fruits from female trees are generally larger and more uniform than

Fig. 14.1a Mature papaya tree bearing fruits

Fig. 14.1b A papaya with male inflorescences

those from hermaphrodite trees; the latter may be extremely variable, producing fruits varying greatly in size. However, some hermaphrodites under specific cultural conditions produce uniform small-sized fruit which have been favoured for marketing. (The 'solo' papaya from Hawaii is a good example.)

The genetic mechanisms of sex inheritance are not clearly understood but are thought to be similar to the sexual balance hypothesis in Drosophila (Hofmeyer, 1967). Teaotia and Singh (1967) report that sexual expression shows seasonal fluctuations. Pure female plants remained sexually stable but hermaphrodites produced a greater proportion of fertile female fruits and fewer pentandrous types in the spring than in the winter in India.

In practice, the crossing of females with hermaphrodites produces a high percentage of fertile trees (females and hermaphrodites) and further, planting of several plants at each planting site and thinning out when their sex can be determined (after a few months under tropical conditions) also increases the proportion of bearing trees (e.g. Agnew, 1968; Ireta Ojeda, 1969). Papayas show a strong tendency to be locally adapted and seed for planting is often best obtained by crossing high yielding or otherwise desirable locally occurring types.

Growth of fruit during its development takes a general sigmoid form and fruit growth is thought to occur following the production of hormones from the integuments and nucellus, as well as from the endosperm and embryo of the seed (Allan, 1969; Kuhne and Allan, 1970).

Sharma and Bajpai (1969) report that pollination in the field (dehiscence of anthers) takes place predominately in day-light hours between 10 a.m. and 3 p.m. Pollen can be stored in a desiccator for a period of one month or more for the purpose of producing desired crosses for seed production.

CLIMATE AND SOILS
Climate
Papayas are grown throughout the tropics and frost-free sub-tropics. As with bananas and pineapples, extension of papaya growing also occurs into the southern temperate zone on the eastern seaboards of the continents of Africa, Australia and South America. In the equatorial lowlands however, growth is much faster and plants may come into bearing in six months and fruits will mature in about 2–3 months from pollination. At the other extreme, papaya cultivation in South Africa (Pietermenitzburg) at an elevation of c. 2,000 ft has shown that fruits take 9–11 months to mature (Allan, 1967). In South Africa, the shape of the sigmoid fruit-growth curve varied with season (and also with variety); mean temperatures below 19°C lengthened the initial and final stages of slow growth rate (Kuhne and Allan, 1970). Low temperatures also cause smaller fruit size and low quality (Hamilton, 1969). However, the importance of local adaptation, particularly in relation to climate, should be stressed in papaya cultivation, and it is necessary to produce locally-adapted strains with the right market qualities for each situation. The demonstration by Allan (loc. cit.) that commercial papaya production is possible in the extreme conditions of inland South Africa indicates the flexibility of this crop.

Under tropical conditions, branching of the main stem of the papaya tree seldom occurs (except in some strains) and cultivation is based on the growing of plants from seeds, but under the conditions of Allan's investigation, vegetative propagation from cuttings is possible (see below). This is consistent with the general finding in plants that the activity of lateral buds is much greater at reduced temperatures.

Under lowland tropical conditions, papayas often suffer excessively from root diseases which suggest that the crop may be better suited to median altitudes in the equatorial regions. Papayas are destroyed by frost conditions and are susceptible to strong winds.

Soils and drainage

A universal requirement for papaya soils is good drainage. Many books on tropical crops recommend that papayas also require a rich humic soil (rather incompatible with good drainage); but they will grow well on sandy soils provided they are fertilized and the rainfall regime or irrigation is adequate. The author has observed that papaya trees always seem to do well on heaps of old builders' rubble, possibly because of enhanced drainage. In Hawaii, papaya cultivation is carried out on coarse volcanic lava rock, and here fertilizers must be applied more frequently than on more normal soils (Ito *et al.* 1968). This also leads to weed control problems using herbicides (see p. 203).

PLANTING

Land preparation, spacing and yield

Papayas are not generally grown on the same scale as the other tropical crops discussed before. Furthermore, because of the short storage life of the fruit (see below) and the predominance of trade based on local consumption, they are often cultivated on land, near the towns, which has been previously been cleared. Also, because of the tendency for the build-up of root diseases in the equatorial tropics, they are often rotated with other crops after two or three years. If large areas of land were to be opened for papaya involving the clearing of jungle or previously planted tree crops, the standard of clearing would have to be the same as for similar-sized trees such as bananas.

The spacing used for papayas is generally between 2 and 3 metres, depending on the size of the tree and the spread of the crown. Carvalho *et al.* compared spacings of 3×2.5, 3×2, 3×1.5, and 3×1 metres in Brazil and found that yield over the first eight months of bearing had increased with density, but the narrower spacings resulted in excessive stem elongation which rendered harvesting and pest control (for mites) difficult.

Yields vary considerably. Generally there is a relationship between fruit size and yield, with large-fruited types being higher yielding. However, market preference is important in selecting varieties to be grown and small-fruited lower yielding types may be more profitable under certain circumstances. For canning purposes, large-fruited types are generally preferred and these may be very high yielders. Under tropical conditions yields from superior large-fruited trees may be as high as sixty tons per acre per year with correct fertilizer use and husbandry. However, yields tend to drop off sharply after the first year of bearing.

Propagation

Commercial papaya growing should use hand-pollinated seed, as discussed before, of appropriate locally adapted strains. Parent trees giving good results in specific crosses should be clearly marked and retained for seed production. In pollination, flower buds should be enclosed in paper bags fixed with pins for fourteen days after pollination (Carvalho, 1966).

Trials carried out in Mexico (Mosqueda Vazquez, 1969) showed that air-dried seeds, after removal of the mucilaginous covering, gave the earliest and most uniform germination. Seeds may be hermetically stored for a year or more.

Seedlings are best raised in polythene bags of size about 20×15 cm laid flat (e.g. Araujo and Bellintani, 1967). These may be transplanted after about twenty to thirty days, depending on the rate of germination. If several seedlings are planted at each point (to eliminate the males) then these should be spaced about 1 ft apart for later thinning out.

In South Africa Allan (1967a) has successfully propagated papayas from soft wood cuttings, which were raised in containers and transplanted after 1 to 3 months depending on the season. Vegetative propagation, of course, secures genetically uniform planting material.

Manuring

Both organic and chemical fertilizers are widely used in papaya growing and are needed for sustained high yields. Ochse *et al.* (1961) reported that in Florida, 0.1 kg of 4:8:5 (N:P:K) fertilizer is applied per tree at two-week intervals for the first 6 months, and thereafter 0.2 kg at the same frequency. Leigh (1969) recommends a pre-planting dressing of 225 kg/ha of 1:3:1:(N:P_2O_3:K_2O) fertilizer and 700 g per plant of this mixture in the first year rising to 1.5 kg/plant at maturity in Australia. Kruger and Menary (1968), also in Australia, and Godoy *et al.* (1968, 1969) in Chile, report yield responses to high fertilizer (up to 700 g of potassium nitrate per month) but Kruger and Menary point out that this leads to a high nitrate content of the fruit which renders it unsuitable for canning. They state that N applications after commencement of fruit development and maturation should not exceed 225 g calcium ammonium nitrate or 112 g urea, if fruit is required for canning. The relationship of N fertilization to yield appears to be related to the formation of the assimilation apparatus (as in bananas—Chapter 2).

Tissue analysis has been used to determine nutrient requirements in papayas, using the most recently matured leaf petiole, identified by the presence of the youngest flowers of the tree in the leaf axil. In Hawaii the standards for N as a percentage of dry weight of the petiole tissue varied between 1.14 and 1.28 with different seasons (Awada *et al.* 1969) and the standard for P is considered to be 0.25 per cent. In Chile the petiole standard for N in *Carica candamarcensis* (the mountain papaya) is considered to be in the range 1.7–2.0, a higher level than above (Hawaii) possibly reflecting a more humid climate (Godoy *et al.* 1968). Standards for K do not appear to have been clearly established.

Weed control

Chemical weed control is most efficient in papayas, and mulching may also be recommended. In Australia Diuron, Paraquat and 2:2-DPA are used when the plants are six months old. The use of arsenical herbicide is forbidden (Leigh, 1969).

In Hawaii, on coarse volcanic soils (mentioned before) petroleum solvent at 150–375 l/ha and Paraquat at 0.5–1.1 kg/ha with surfactant 2.5–5 l/ha are used for weed control, and Diuron is reported as showing promise as a suitable herbicide (Ito *et al.* 1968). Papayas are sensitive to hormonal herbicides such as 2:4-D.

Papain and pectin from papayas

These two products are also obtainable from papayas on a commercial scale. Papayas contain pectin levels of 2 per cent or more (Medina Ramirez, 1968) and there is a demand for pectins in the jam industry. Pectins may be extracted from waste material in papaya canning. Papain is obtained by 'tapping' the fruits which is done by making shallow vertical incisions in the green fruits and later scraping off the exuded sap which is high in papain. Tapping is best done in the early morning when the turgor pressure is high (analogous to rubber). Northwood (1970) reports a yield of 169 kg/ha of papain over an 11-month period.

MAJOR PESTS AND DISEASES

Mites and nematodes are generally the most destructive pests in papayas (e.g. Guerout, 1969a); in addition aphids and other insect species are responsible for the spread of the papayas' numerous virus diseases. Chemical control and hygiene procedures are necessary (e.g. Kralovic, 1967) to secure good papaya production on a long-term basis.

Formalin (25 ml of 4 per cent) per hole and 'Nemagon' have been used to control nematodes (Singh and Shanker, 1968).

Among the fungi, *Phytopthora* affects both roots and fruits (Hunter, 1969; Ko, 1971). Soil fumigation is used in Hawaii (Murashige and Nakano, 1966).

Krochmal (1968) reports that anthracnose (*Colletotrichum gleosporioides*) can be controlled by spraying with 'mancozeb' or copper fungicides and 'papaya decline' caused by a complex of fungi by spraying with zinc salts or 'maneb' at intervals of 10 to 14 days. Greasy spot-disease caused by *Corynespora* (*Helminthosporium*) *cassiicola* and possibly *Botryodiplodia sp.* is described by Bird *et al.* (1966) from the Virgin Islands, and a disease caused by *Calonectria sp.* causing stunting, loss of leaves and decay of the stem at the soil line is described by Laemmlen and Aragaki (1971).

Papayas suffer from a wide range of virus diseases which have been given various names; papaya mosaic (e.g. Kralovic, 1967), distortion ringspot or distortion virus (e.g. Story and Halliwell, 1969), distortion mosaic (e.g. Schaefers, 1969), yellow crinkle (e.g. Greber, 1966), virus debility characterized by stunted growth and leaf chlorosis leading to the formation of small (pencil point) crowns (e.g. Peregrine, 1968) and bunchy top (similar to above—e.g. Naylor, 1966). The only satisfactory control measures, apart from breeding for resistance mentioned before, is through hygiene and quarantine procedures and vector control. The latter, to be effective, must involve study of the biology of the vectors which are chiefly aphids, and also leaf hoppers, particularly on seasonal movements and fluctuation in the vector populations.

BIBLIOGRAPHY

Abeywardena, V. (1969) *Ceylon Coconut Planters' Rev.* 5, p. 173.

———— and Fernando, J.K.T. (1963) *Ceylon Coconut Quart.* 14, p. 74.

Abraham, A. (1970) In: *Proc. 2nd Internat. Symp. on Tropical Root Crops* (Hawaii) 2, p. 76.

Abraham, P.D. *et al.* (1968) *J. Rubber Res. Inst. (Malaya)* 20, p. 291.

Abraham, P.D. *et al.* (1971a) *J. Rubber Res. Inst. (Malaya)* 23, p. 85.

———— (1971b) *J. Rubber Res. Inst. (Malaya)* 23, p. 90.

———— (1971c) *J. Rubber Res. Inst. (Malaya)* 23, p. 114.

———— and Tayler, R.S. (1967) *Exp. Agr.* 3, p. 1.

Adams, S.N. (1962) In: *Agriculture and Land Use in Ghana.* Ed. J.B. Wills. Oxford U. Press.

Adijuwana, H. and Soerianegara, I. (1970) *Menara Perkebunan*, 26, p. 1.

Adsuar, J. (1971) *J. Agr. Univ. Puerto Rico* 55(2), p. 265.

Agnew, G.W.J. (1968) *Greensland Agr. J.* 94(1), p. 24.

Ahenkorah, Y. (1968) *Soil Sci.* 105, p. 24.

Akbar, V. *et al.* (1971) *Oléagineux* 26, p. 527.

Akhmetov, G.S. and Bairamov, B.I. (1967/8) *Fertilité* 30, p. 65.

Alers, S. and Samuels, G. (1962) *J. Agr. Univ. Puerto Rico* 47, p. 24.

Ali, F.M. (1969) *Exp. Agr.* 5, p. 209.

Ali, M.K. and Talukdar, M.R. (1965) *Agr. Pakistan* 16, p. 145.

Alink, J.M. and Wolf, J. (1964) *Tyd. Heidmeij* 75, p. 338.

Allan, P. (1967) *Farming S. Afr.* 42 (10), pp. 25, 29, 31.

———— (1967a) *Farming S. Afr.* 42(11), p. 15.

———— (1969) *Agr. Sci. S. Afr., Agroplantae* 1(3 & 4), p. 163.

Allbrook, R.F. (1967) *Exp. Agr.* 3(3), p. 215.

Allen, A.W. (1966) *2nd Malaysia Soil Scientists Confr.* (Kuala Lumpur).

Allen, E.F. (1953) *Malayan Agric. J.* 15, p. 43.

Alvarez, R. and Friere, E.S. (1962) *Bragantia* 21, p. 31.

———— *et al.* (1963) *Bragantia* 22, p. 657.

Alvim, P. de T. (1958) *Coffee and Tea Industr.* 81, p. 17.

———— (1966) *Cocoa Growers Bull.* 7, p. 15.

———— (1968) *Comm. Techn. CEPLAC* 20, p. 5.

Americanos, P.G. (1971) *Techn. Bull. Cyprus Agr. Res. Inst.* 8, p. 1.

Amon, B.E. (1967) *Publn. Commit. Tech. Co-oper. Afr.* 98, p. 339.

Amorim, H.V. de *et al.* (1968) *Anais Escola Super. Agr.* (Brazil) 25, p. 121.

Ampuero, E.C. (1967) *Cocoa Growers Bull.* 9, p. 15.

Ananth, B.R. and Hanumantha Rao, H. (1970) *Ind. Coffee* 34, p. 15.

Anderson, C.A. (1970) *Proc. Soil and Crop Sci.* (Florida) 30, p. 150.

Anderson, I.C., Greer, H.A.L. and Tanner, J.W. (1965) In: *Genes to Genus*, Illinois, Internat. Min. & Chem. Corpn.

Angkapradipta, P. *et al.* (1972) *Menara Perkebunan* 40, p. 47.

Angladette, A. (1964) In: *Mineral Nutrition of the Rice Plant*, p. 355, Johns Hopkins Press.

Antoni, H. (1965) *Rev. Ind. Agr. Tucuman* 43, p. 47.

Arasu, N.T. (1970) *Malaysian Agr. J.* 47, p. 409.

Araujo, C.M. and Bellintani, V. (1967) *Agronomia* (Brazil) 25(3–4), p. 33.

Are, L. and Afonja, B. (1966) *Tech. Working Party on Cocoa Prod.* (Rome), Paper (Ca) 66/13, p. 1.

Arisz, W.H. (1918) *Arch. v.d. Rubbercult. in Ned. Indie.* 2, p. 2 (cited by Dijkman, 1951).

———— (1919–20) *Arch. v.d. Rubbercult. in Ned. Indie.* 3–4, p. 321 (cited by Dijkman, 1951).

Arnold, G.W., Bennett, D. and Williams, C.N. (1967) *Aust. J. Agr. Res.* 18, p. 245.

Arraudeau, M. (1967) In: *Proc. 1st Internat. Symp. on Trop. Root Crops* (Trinidad) 1(3), 180.

Arscott, T.G. *et al.* (1965a & b) *Trop. Agr.* 42, pp. 215 and 210.

Artschwager, E. (1940) *Jour. Agric. Res.* 60, p. 49 (cited by Dillewijn, *loc. cit.*).

Ascenso, J.C. (1968) *Trop. Agr.* (Trinidad) 45, p. 323.

———— and Bartley, B.G.D. (1966) *Euphytica* 15, p. 211.

Ascomaning, E.J.A. and Kwakwa, R.S. (1967) *Confr. Int. Res. Agron. Cacaoyeres* (Abidjan, 1965), p. 39.

Aso, P.J. and Dantur, N.C. (1969) *Rev. Ind. Agr. Tucuman* 46(1), p. 35.

———— and Dantur, N.C. (1970) *Rev. Ind. Agr. Tucuman* 47(1), p. 1.

———— and Dantur, N.C. (1971) *Rev. Ind. Agr. Tucuman* 48(2), p. 31.

———— and Stein, E. (1967) *Rev. Ind. Agr. Tucuman* 45(2), p. 107.

Aspiras, M. (1964) Unpublished, IRRI (cited in Mooman and Vergara, *loc. cit.*).

Awada, M. *et al.* (1969) *Misc. Publ. Co-op. Ext. Serv.* (Univ. Hawaii) 64, p. 30.

Ayala, A. (1968) In: *Crop Nematology*, Eds. Smart, G.C. & Perry, p. 49.

Azeri, T. and Herper, E. (1972) *Plt. Disease Reporter*, USDA 56, p. 352.

Bachy, A. (1969) *Oléagineux* 24, p. 533.

———— and Martin, G. (1967) *Oléagineux* 22, p. 533.

———— and Regaud, J.N. (1968) *Oléagineux* 23, p. 701.

Baker, C.J. (1970) *Kenya Coffee* 34, p. 226.

Baker, K. and Collins, J.L. (1939) *Amer. J. Bot.* 26, p. 697.

Baker, R.E.D. (1961) In: *Cocoa* by Urquhart, *loc. cit.*

Balasubramaniam, M.A. (1970) *Ind. Coffee* 34, p. 47.

Baldwin, J.T. Jr. (1947) *J. Heredit.* 38, p. 54.

Bar-Akiva, A. *et al.* (1969) *Exp. Agr.* 5(4), p. 339.

———— and Kaplan, M. (1967) *Fruits, France* 22(3), p. 153.

———— and Lavon, R. (1967) *Israel J. Agr. Res.* 17(1), p. 7.

Barber, C.A. (1918) *Mem. Dept. Agric. India, Bot. Sci.* 9, p. 133 (from Dillewijn, *loc. cit.*).

Barbier, E. (1967) *Agrumes Presse* 26, p. 19.

Barbosa, B.C. (1970) *Two and a Bud* 17, p. 71.

Barlow, C.H. and Lim, S.C. (1967) *J. Rubber Res. Inst.* (Malaya) 20, p. 44.

Barnes, A.C. (1964) *The Sugar Cane*, London, Leonard Hill.

Barnes, D.E. and Tan, W.S. (1969) In: *Prog. in Oil Palm*, p. 185.

Bartley, B.G.D. (1968) *Cocoa Growers Bull.* 10, p. 9.

———— and Amponsah, J.D. (1966) *Ann. Rep. Cacao Res.*, p. 50.

Barua, D.N. (1966) *Two and a Bud* 13, p. 58.

———— (1968) *Two and a Bud* 15, p. 103.

———— (1969) *Nature* (London) 224, p. 514.

———— and Dutta, K.N. (1971) *Two and a Bud* 18, p. 8.

Barua, S.C. and Grice, W.J. (1968) *Two and a Bud* 15, p. 20.

Batchelor, L.D. and Webber, H.J. (1948) *The Citrus Industry* 2 vols., California.

Beachall, H.M. and Jennings, P.R. (1964) In: *Mineral Nutrition of Rice Plant*, p. 29, Johns Hopkins Press.

Bealing, F.J. (1969) *J. Rubber Res. Inst.* (Malaya) 21, p. 445.

Beak, B.D.A. and Chan, S.R. (1958) *Trop. Agr.* (Trin.) 35, p. 59.

Beeny, J.M. (1970) In: *Crop Diversification in Malaysia*, Eds. E.K. and J.W. Bonkers, Malaysia.

———— *et al.* (1967) *The Planter* (Malaya) 18(5), p. 1.

Benac, R. (1969) *Café Cacao Thé* 13, p. 187.

Ben-Yehoshua, S. (1967) *Israel J. Agr. Res.* 17, p. 17.

Bernard, G. and Noiret, J.M. (1970) *Oléagineux* 25, p. 67.

Bertin, A. (1968) *Agrumes Presse* 42, p. 1.

Bettencourt, A.J. and Carvalho, A. (1968) *Bragantia* 27, p. 35.

Bevan, J.W.L. *et al.* (1966) *Planting Technique for Oil Palm in Malaysia*, Kuala Lumpur.

Bhambota, J.R. and Kanwar, J.S. (1970) *Ind. J. Agr. Sci.* 40(6), p. 485.

Bhat-Shama, K. and Leela, M. (1968) *Indian Farming* 18, p. 19.

Bhavanandan, V.P. and Fernando, V. (1970) *Tea Quart.* (Ceylon) 41, p. 94.

Bicudo, L. (1969) *Coopercotia* 26, p. 47.

Bielorai, H. and Levy, J. (1971) *Israel J. Agr. Res.* 21(1), p. 3.

Bierhuizen, J.F. *et al.* (1969) *Acta Bot. Neerlandica Rep.* 18, p. 367.

Bigornia, A.E. and Infante, N.A. (1965) *Philippine J. Sci.* 94, p. 459.

Bird, J. *et al.* (1966) *J. Agr. Univ. Puerto Rico* 50(3), p. 186.

Blaak, G. (1969) *Euphytica* 18, p. 153.

Black, R.F. (1962) *Qld. J. Agr. Sci.* 19, p. 435.

Blommenstein, J.A. van and Oothuysen, C.M. (1970) *Agric. Res. Dep. Agr. Techn. Serv. Rep. S. Afr.*

Blondel, L. (1967) *Fruits, France* 22(1), p. 19.

Blow, R. (1968) *Cocoa Growers Bull.* 11, p. 10.

Blummenstock, D.A. and Thornthwaite, A.V. (1941) *Yearbook*, U.S. Dept. Agr.

Bobilioff, W. (1918) *Arch. v.d. Rubbercult in Ned. Indie* 2, p. 735.

———— (1923) *Anatomy and Physiology of Hevea brasiliensis.* Part I, Anatomy of *Hevea brasiliensis*, Zurich, *Art. Inst. Orell Fussli* (cited by Dijkman, 1951).

Bolhuis, G.G. (1967) In: *Proc. 1st Internat. Symp. on Tropical Root Crops* (Trinidad) 1(1), p. 81.

Bond, T.E.T. (1942) *Ann. Bot.* 6, p. 607.

———— (1944) *Ann. Appl. Biol.* 31, p. 300.

———— (1945) *Ann. Bot.* 9, p. 1183.

Boonsirat, Ch. and Paardekooper, E.C. (1971) *Rep. Rubb. Centre, Thailand* 19, p. 15.

Borden, R.J. (1941) *Hawaiian Plant. Rec.* 45, p. 241.

———— (1942) *Hawaiian Plant. Rec.* 46, p. 39.

Bouchard, L. and Damour, M. (1971) *Agron. Trop.* 26, p. 256.

Bouharmont, P. (1971) *Café Cacao Thé* 15, p. 202.

Bould, C. *et al.* (1971) *Kenya Coffee* 36, p. 37.

Boulle, P.F. (1961) *Pro. Ann. Congr. S. Afr. Sugar Tech. Assn.* 35, p. 129.

Bourne, B.A. (1948) *Sug. Jour.* 10, p. 3.

———— (1963) *Sug. Jour.* 26, p. 41.

Bowden, B.N. (1964) *J. Trop. Geog.* 19, p. 1.

Boyer, J. (1965) *Oléagineux* 20, p. 437.

———— (1969) *Café Cacao Thé* 13, p. 187.

Boysen-Jensen, P. (1932) *Die stoffproduktion dér Pflanzen* Fisher, Jena.

Bramley, W. (1968) *Farming S. Afr.* 43, p. 39.

Braudeau, J. (1957) *Cocoa Confr.* (Lond.), p. 59.

Bravo, C.M. (1969) *Turrialba, Rev. Interam. Ciencias Agr.* 19, p. 522.

Bredell, G.S. (1970) *Farming S. Afr.* 46, p. 11.

Briant, A.K. and Johns, R. (1940) *East Afr. Agr. J.* 6, p. 404.

Briolle, C.E. (1969) *Oléagineux* 24, p. 545.

Broadbent, P. (1970) *Agr. Gaz.* NSW. 81, p. 87.

Broekmans, A.F.M. (1957) *West Afr. Inst. for Oil Palm Res. J.* 2(7), p. 187.

Brosehart, H. *et al.* (1957) *Plant and Soil* 8, p. 289.

Brown, D.A.La. *et al.* (1967) *Cocoa Growers Bull.* 9, p. 9.

———— *et al.* (1968) *Cocoa Growers Bull.* 10, p. 13.

Brown, W.R. (1920) *Pusa Agr. Res. Inst. Bull.* 93, p. 7 (cited by Batchelor and Webber, *loc. cit.*).

Brun, J. (1971) *Fruits, France* 26, p. 533.

Brunin, C. (1968) *Oléagineux* 23, p. 303.

———— (1970) *Oléagineux* 25, p. 269.

Bull, R.A. (1966a) *The Planter* (Malaya) 42, p. 248.

———— (1966b) *Malayan Agriculturist* 6, p. 27.

Burg, B. van der. (1969) *World Crops* 21, p. 105.

Burger, W.P. (1967) *Farming, S. Afr.* 41(12), p. 9.

Burle, L. (1957) *Cocoa Confr.* (Lond.), p. 52.

Burns, R.M. (1971) *California Agr.* 25(2), p. 10.

———— and Coggins, C. (1969) *California Agr.* 23, p. 18.

———— *et al.* (1970) *California Agr.* 24(5), p. 8.

Burton, G.W. *et al.* (1957) *Agron. J.* 49, p. 498.

Buttery, B.R. (1961) *J. Rubb. Res. Inst. Malaya* 17, p. 46.

Buttery, B.R. and Boatman, S.G. (1964) *Science* 145, p. 285.

Cabala-Rosand, F.P. *et al.* (1967) *Cacau Atualidades* 4, p. 52.

Cabanilla, H.L. and King, J.R. (1966) *Trop. Agr.* (Trinidad) 43, p. 187.

Calavan, E.C. and Christiansen, D.W. (1966) *Israel J. Bot.* 15, p. 121.

Calzada-Benza, J. *et al.* (1967) *Agron. Trop. Venezuela* 17(4), p. 381.

Camacho, E.S. *et al.* (1968) *Proc. Trop. Region Amer. Soc. Hort. Sci.* 12, p. 163.

Camargo, A. *et al.* (1967) *Bragantia* 26, p. 1.

Campbell, C.W. (1968) *Proc. Trop. Region Amer. Soc. Hort. Sci.* 12(16), p. 184.

Campbell, J.S. and Gooding, H.J. (1962) *Trop. Agr.* 39, p. 261.

Cannell, M.G.R. (1969) *Kenya Coffee* 34, p. 204.

———— (1971a) *Ann. Appl. Biol. Reps.* 67, p. 99.

———— (1971b) *Exp. Agr.* 7, p. 63.

———— (1971c) *Kenya Coffee* 36, p. 176.

———— (1971d) *Kenya Coffee* 36, p. 68.

———— and Huxley, R.A. (1970) *Kenya Coffee* 35, p. 139.

Capoor, S.P. *et al.* (1967) *Ind. J. Agr. Sci.* 37, p. 572.

Capriles de reys, L. *et al.* (1966) *Techn. Working Party on Cocoa Prod.* (Rome) (Ca) 66/7, p. 1.

Carlos, J.T. and Krezdorn, A.H. (1968) *Proc. Trop. Region Amer. Soc. Hort. Sci.* 12, p. 99.

Caro Costas, R. (1968) *J. Agr. U. Puerto Rico* 52(3), p. 256.

Carpenter, P.H. (1950) *The Wealth of India*, CSIRI, Delhi.

Carr, M.K.V. (1972) *Exp. Agr.* 8, p. 1.

Carriere de Belgarric, R. (1967) *Oléagineux* 22, p. 215.

Carter, W. (1952) *Bot. Rev.* 18, p. 680.

———— (1966) *Nature* (London) 212, p. 320.

Carvajal, J.F. (1969) *Rev. Cafetalera* 94, pp. 9, 17, 24.

———— *et al.* (1969) *Turr. Rev. Interam Ciencias Agr.* 19, p. 13.

Carvalho, A.M. (1966) *Agronomica* (Brazil) 18(1–2), p. 9.

———— *et al.* (1966) *Bragantia* 25, Nota 13, p. LIX.

Cary, P.R. (1971a) *Farmers Newsletter* (CSIRO) 112, p. 10.

———— (1971b) *Farmers Newsletter* (CSIRO) 112, p. 25.

Casem, E.O. (1964) *Paper at P.S.A.E. Convention*, Manila (cited in IRRI Rice Prod. Manual).

Cassidy, N.G. (1968) *Trop. Agr.* (Trinidad) 45, p. 247.

Cassin, J. (1967) *Agrumes Presse* 31, p. 1.

Castro, V.G. and Cortez, R.G. (1965) *Philippine Sugar Inst. Quart.* 11, p. 21.

Catling, H.D. (1970) *FAO Plant Proct. Bull.* 18, p. 8.

Chadah, Y.R. (1962) *Trop. Sci.* 4(1).

Chalker, F.C. and Turner, D.W. (1969) *Agr. Gaz.* (NSW) 80, p. 474.

Chambers, G.M. (1970) *World Crops* 22, p. 80.

Champion, J. (1961) *Fruits, France* 16, p. 191.

———— (1970) *Fruits, France* 25, p. 369.

———— and Mounet, J. (1962) *Fruits D'outre Mer.* 17, p. 147.

Chan, S.K. (1970) In: *Crop Diversification in Malaysia*, Eds. E.K. and J.W. Bonkers, Malaysia.

Chandler, R.F. (1963) *IRRI Newsletter* 41, p. 4.

Chandler, W.H. (1958) *Evergreen Orchards*, Lea and Febiger, Philadelphia.

Chapman, T. and Cowling, D.J. (1965) *Trop. Agr.* 42, p. 189.

Charles, A.E. (1965) *Papua New Guinea Agr. J.* 17, p. 65.

———— (1969) *FAO Report* PL: CNP/68/19, p. 1.

———— and Douglas, L.A. (1965) *Papua New Guinea Agr. J.* 17, p. 76.

Charrier, A. (1971) *Café Cacao Thé* 15, p. 181.

Cheesman, E.E. (1948) *Kew Bull.*, p. 145.

Chen, R.A. (1966) *Trop. Agr.* (Trinidad) 43, p. 211.

Chenery, E.M. (1955) *Plant and Soil* 6, p. 174.

Chew, W.Y. (1970) *Malaysian Agr. J.* 47(4), p. 483.

Child, R. (1964) *Coconuts*, Longmans.

Childs, A.H.B. (1961) *Cassava Bull. No. 15.* Min. of Agr., Tanganyika.

Choussey, F. (1963) *Agricultura el Salvidor* 4, p. 9, 11.

Christoi, R. and Souza, F.E. de (1966) *Bol. Recursos Nature Sudena* 4, p. 23.

Chuah, S.E. (1967) *J. Rubber Res. Inst.* (Malaya) 20, p. 100.

Church, C.G. (1933) *California Citrogr.* 18, p. 348.

Clark, C. and Haswell, M.R. (1964) *The Economics of Subsistence Agriculture*, London, Macmillan.

Clements, H.F. (1940) *Hawaiian Plant. Rec.* 44, p. 201.

———— (1943) *Hawaiian Plant. Rec.* 47, p. 257 (from Dillewijn, *loc. cit.*).

————, Martin, J.P. and Moriguchi, S. (1941) *Hawaiian Plant. Rec.* 45, p. 227 (from Dillewijn, *loc. cit.*).

Clements, R.F. (1968) In: *Weed Control Handbook I.*, Eds. J.D. Fryer & S.A. Evan, London, Blackwell Scientific Publication.

Cleves, S. (1970) *Biol. Techn. Oficinia Cafe, Costa Rica* 3, p. 1.

Cobley, L.S. (1956) *An Introduction to the Botany of Tropical Crops*, Longmans.

Cody, R.S. (1969) *Hacienda* 64(2), p. 44.

Cohen, A. (1967) *Israel J. Agr. Res.* 17(1), p. 29.

Collins, J.L. (1933) *Cytologia* 4, p. 248.

———— (1968) *The Pineapple*, London, Leonard Hill.

Colonna, J.P. (1970) *Cah. O.R.S.T.O.K.* (Ser. Biol.) 13, p. 67.

Comas, J.O. (1966) *Bull. Doc.* 45, p. 31.

Commoner, B. *et al.* (1943) *Amer. J. Bot.* 30, p. 23.

Compagnon, P. and Tixier, P. (1950) *Rev. Gen. Caoutch.* 27, p. 525 (cited by Abraham *et al.* 1967a).

Cook, O.F. (1901) *Contrib. U.S. Nat. Herb.* 7, p. 237.

———— (1910) *Contrib. U.S. Nat. Herb.* 14, p. 271.

Cook, V.A. (1966) *Cocoa Growers Bull.* 7, p. 20.

Coomans, P. (1971) *Oléagineux* 26, p. 295.

Coomber, B.B. and Kaye, G.R. (1971) *Farmers Newsletter* (CSIRO) 110, p. 16.

Cooper, H.R. (1931) *Ind. Tea Assn. Quart. Jour.*, p. 120.

Corley, R.H.V. *et al.* (1971) *Exp. Agr.* 7, p. 129.

———— *et al.* (1972) In: *International Oil Palm Conference*, Malaysia.

———— et al. (1973) *Euphytica* 22, p. 48.

Cornelison, A.H. and Humbert, R.P. (1960) *Hawaiian Plant. Rec.* 55, p. 331.

Corns, J.B. (1966) *Proc. Trop. Region Amer. Soc. Hort. Sci.* 9, p. 16.

Coste, R. (1956) *Les cafeiers et les cafes dans le monde*, Paris.

Cours, G. (1951) *Mem. Inst. Sci. Madagascar* 3, p. 1.

Courtois, G. (1968) *Oléagineux* 23, p. 641.

Coussement, S. *et al.* (1970) *Café Cacao Thé* 14, p. 105.

Cox, J.E. (1967) *Agr. Gaz.* (NSW) 78(1), p. 26.

Cramer (1926)—cited by Dijkman, 1951.

Crope, S.M.S. *et al.* (1970) *Rev. Ceres* 17, p. 217.

Crosbie, A.J. and Brammer, H. (1957) *Cocoa Confr.* (London), p. 266.

Croucher, H.H. and Mitchell, W.K. (1940) *Dep. Sci. Agri. Jamaica*, Bull. 19, p. 30.

Cruickshank, A.M. (1970) *Cocoa Growers Bull.* 15, p. 4.

———— and Murray, D.B. (1966) *Ann. Rep. Cacao Res.*, p. 23.

Cunningham, R.K. *et al.* (1961) *J. Hort. Sci.* 36, p. 116.

Cutuli, G. (1968) *Potash Rev.* 8/19, Feb., p. 1.

Dagg, M.A. (1970) *Agr. Metrology* 7, p. 303.

———— (1971) *Kenya Coffee* 36, p. 149.

Daito, H. and Hiros, K. (1970) *Bull. Hort. Res. Sta.* (Ser. B.) 10, p. 35.

Dakwa, J.T. (1966/7) *Ann. Rep. Cocoa Res. Inst.* (Ghana), p. 33.

Daniel, P.M. (1951) *Tea Quarterly* 21, p. 1.

Daniels, J. and Krishnamurthi, M. (1965) *Internat. Sug. J.* 67, p. 195.

Dantur, N.C. and Aso, P.J. (1969) *Rev. Ind. Agricola Tucuman* 46, p. 97.

Das, N. and Baruah, A. (1967) *Ind. J. Agr. Sci.* 37, p. 27.

Das, V.K. (1931) *Repts. Assn. Hawaiian Sug. Tech.* 15, p. 43.

———— (1936) *Hawaiian Plant Rec.* 40, p. 103.

Daudin, J. (1955) *IFAC Bull.* 13, p. 55.

Davis, R.M. (1966) *Proc. Caribbean Region. Amer. Soc. Hort. Sci.* 9, p. 188.

Davis, T.A. (1961) *Nature* (London) 192, p. 277.

———— (1968) *Ceylon Coconut Quart.* 19, p. 116.

———— (1969a) *World Crops* 21, p. 253.

———— (1969b) *Ceylon Coconut Planters Rev.* 6, p. 1.

———— and Pillai, N.G. (1966) *Oléagineux* 21,
p. 669.

Davies, W.N.L. *et al.* (1966) *Ann. Rep. Tate and Lyle Centr. Agr. Res. Sta.* 1965, p. 199.

DeLange, J.H. and Vincent, A.P. (1970) *Agr. Sci. S. Afr. Agro-plantae* 2(4), p. 121.

Delorme, M. (1966) *Oléagineux* 21, p. 1.

Delvaux, R. (1967) *Oléagineux* 22, p. 75.

Dennett, J.H. (1932) *Malayan Agr. J.* 20, p. 518.

Deuss, J. (1969) *Café Cacao Thé* 13, p. 283.

———— (1971) *Café Cacao Thé* 15, p. 115.

Dickson, A.G. and Samuels, E.W. (1956) *J. Arnd. Arbor.* 37, p. 307.

Dierendonck, F.J.E. van (1959) *The Manuring of Coffee, Cocoa, Tea and Tobacco*, Geneva.

Dijkman, M.J. (1951) *Hevea.* U. of Miami Press.

Dillewijn, C. van (1952) *The Botany of Sugar Cane.* Chron. Botanica, USA.

Dodson, P.G.C. (1968) *Exp. Agr.* 4, p. 103.

Doku, E.V. (1969) *Tapioca in Ghana*, Ghana U. Press.

Donadio, L.C. and Moreiro, C.S. (1969) *Rev. Agr. Brasil* 44(4), p. 123.

Donald, C.M. (1961) *Symp. Soc. Exp. Biol.* No. 15.

Dore, J. (1960) *Malayan Agr. J.* 43, p. 28.

Douglas, L.A. (1965) *Papua New Guinea Agr. J.* 17, p. 87.

Drescher, R.W. (1970) *Ser. Techn. Estac. Exp. Agropec. Concordia, INTA* 36, p. 1.

D'Souza, G.I. (1971) *Turr. Rev. Interam. Ciencias Agr.* 21, p. 146.

DuCharme, E.P. (1968) In: *Trop. Nematology*, Eds. G.C. Smart & V.G. Parry, p. 20.

Duff, A.D.S. (1968) *Oléagineux* 23, p. 311.

Dunsmore, J.R. (1957) *Malayan Agri. J.* 40, p. 159.

Dutta, K.N. (1968) *Two and a Bud* 15, p. 1.

Dutta, S.K. (1967) *Two and a Bud* 14, p. 86.

———— (1968) *Investors Guardian* 211, p. 698.

———— and Grice, W.J. (1966) *Two and a Bud* 13, p. 70.

———— and Sharma, K.N. (1967) *Two and a Bud* 14, p. 76.

Eden, T. (1940) *Emp. J. Exp. Agr.* 8, p. 269.

———— (1965) *Tea.* 2nd Edition, Longmans.

Edwards, D.F. (1969) *Cocoa Growers Bull.* 13, p. 14.

Eid, S.A. (1972) *Plt. Disease Rep.* (USDA) 56, p. 318.

Ekandem, M.J. (1964) *Fed. Dep. Agr. Res. Memo. Nigeria*, No. 55.

Ekern, P.C. (1965) *Plant Physiol.* 40, p. 736.

El-Nokrashy, M.A. *et al.* (1966) *Agr. Res. Rev.* (UAR) 44(3), p. 26.

El-Wakeel, A.T. and Soliman, A.F. (1969) *Agr. Res. Rev.* (UAR) 47(2), p. 61.

El-Zeftawi, B.M. (1970) *J. Aust. Inst. Agr. Sci.* 36, p. 139.

———— (1971) *J. Aust. Inst. Agr. Sci.* 37, p. 151.

Enyi, B.A.C. (1970) *Beitr. Trop. Sub-Trop. Landwirtsh U. Tropenveterinarmed* 8(1), p. 71.

Espinosa, F.M. (1969) *Agricultura el Salvador* 9, p. 61.

———— (1970) *Cafe Nicaragua* 223, p. 3.

Ettori, O.J. (1970) *Agricultura em Sao Paulo* 17, p. 43.

Evans, H. (1934) *Sugar Cane Res. Sta., Mauritius, Bull.* 5.

———— (1935a) *Emp. Jour. Exp. Agric.* 3, p. 351.

———— (1935b) *Sugar Cane Res. Sta., Mauritius, Bull.* 7, p. 36.

———— and Wieche, P.O. (1947) *Sugar Cane Res. Sta., Mauritius Bull.* No. 19, p. 36.

Evans, L.T., Wardlaw, L.F. and Williams, C.N. (1964) In: *Grasses & Grassland*, London, Macmillan.

Evans, W.F. *et al.* (1971) In: *3rd Asian Pacific Weed Confr.*, Malaysia.

Evatt, N.S. (1958) *Texas Agr. Exp. Sta. Prog. Rep.* 2006.

Ewart, G.Y. (1949) *Hawaii Sug. Tech.* 8, p. 6.

Fagan, H.L. (1971) *Bull. Citrus Res.* 18, p. 1.

Fahn, A. *et al.* (1961) *Bot. Gaz.* 123, p. 116.

FAO of United Nations (1964) *Production Yearbook* 18.

Far East Trade and Dev. 24(4), 238.

Fawcett, W. (1913) *The Banana*, London.

Fennah, R.G. and Murray, D.B. (1957) *Cocoa Confr.* (London), p. 222.

Fenwick, D.W. (1968) *Ann. Rep. Res. Sta. Trin. & Tobago Coconut Res. Ltd.*, p. 37.

———— and Maharaj, S. (1967) *J. Agr. Soc.* (*Trin. & Tobago*) 68, p. 60.

Fernandez Caldas, E. and Garcia, V. (1972) *Fruits, France* 27, p. 509.

Fernie, L.M. (1964) In: *Handbook of Arabica Coffee in Tanganyika*, Tanganyika Coffee Board.

———— (1970) *Kenya Coffee* 35, p. 49.

Fernando, L.H. (1969) *Tea Quart.* (Ceylon) 40, p. 53.

Fernando, P.M. and Tambiah, M.S. (1970) *J. Rubb. Res. Inst.* (*Ceylon*) 46, pp. 69 & 88.

Fernando, D.M. and Silva, M.S.C. (1971) *Quart. J. Rubb. Res. Inst.* (*Ceylon*) 48, p. 19.

Fernando, T.M. (1971) *Quart. J. Rubb. Res. Inst.* (*Ceylon*) 48, p. 100.

Ferreira, J.J. (1970) *Rev. Fac. Agron. Vet.* (Argentina) 18(1), p. 59.

Feucht, R.W. and Arancibia, L.M. (1967) *Bol. Tech. Estac. Exp. Agron.* U. Chile.

Figueroa, Z.R. (1970) *Invest. Agropec Peru* 1, p. 43.

Fischer, R.A. (1923) *Phil. Trans. Roy. Soc. B.* 213, p. 89.

Fletcher, W.W. *et al.* (1968) In: *Weed Control Handbook I*, Eds. J.D. Fryer & S.A. Evans, London, Blackwell Scientific Publication.

Foale, M.A. (1965) *Oléagineux* 20, p. 585.

———— (1967) *Ceylon Coconut Quart.* 18, p. 31.

———— (1968a) *Aust. J. Agr. Res.* 19, p. 927.

———— (1968b) *Oléagineux* 23, p. 721.

———— (1968c) *Aust. J. Agr. Res.* 19, p. 781.

———— (1969) *S. Pac. Bull.* 19, p. 23.

Fogliata, F.A. (1965) *Rev. Industr. Agricola Tucuman* 43, p. 1.

———— (1965a) *Rev. Industr. Agricola Tucuman* 43, p. 25.

Foguet, J.L. (1967) *Bot. Estac. Exp. Agr. Prov. Tucuman* 106, p. 1.

Fordham, R. (1971) *Exp. Agr.* 7, p. 171.

Forsyth, J. (1969) *Farmers Newsletter* (CSIRO) 103, p. 6.

Foster, D.H. (1969) *Proc. Qld. Soc. Sug. Cane Technol.*, 36th confr., p. 21.

Franciosi, T.R. and Ponce, A.M. (1970) *Proc. Trop. Region Amer. Soc. Hort. Sci.* 14, p. 101.

Franco, C.M. and Mardes, C.H. (1953) *Biol. Superintend. Serv. Cafe* 28, p. 318.

Frederico, D. and Maestri, M. (1970) *Rev. Cores.* 17, p. 171.

Freiberg, S.R. (1965) In: *Fruit Nutrition*, Ed. Childers., N.Y.

Fremond, Y. and Nuce de Lamothe, M. de (1968) *Oléagineux* 23, p. 93.

———— and Nuce de Lamothe, M. de (1971) *Oléagineux* 26, p. 459.

———— and Brunin, C. (1966) *Oléagineux* 21, p. 213, p. 361.

———— and Goncalves, A.J.L. (1967) *Oléagineux* 22, p. 601.

———— and Orgias, A. (1952) *Oléagineux* 7, p. 345.

———— and Ouvrier, M. (1971) *Oléagineux* 26, p. 609.

Frey-Wyssling, A. (1932) *Arch. v.d. Rubbercult in Ned. Indie.* 16, p. 241 (cited by Dijkman, 1951).

———— (1933) *Ber. Schweiz. Bot. Geo.* v. 42

(cited by Dijkman, 1951).

Gaastra, P. (1963) In: *Environmental Control of Plant Growth*, Ed. L.T. Evans, N.Y. Acad. Press.

Gaillard, J.P. (1969) *Fruits, France* 24, p. 75.

———— (1970) *Fruits, France* 25, p. 11.

Gallo, J.R. *et al.* (1966) *Bragantia* 25(7), p. 77.

———— *et al.* (1970) *Bragantia* 29, p. 237.

Gangwar, R.P. and Singh, S.N. (1965) *Trop. Agr.* (Ceylon) 121, p. 55.

Garcia, A.N. (1969) *Agronimia Venezuela* 11, p. 45.

Garcia-Benavides, J. (1971) *Agron. Trop.* (Venezuela) 21(2), p. 77.

Garcia, V. (1968) *Fruits, France* 23, p. 480.

Gascon, J.P. *et al.* (1965) In: *The Oil Palm*, London, Trop. Prod. Inst.

Gaspar, A.M. and Diniz, A.C. (1965) *Agron. Angolana* 22, p. 121.

Gavande, S.A. (1968) *Cacao Inst. Interam. Ciencias Agr.* 13, p. 5.

Geerts, J.M. (1917) *Plantkunde van het suikerriet* Weltevreden, p. 151 (from Dillewijn, *loc. cit.*).

Gestin, A.J. (1971) *Café Cacao Thé* 15, p. 13.

Ghosh, S.K. *et al.* (1971) *Current Sci.* 40, p. 299.

Ghosh, T.K. (1968) *Two and a Bud* 15, p. 3.

Gilbert, S.M. (1938) *Trop. Agr.* (Trinidad) 15, p. 52.

Glasziou, K.T. *et al.* (1960) *Amer. J. Bot.* 47, p. 743.

Glendenning, D.R. (1966) *Euphytica* 15, p. 116.

———— (1967) *Euphytica* 16, p. 76.

Glover, P.M. (1949) *Ind. Tea Assn.*, Memorandum No. 21.

Godfrey-Sam Aggrey, W. (1969) *World Crops* 21, p. 33.

Godoy, H.J.D. *et al.* (1968) *Agricultura Tecn.* (Chile) 28(4), p. 149.

———— *et al.* (1969) *Agricultura Tecn.* (Chile) 29(1), p. 9.

Golding, F.D. (1936) *Bull. Agr. Dept. Nigeria* No. 11, p. 1.

Goncalves, J.R.C. (1968) *Trop. Agr. Trinidad* 45, p. 331.

Gosnell, J.M. and Lang, A.C. (1969) *Proc. Ann. Congr. S. Afr. Sugar Technical Assn.* 43, p. 26.

Goujon, M. (1971) *Café Cacao Thé* 15, p. 308.

Granada, C.G.A. and Sanchez, P.A. (1969) *Acta Agron.* 19, p. 107.

Grant, C.J. (1964) In: *Mineral Nutrition of the Rice Plant*, p. 15, Johns Hopkins Press.

Grant, P.M. and Shaxson, T.F. (1970) *Trop. Agr.* (Trinidad) 47, p. 31.

Greber, R.S. (1966) *Queensland J. Agr. Anim. Sci.* 23(2), p. 147.

Green, G.C. and Kuhne, F.A. (1969) *Agr. Sci S. Afr., Agroplantae* 1, p. 157.

Green, M.J. (1968) *Tea Jour. Tea Boards of E. Afr.* 9, p. 22.

———— (1970) *J. Tea Boards of E. Afr.* 10, p. 11.

———— (1971) *J. Agr. Sci. Camb., Repr.* 76, p. 143.

Green, R. (1970) *Q. newsletter*, Malawi, Tea Res. Foundation.

Green, G.C. (1963) *Tech. Note World Met. Organ.* 53, p. 113.

Greenwood, J.M.F. and Promdej, S. (1971) *Rep. Rubb. Res. Centre, Thailand* 19, p. 27.

Greenwood, M. and Posnette, A.F. (1950) *J. Hort. Sci.* 25, p. 164.

Grey, B.S. (1965) *The Requirement for Assisted Pollination in Malaya*, 2nd Oil Palm Confr. (Malaya).

Grice, D.S.G. and Proudman, G.D. (1968) *Farming S. Afr.* 44, p. 13.

Grice, W.J. (1968) *Two and a Bud* 15, p. 61.

Griffiths, E. (1969) *Span* 12, p. 92.

Grist, D.H. (1955) *Rice*, London, Longmans.

Groszman, H.M. (1948) *Qld. Agr. J.* 67, p. 78.

Guerout, (1969) *Fruits, France* 24, p. 436.

———— (1969a) *Fruits, France* 24(6), p. 325.

Guidian, R. (1970) *Agricultor Costarricense* 28, p. 136.

Gutiérrez, Z.G. and Jiménez, Q.M.F. (1970) *Rev. Cafetalera* 98, p. 35.

Haarer, A.E. (1962) *Modern Coffee Production*, London, Leonard Hill.

———— (1963) *Coffee Growing*, Oxford U. Press.

Hadfield, W. (1968) *Nature* (London) 219, p. 282.

Hainsworth, E.A. (1969) *Tea Jour. of Tea Boards of E. Afr.* 10, p. 14.

Halais, P. (1935) *Rev. Agric. Maurice* No. 80, p. 44.

———— (1950) *Proc. Int. Soc. Sug. Cane Tech.*, 7th Congr., p. 218.

Halma, F.F. (1934) *J. Pomol. and Hort. Sci.* 12, p. 99.

Ham, J. (1940) Plantverband en uitdunning. In: *Dictaat van den cursus over de rubbercultuur*, Buitenzorg, Den Dienst van den Landbouw, p. 91 (cited by Dijkman, 1951).

Hamilton, R. (1969) *Misc. Publ. Co-op. Ext. Serv.* (Univ. Hawaii) 64, p. 25.

Hammond, P.S. (1962) In: *Agriculture and Land Use in Ghana*, Oxford U. Press.

Hampden, C. (1973) *Industr. Britain* 14, p. 5.

Hanna, A.D. (1967) *PANS.*, Sect. A. 13, p. 214.

———— (1969) *PANS.*, Sect. A. 15, p. 340.

Hardon, J.J. (1969) *Progress in Oil Palm*, p. 13.

———— (1970) *Oléagineux* 25, p. 449.

———— *et al.* (1969) *Exp. Agr.* 5, p. 25.

———— *et al.* (1972) *Euphytica* 21, p. 257.

———— and Turner, P.D. (1967) *Exp. Agr.* 3, p. 105.

Hardy, M. (1967) *Ann. Rep. Mauritius Sug. Ind. Res. Inst.* 1966, p. 95.

Harler, C.R. (1966) *Investors Guardian* 208, p. 784.

———— (1971) *World Crops* 23, p. 275.

Harries, H.C. (1970) *Oléagineux* 25, p. 927.

———— *et al.* (1969) *FAO paper* PL/CNP/68/33, p. 1.

Hartley, C.W.S. (1967) *The Oil Palm*, Longmans.

Hartley, R. (1969) *Tea Jour. Tea Boards of E. Afr.* 10, p. 13, 15, 17.

Hasida, T. (1969) In: *Weed Control Basic to Agricultural Development*, p. 107.

Have, H. Ten (1967) *Research on Breeding for Mechanical Culture of Rice in Surinam*, Pudoc.

Hatch, M.D. (1971) In: *12th Pacific Science Congress*, Canberra.

Havord, G. and Hardy, F. (1957) *Cocoa Confr.* (Lond.), p. 272.

Henderson, T.J. *et al.* (1970) *Report of Atmospherics Inc., California, to Natural Science Foundation,* Washington.

Hernandez Medina, E. and Lugo Lopez, M.A. (1969) *J. Agr. U. Puerto Rico* 53, p. 33.

Hilton, R.N. (1963) *Maladies of Hevea*, Rubber Res. Inst., Malaysia.

Himme, M. and Petit, J. (1957) *Cocoa Confr.* (Lond.), p. 227.

Hirose, K. (1970) *Jap. Agr. Res. Quart.* 5, p. 31.

———— *et al.* (1970) *Bull. Hort. Res. Sta.* (Ser. B.) 10, p. 17.

Hislop, E.C. (1964) *Cocoa Growers Bull.* 2, p. 4.

Ho, C.T. (1969) *Fertilité* 33, p. 19.

———— (1970) *Soils and Ferts.*, (Taiwan) 1969, p. 57.

Hofmeyer, J.A.J. (1967) *Agron. Trop.* (Venezuela) 17(4), p. 345.

Holland, D.A. (1967) *Oléagineux* 22, p. 307.

Holly, K. and Steele, B. (1968) In: *Weed Control Handbook I.*, Eds. J.D. Fryer & S.A. Evans, London, Blackwell Scientific Publication.

Holmes, E. (1930) *Trop. Agr.* (Trin.) 7, p. 320.

Homes, M.V. (1957) *Cocoa Confr.* (Lond.), p. 257.

Hosein, I. (1967) *Bull. Citrus Res.* (Trinidad) 7, p. 1.

———— (1969) *Bull. Citrus Res.* 15, p. 1.

Houtman, P.W. (1916) *Archief Suikerind Ned. Indie* 24, p. 637 (from Dillewijn, *loc. cit.*).

Hoyle, J.C. (1968) *J. Agr. Soc.* (Trin. & Tobago) 68, p. 300.

———— (1971) *Exp. Agr.* 7, p. 1.

Huertao, A.S. (1940) *Philippine Agr.* 45, p. 258.

Humbert, R.P. (1968) The Growing of Sugar Cane, N.Y. & Lond., Elsevier Co. Amst.

Hunter, J.E. (1969) *Hawaii Farm Sci.* 18(2), p. 4.

Hurov, H.R. (1960) *Proc. Nat. Rubber Res. Confr.* (Malaysia), p. 419.

Hurtado, M.J.R. and Nunez, G.R. (1970) *Acta Agronomica* (Colombia) 20, p. 179.

Huxley, P.A. (1970) *Span* 13, p. 24.

———— and Cannell, M.G.R. (1970) *Kenya Coffee* 35, p. 176.

———— and Ismail, S.A.H. (1970) *Kenya Coffee* 35, p. 76, p. 87.

Ikegaya, K. (1970) *Japan Agr. Res. Quart.* 5, p. 38.

Inada, K. (1967) *Jap. Agr. Res. Quart.* 2, p. 6.

Inoue, H. (1971) *Gakuzyutu Hokoku* (*Tech. Bull. Fac. Agr., Kagawa U.*) 22, p. 83.

———— *et al.* (1971) *Gakuzyutu Hokoku* (Tech. Bull. Fac. Agr., Kagawa U.) 23(1), p. 27.

Ippei Suzuki (1971) In: *3rd Asian Pacific Weed Confr.*, Malaysia.

Ireta Ojeda, A. (1969) *Agron Sinaloa* 3(3), p. 8.

Ishiguka, Y. and Tanaka, A. (1961) *Working Party on Rice Soils No. 8*, Paper No. 29 (cited by Moomaw & Vergara, *loc. cit.*).

Ito, P.J. *et al.* (1968) *Proc. Trop. Region Amer. Soc. Hort. Sci.*

Iwamasa, M. (1967) *Jap. Agr. Res. Quart.* 2(3), p. 19.

Jagoe, R.B. (1952) *Molog. Agric. J.* 35, p. 12.

Janardhan, K.V. *et al.* (1971) *Ind. Coffee* 35, p. 145.

Jaunet, J.P. (1968) *Oléagineux* 23, p. 243.

Jayasekera, N.E.M. and Senanayake, Y.A.D (1971) *J. Rubb. Res. Inst.* (Ceylon) 48, p. 66.

Jennings, D.L. (1960) *Emp. J. Exp. Agr.* 28, p. 261.

———— (1970) In: *Proc. 2nd Internat. Symp. on Tropical Root Crops*, Hawaii (2), p. 64.

Jensen, C.A. *et al.* (1927) *California Citrogr.* 12(234), p. 266.

Jesinger, R., Rashid, S. and Tajuddin, S. (1971) In: *3rd Asian Pacific Weed Confr.*, Malaysia.

Jiménez-Saénz, E. (1967) *Turrialba Rev. Interam, Ciencias Agr.*

Jonge de, P. (1969) *J. Rubber Res. Inst.* (Malaya) 21, p. 283.

Jordan, D. and Opoku, A.A. (1966) *Trop. Agr.*

(Trinidad) 43, p. 135.

Joseph, K.T. (1971) *The Planter* 47(538), p. 7.

Joshi, M.C. *et al.* (1965) *Bot. Gaz.* 126, p. 174.

Kalma, J.D. (1968) *Israel J. Agr. Res.* 18(1), p. 20.

Kanapathy, K. (1958) *Malayan Agr. J.* 41, p. 18.

Kasasian, L. (1965) *Cocoa Growers Bull.* 5, p. 10.

————— (1967) *Bull. Citrus Res.* 3, p. 1.

Kasasian, L. and Seeyave, J. (1968) *Proc. 9th. Brit. Weed Control Confr.* 2, p. 768.

————— and Smith, R.W. (1968) In: *Techn. meeting on herbicides, Coconut Industry Board*, p. 20.

Katsuo, K. (1969) *Japan Agr. Res. Quart.* 4, p. 28.

Kawaguchi, K. (1966) *Jap. Agr. Res. Quart.* 1, p. 7.

Kawamura, A. *et al.* (1970) *Bull. Shikoku Agr. Exp. Sta.* 22, p. 1.

Kaye, G.R. (1970) *Farmers Newsletter* (CSIRO) 106, p. 23.

Kelang, G.P. and Haaren, A.J.H. van (1967) *Papua New Guinea Agr. J.* 19, p. 72.

Kenji Noda (1971) In: *3rd Asian Pacific Weed Confr.*, Malaysia.

————— and Keio Ozawa (1971) In: *3rd Asian Pacific Weed Confr.*, Malaysia.

Kentin, R.H. and Asante, E.A.T. (1969) *Rep. Cocoa Res. Inst.* (Ghana), p. 100, p. 102.

Kerns, K.R. and Collins, J.L. (1947) *J. Hered.* 38, p. 322.

————— *et al.* (1936) *New Phytol.* 35, p. 305 (cited by Collins, 1968).

Kerr, A. (1966) *Tea Quart.* (Ceylon) 37, p. 168.

Khan, N.A. (1967) *Pakistan Rev.* 15(3), p. 27.

Khanna, S.S. *et al.* (1969) *Ind. J. Agr. Sci.* 39(8), p. 830.

Khosla, D.V. (1965) *Foreign Agr., V.S. Dep. Agr.* 3, p. 8.

Kiely, T.B. *et al.* (1972) *Agr. Gaz. NSW.* 83(1), p. 20.

King, N.J., Mungomery, R.W. and Hughes, C.G. (1965) *Manual of cane Growing*, Elsevier, N.Y.

Ko, W.H. (1971) *Hawaii Farm Sci.* 20(4), p. 9.

Kobus, J.D. (1900) *Archief Java—Suikerind* 8, p. 453 (from Dillewijn, *loc. cit.*).

Kortleve, A. (1928) *Arch. v.d. Rubbercult in Ned. Indie* 12, p. 605.

Kotalawala, J. (1968) *Ceylon Coconut Planters' Rev.* 5, p. 112.

Kowal, J.M.L. (1959) *Emp. J. Exp. Agr.* 27.

Kralovic, J. (1967) *Rev. Agr. Cuba* 1(1), p. 53.

Kratochvil, F. (1956) *World Crops* 8, p. 146.

Krause, B.H. (1948/9) *Bot. Gaz.* 110, pp. 159, 333, 551.

Krekule, J. (1969) *Trop. Agr.* (Trinidad) 46, p. 69.

Krezdorn, A.H. (1966) *Proc. Caribb. Region Amer. Soc. Hort. Sci.* 9, p. 55.

————— (1970) *Proc. Trop. Region Amer. Soc. Hort. Sci.* 14, p. 34.

Krochmal, A. (1968) *World Crops* 20(3), p. 40.

————— and Samuels, G. (1967) In: *Proc. 1st Internat. Symp. on Trop. Root Crops* (Trinidad) 1(2), p. 97.

Kroon, A.H.J. (1970) *Trop. Abstracts* No. 633–855-36.

Kruger, N.S. and Menary, R.C. (1968) *Queensland Agr. J.* 94(4), p. 234.

Kuhne, F.A. and Allan, P. (1970) *Agr. Sci. S. Africa, Agroplantae* 2(3), p. 99.

Kulasegaram, S. and Kathiravetpillai, A. (1971) *Tea Quart.* (Ceylon) 42, p. 57.

Kunhi-Muliyar, M. and Nelliat, E.V. (1971) *Oléagineux* 26, p. 687.

Kurz, S. (1865) *J. Agric. Soc. India* 14, p. 295.

Lacoeuilhe, J.J. and Martin-Prevel, P. (1971) *Fruits, France*, 26, p. 161.

Laemmlen, F.F. and Aragaki, M. (1971) *Plt. Disease Reporter, USDA* 55(8), p. 743.

Lassoudiere, A. (1971) *Fruits, France*, 26, p. 501.

————— (1972) *Fruits, France* 27, p. 5.

————— and Maubert, P. (1971) *Fruits, France* 26, p. 321.

————— and Pinon, A. (1971) *Fruits, France* 26, p. 333.

Lavabre, E.M. (1971) In: *Symp. sur le desher bage dés cultures tropicales*, Antibes.

Laville, E. (1971) *Fruits, France* 26, p. 301.

Laycock, D.H. (1958) In: *Proc. 1st Meeting in Appl. Metrol.* (Nairobi).

————— (1961) *Trop. Agric.* 38, p. 195.

————— (1967) *Two and a Bud* 14, p. 16.

Leach, J.R., *et al.* (1971) *Cocoa Growers Bull.* 16, p. 21.

Leake, H.M. (1928) *Proc. Roy. Soc. Lond. B.* 103, p. 82.

————— (1929) *Trop. Agr.* 6, p. 43.

————— (1934) *Int. Sug. Jour.* 36, pp. 147, 183, 227.

Le Bourdelles, J. (1968) *Fruits, France* 23, p. 221.

Leffingwell, R.J. (1963) *Sugary Azucar* 58, p. 78 (cited by Salter & Goode, *loc. cit.*).

Leigh, D.S. *et al.* (1966) *Agr. Gaz.* (NSW) 77, p. 422.

————— (1969) *Agr. Gaz.* (NSW) 80, p. 6, p. 369.

Lelyveld, L. van (1957) Report on Official Visit to Principal Pineapple-Producing Countries of the

World, April–Aug., 1957 (unpublished, S. Afr. Dep. Agr., cited by Green, 1963).

Lent, R. (1966) *Turrialba Rev. Interam. Ciencias Agr.* 16, p. 352.

Liefstingh, G. (1966) *Cocoa Growers Bull.* 6, p. 12.

Light, de, N.M. (1937) *Bergcultures* 11, p. 1766.

Lin, C.F. (1965) *Soils & Ferts.* (Taiwan), p. 64.

Linford, M.B., King, N. and Magistad, O.C. (1934) *Pineapple Quart.* 4, p. 176.

List, G. (1969) *Kaffee & Tea Markt.* 19, p. 25.

Liyanage, D.V. (1953) *Ceylon Coconut Quart.* 4, p. 127.

———— (1955) *Rep. Coconut Conf.* (Ceylon), p. 11.

———— (1963) *Ceylon Coconut Quart.* 14, p. 89.

———— (1966) *Ceylon Coconut Planters' Rev.* 4, p. 27.

———— (1969) *Ceylon Coconut Quart.* 20, p. 161.

———— and Abeywardena, V. (1957) *Trop. Agr. Mag. Ceylon Agr. Soc.* 113, p. 1.

Lockard, R.G. (1959) *Min. of Agric.* (*Malaysia*) *Bull.* No. 108.

Loh, C.S. and Chen, K.C. (1948) *Jour. Sug. Cane Res.* 2, p. 1.

Lomas, J. and Gat, Z. (1967) *Agr. Meteorol.* 4(6), p. 415.

Longworth, J.F. (1963) *Cocoa Growers Bull.* 1, p. 17.

Lopez Hamandez, J.A. (1963) *Sugary Azucar* 58, p. 31.

Lopez, M.B. *et al.* (1965) *Philippine Sugar Inst. Quart.* 11, p. 86.

Lowe, J.S. (1971) In: *3rd Asian Pacific Weed Confr.*, Malaysia.

Maas, P.W. Th. and Imambaks, M.J. (1971) *Surinaamse Landb.* 19, p. 32.

———— and Wondenberg, J. van (1970) *Surinaamse Landb.* 18, p. 87.

Mabey, S.E. (1967) *E. Afr. Agr. For. J.* 33, p. 14.

Maestri, M. *et al.* (1970) *Rev. Ceres* 17, p. 227.

Magloblisvili, S.V. and Arobelidze, S.M. (1970) *Subtropiceskie Kul'tury* 5, p. 73.

Maitin, P.E. (1967) *Rhod. Agr. J.* 64(3), p. 61.

Malavolta, E. *et al.* (1955) *Plant. Phys.* 30, p. 81.

Maliphant, G.K. (1970) *Citrus Res.* (Trinidad), p. 21.

Manciot, R. Le (1968) *Oléagineux* 23, p. 169.

Mangelsdorf, A.F. (1950) *Econ. Bot.* 4, p. 150.

Mangoensoekardjo, S. and Nurdin (1970) *Perkebunan Medan* 1, p. 13.

Manica, I. and Anderson, O. (1969) *Rev. Ceres* 16(88), p. 121.

Manipura, W.B. (1971) *Tea Quart.* (Ceylon) 42, p. 30.

Mann, C. (1970) *Tea & Coffee Trade Jour.* 139, p. 30.

Manning, H.L. (1956) *Proc. Roy. Soc. B.* 144, p. 460.

Mao, Y.T. (1959) FAO publication No. 59/9/7101.

Marchal, J. and Lacoeuilhe, J.J. (1969) *Fruits, France* 24(6), p. 229.

Marques, M.A.P. (1971) *Rev. Agr., Mocambique* 136, p. 29.

Martin, F.W. (1970) In: *Proc. 2nd Internat. Symp. on Tropical Root Crops* (Hawaii) 1(2), p. 53.

Martin, J.P. (1938) *Exp. Sta. Hawaii Sug. Plant. Assoc.*, p. 295.

Martin, W.S. (1936) *Rep. Dep. Agr.* (Uganda), p. 53 (cited by Wellman, *loc. cit.*).

Martin–Prevel *et al.* (1966) *Fruits, France* 21, p. 577.

Martineaux, P.G. *et al.* (1969) In: *Prog. in Oil Palm*, p. 82.

Martinez, A.P. (1965) *FAO Plant Proct. Bull.* 13, p. 25.

Martinez, A.L. and Wallace, J.M. (1968) *Philippine J. Plant. Ind.* 33, p. 119.

Mascaro, M.A. (1962) *Sugary Azucar* 57, p. 34.

Matos, J.R. de (1970) *Solo, Centro, Acad.* (*Luiz de Queiroz*), *Brasil* 62, p. 51.

Matsubayashi, M. (1963) *Theory and Practice of Growing Rice*, Tokyo, Fuji Publishing C.

Matsushima, S. (1962) *Min. of Agri.* (*Malaya*) *Bull.* No. 112.

———— (1964) In: *Mineral Nutrition of the Rice Plant*, Johns Hopkins Press.

———— and Tsunoda, K. (1958) *Proc. Crop Sci. Soc. Japan* 26, p. 243.

May, E.B. (1956) *West Afr. Inst. Oil Palm Res. Jour.* 2(5), p. 6.

McClelland, T.B. (1917) *Rep. Puerto Rican Agr. Exp. Sta.* 22.

———— (1926) *Exp. Sta. Bull.* (Puerto Rico), No. 31.

McFadyean, A. (1944) *The History of Rubber Regulation*, London, George Allen & Unwin.

McKelvie, A.D. (1955) *Rep. W. Afr. Cocoa Res. Inst.* 1954–5, p. 89.

———— (1957) *Cocoa Confr.* (London), p. 51.

———— (1962) In: *Agriculture and Land Use in Ghana*, Ed. J.B. Wills, Oxford U. Press.

McMartin, A. (1946) *S. Afr. Sug. J.* 30, p. 71.

———— (1946a) *S. Afr. J. Sci.* 42, p. 122.

———— (1949) *Proc. S. Afr. Sug. Tech. Assn.* 23, p. 131.

Medina Ramirez, G. (1968) *Technologia* (Colombia) 10(54), p. 16.

Melin, Ph. and Marseault, J. (1972) *Fruits, France* 27, p. 495.

Mello, F.A.F. *et al.* (1966) *Anais Escola Super. Agr. U. Sao Paulo* 23, p. 95.

———— *et al.* (1969) *Rev. Agr.* (Brazil) 44, p. 103.

Menuel, R. and Gerard, Ph. (1967) *Oléagineux* 22, p. 235.

Micheletti deZerpa, D. (1967) *Agron. Trop.* (Venezuela) 17(4), p. 361.

Migvar, L. (1965) *Ag. Extension Circ. Trust Terr.* (Pacific Isl.) 6, p. 1.

Mikkelsen, D.S. *et al.* (1958) *Rice Fertilization California Agr. Esp. Sta. Bull.,* U. of California.

Milford, G.F.J. *et al.* (1969) *J. Rubber Res. Inst.* (Malaya) 21, p. 274.

Minessy, F.A. *et al.* (1970) In: *Proc. Confr. Subtrop. Fruits* (Lond.) 1969, p. 245.

———— *et al.* (1971) *Plant & Soil* 34(1), p. 1.

Mitchell, A.R. and Nicholson, M.E. (1965) *Qld. J. Agr. Anim. Sci.* 22(2), p. 409.

Mitchell, H.W. (1964) In: *Handbook of Arabica Coffee in Tanganyika,* Tanganyika Coffee Board.

Mitchell, H.W. (1969) *Kenya Coffee* 34, p. 130.

Mitchell, P. and Cannon, R.C. (1953) *Adv. Leafl. No. 286. Division,* Pl. Ind. Dep. Agric. Qld., pp. 1–36.

Mok, C.K. (1970) *C.R. Assoc. Int. Essais Semences* 35, p. 243.

Mollegaard, H. (1971) *Oléagineux* 26, p. 449.

Monaco, L.C. and Carvalho, A. (1969) *Span* 12, p. 74.

Mongelard, C. (1969) *Ann. Rep. Mauritius Sug. Ind. Res. Inst.* 1968, p. 83.

Montieth, J.L. (1965) *Exp. Agr.* 4, p. 241.

Montieth, N.H. (1966) *Trop. Agri.* (Trin.) 43, p. 75.

Moomaw, C. and Vergara, S. (1964) In: *Mineral Nutrition of the Rice Plant,* p. 3, John Hopkins Press.

Moreau, B. (1971) *Fruits, France* 26, p. 349.

Moreira, R. (1969) *Coopercotia* (Brazil) 26, p. 21.

Moreira, C.S. (1969) *Rev. Agr.* (Brazil) 44, p. 41.

———— *et al.* (1969) *Sole, Centro Acad. 'Luiz de Queiroz'* (Brazil) 61, p. 71.

Morgan, W.B. (1959) *J. Trop. Geog.* 13, p. 58.

Mortensen, E. and Bullard, E.T. (1964) *Handbook of Tropical and Sub-Tropical Horticulture,* Washington, D.C.

Mortimer, C.H. (1941) *J. Ecol.* 29, p. 280.

Moscoso, G.G. (1970) In: *Proc. Confr. Trop. Subtrop. Fruits* (Lond.) 1969, p. 193.

Mosqueda Vazquez, R. (1969) *Agr. Techn.* (Mexico) 2(11), p. 487.

Moss, D.N. (1962) *Nature* (London) 193, p. 587.

Moss, G.I. (1970a) *Aust. J. Agr. Res.* 21(2), p. 233.

———— (1970b) *Farmers Newsletter* 106, p. 17.

———— (1970c) *Phyton. Argentina* 27(2), p. 141.

———— (1971) *Aust. J. Agr. Res.* 22(4), p. 625.

———— (1972a) *Aust. J. Exp. Agr. Anim. Husbd.* 12(54), p. 96.

———— (1972b) *Aust. J. Exp. Agr. Anim. Husbd.* 12(55), p. 195.

Moy, E.B. (1956) *West Afr. Inst. for Oil Palm Res. Jour.* 2(5), p. 6.

Mulinge, S.K. (1971) *Kenya Coffee* 36, p. 9, p. 11.

Muller, H.R.A. (1931) *Bull. Inst. Plzeikt. Buiteny* 24, p. 1.

Mungomery, W.V. *et al.* (1966) *Queensland J. Agr. Anim. Sci.* 23(2), p. 103.

Murashige, T. and Nakano, R.T. (1966) *Proc. Caribbean Region Amer. Soc. Hort. Sci.* 9, p. 235.

Murray, D.B. (1964) *Cocoa Growers Bull.* 3, p. 8.

———— (1966) *Ann. Rep. Cacao Res.* p. 34.,

———— (1967) *Cocoa Growers Bull.* 9, p. 25.

———— (1967) *Confr. Int. Rev. Agron. Cacaoyeres* (Abidjan) 1965, p. 109.

———— and Maliphant, G.K. (1964) *Ann. Rep. Cocoa Res.,* p. 35.

Muthappa, B.N. (1970) *Ind. Coffee* 34, p. 263.

Nadir, M. (1965) *Al Awamina* 16, p. 123.

Nagai, I. (1959) *Japonica Rice, Its Breeding and Culture,* Tokyo, Yokendo.

Nambiar, M.C. *et al.* (1969) *Ind. J. Agr. Sci.* 39, p. 455.

Namekata, T. and Rossetti, V. (1969) *Biologico* (Brazil) 35, p. 289.

Nanayakkara, A. (1967) *Ann. Rep. Tea Res. Inst.* (Ceylon) 2, p. 53.

———— (1968) *Ann Rep. Tea Res. Inst.* (Ceylon) 2, p. 53.

Narayanan Potty, S. *et al.* (1968) *Rubb. Board Bull.,* India 10, p. 83.

Nasharty, A.H. *et al.* (1969) *Agr. Res. Rev. UAR* 47, p. 71.

Nathanael, W.R.N. (1966) *Ceylon Coconut Planters' Rev.* 4, p. 39.

———— (1969) *Fertilité* 35, p. 11.

Nature (1970) 227 (5260), p. 778.

Navasero, S.A. and Tanaka, A., (1966) *Plant and Soil* 24, p. 17.

Nayar, T.C. and Bakthavathsalu, C.M. (1955) *Ind. J. Hort.* 12, p. 22.

Naylor, A.G. (1966) *Proc. Caribbean Reg. Amer. Soc. Hort. Sci.* 9, p. 30.

Nelliat, E.V. (1968) *Ind. J. Agr. Sci.* 38, p. 737.

Nethsinghe, D.A. (1966a) *Ceylon Coconut Planters' Rev.* 4, p. 30.

————— (1966b) *Ceylon Coconut Planters' Rev.* 4, p. 55.

Newhall, A.G. and Diaz, F. (1967) *Cacao Inst. Interam. Ciencias Agr.* 12, p. 12.

Newing, G.G. (1971) *Tea Jour. of Tea Boards of E. Afr.* 12, p. 24.

Nforzato, R. *et al.* (1968) *Bragantia* 27, p. 135.

Ng, E.K. *et al.* (1969) *J. Rubber Res. Inst.* (Malaya) 21, p. 292.

Ng, K.T. and Goh, K.H. (1970) *Chemara Comm.* (*Malaysia*) *Limited Distrib.* No. 4.

Ng, S.K. (1970) *Oléagineux* 25, p. 637.

————— *et al.* (1969) *Malaysian Agr. J.*, p. 41.

————— Thamboo, S. and de Souza, P. (1965) *Malaysian Agr. J.* 46, p. 332.

————— and Thamboo, S. (1967) *Malaysian Agr. J.* 46, p. 3.

————— and Walters, E. (1969) In: *Progress in Oil Palm*, p. 67.

Nichols, R. (1964) *Ann. Bot.* 28, p. 619.

————— (1965a) *Ann. Bot.* 29, p. 181, p. 197.

————— (1965b) *Cocoa Growers Bull.* 4, p. 10.

————— and Walmsley, D. (1965) *Plant and Soil* 23, p. 149.

Nichols, R.F.W. (1947) *East Afr. Agr. J.* 12, p. 184.

Nieuwolt, S. (1965) *J. Trop. Geog.* 20, p. 34.

Nightingale, G.T. (1942) *Bot. Gaz.* 103, p. 409.

Ninan, C.A. *et al.* (1966) *Ind. Coconut J.* 18, p. 12.

Nizwar Hakim Harahap and Soerianegara, I. (1970) *Menara Perkebunan* 39, p. 47.

Nobuo Onodera and Takayuki Isokawa (1971) In: *3rd Asian Pacific Weed Confr.*, Malaysia.

Norman, G.G. (1966) *Proc. Caribbean Region Amer. Soc. Hort. Sci.* 9, p. 128.

Norman, P.A. *et al.* (1968) *J. Econ. Entomol.* 61, p. 238.

Norman, S. *et al.* (1971) *Market. Res. Rep.* USDA. 896, p. 1.

Normanha, E.E. and Pereira, A.S. (1949) In: *First Congr. Agronomic Res. in S. Amer.* (Uruguay) (cited by Krochmal & Samuels, *loc. cit.*).

Normanha, E.S. *et al.* (1968) *Bragantia* 27, p. 149

(from *Trop. Abstr.* 1970).

Northwood, P.J. (1970) *E. Afr. Agr. Forestry J.* 36(1), p. 45.

Nour-El-Din, F. and El-Bauna, M.T. (1967) *Agr. Res. Rev.* (UAR) 45, p. 107.

Nugent, V. *et al* (1967) *Trop. Sci.* 9, p. 182.

Nunes, M.A. *et al.* (1969) *Acta Bot. Neerlandica Rep.* 18, p. 420.

Nutman, F.J. (1970) *PANS* 16, p. 277.

————— and Roberts, F.M. (1969) *Café Cacao Thé* 13, pp. 131, 221.

————— and Roberts, F.M. (1971) *Inst. Pest Control* 13, p. 22.

Nyenhuis, E.M. (1965) *Farming S. Afr.* 40, p. 52.

————— (1967) *Farming S. Afr.* 43, p. 31.

————— (1967a) *Farming S. Afr.* 43, p. 51.

————— (1967b) *Farming S. Afr.* 43, p. 61.

Obi, J.K. (1967) *Publn. Comm. Tech. Co-oper. Afr.* 98, p. 434.

Ochs, R. (1965) *Oléagineux* 20, pp. 365, 433, 493.

Ochse, J.J. *et al.* (1961) *Trop. and Sub-trop. Agr.*, London, Macmillan.

Oguntoyinbo, J.S. (1966) *J. Trop. Geog.* 22, p. 38.

Okuda, A. and Takahashi, E. (1962) *J. Soil Sci. and Manure* (Japan) 33, p. 1.

Ollagnier, M. *et al.* (1970) *Fertilité* 36, p. 3.

————— and Gascon, J.P. (1965) In: *The Oil Palm*, Trop. Prod. Inst., p. 3.

————— and Martin, G. (1967) *Oléagineux* 22, p. 579.

————— and Ochs, R. (1971) *Oléagineux* 26, p. 1, p. 367.

Opake, L.K. and Toxopeus, H. (1967) *Nigerian Agr. J.* 4, p. 22.

Oppenheim, C. (1956) *Min. Agric. Israel, Spec. Bull.* 3, p. 60.

Osorio, B.J. and Castillo, Z. (1969) *Conicafé* 20, p. 20.

Othieno, C.O. (1968) *Tea Jour. Tea Boards of E. Afr.* 9, pp. 19, 21.

Ou, S.H. *et al.* (1965) *Plt. Disease Rep.* 49, p. 778.

Overbeek, J. van (1946) *Bot. Gaz.* 108, p. 64.

Owen-Jones, J.B. (1967) *The Planter* (Malaysia) 43, p. 95.

Paardekooper, E.C. (1965) *Planting Manual No. 11*, Rubber Res. Inst. (Malaya).

Panje, R.R. *et al.* (1965) *Ind. Sug. J.* 10, p. 1.

Pantastico, E.B. *et al.* (1968) *Proc. Trop. Region Amer. Soc. Hort. Sci.* 12, p. 171.

Papadakis, J. (1965) *Crop Ecology Survey of West Africa*, FAO.

Parham, J.W. (1968) *Oléagineux* 23, p. 381.

Park, M. (1928) *Trop. Agr.* 70, p. 171.

——— (1934) *Trop. Agr.* 83, p. 141.

Patel, J.S. (1938) *The Coconut: A Monograph*, Madras Govt. Press.

——— and Anandan, A.P. (1936) *Madras Agr. J.* 24, p. 5.

Pax, F. (1910) In: *Das Pflanzenreich*. Engler, A. Heft. 44, p. 21.

Pearsall, W.H. (1938) *J. Ecol.* 26, p. 180.

Pearse, T.L. and Gosnel, J.M. (1969) *Proc. Ann. Congr. S. Afr. Sug. Technol. Assoc.* 43, p. 50.

Peregrine, W.T.H. (1968) *E. Afr. Agr. and Forestry J.* 33, p. 316.

Pereira, A.L.G. and Zagatto, A.G. (1967) *Arq. Inst. Biol.* 34, p. 153 (from *Trop. Abstr.* 1968).

Pereira, H.C. and Jones, P.A. (1950) *East Afr. Agr. Jour.* 15, p. 174.

——— and Jones, P.A. (1954) *Field Responses by Kenya Coffee to Fertilizers, Manures and Mulches.*

Pethiyagoda, U. (1970) *Tea Quart.* (Ceylon) 41, p. 69.

Petinov, N.S. (1961) *Izv. Akad*, Nauk (USSR) Ser. Biol. No. 5.

Petrov, I.A. (1967) *Bietr. Trop. Sub-trop. Landwirtsch. U. Tropenveterinarmed* 5(2), p. 119.

Phelps, R.H. (1969) *Bull. Citrus Res.* 16, p. 1.

Phillips, A.L. (1970) *J. Agric. Univ. Puerto Rico* 54, p. 504.

——— and Rest, D.J. (1969) *Rev. Café Puerto Rico* 25, p. 27.

Pilot, J. (1968) *Rev. Agr. Sucr. Ile Maurice* 47, p. 29.

Planters Bull. Rubber Res. Inst. (Malaya) (1957), No. 28, p. 20.

——— Rubber Res. Inst. (Malaya) (1959), No. 43, p. 52.

——— Rubber Res. Inst. (Malaya) (1960), No. 47, p. 39.

——— Rubber Res. Inst. (Malaya) (1960), No. 50, p. 98.

——— Rubber Res. Inst. (Malaya) (1961), No. 57, p. 183.

——— Rubber Res. Inst. (Malaya) Conference (1964), No. 74.

——— Rubber Res. Inst. (Malaya) (1967), No. 91, p. 139.

——— Rubber Res. Inst. (Malaya) (1967), No. 91, p. 147.

Poedjowardojo, S. and Darjadi, N. (1969) *Menara Perkebunan* (Indonesia) 38, p. 14.

Pomier, M. (1967) *Tech. Paper. S. Pac. Comm.* 153, p. 1.

——— (1968) *Oléagineux* 23, p. 241.

——— (1969) *Oléagineux* 24, p. 13.

Poncin, L. and Bellefroid, de V. (1957) *Cocoa Confr.* (Lond.), p. 281.

Ponnamperuma, F.N. (1955a) *Trop. Agriculturist* 111, p. 92.

——— (1955b) Ph. D. Thesis, Cornell University.

——— (1964a) In: *Mineral Nutrition of the Rice Plant*, p. 295, John Hopkins Press.

——— (1964b) In: *Mineral Nutrition of the Rice Plant*, p. 461, John Hopkins Press.

Poon, Y.C. (1970) *Exp. Agr.* 6, p. 113.

Portmouth, G.B. (1957) *Tea Quarterly* 28, p. 8.

Posnette, A.F. and Entwistle, H.M. (1957) *Cocoa Confr.* (Lond.), p. 66.

Praedo, E.P. de and Miranda, E.R. de (1967) *Cacau Atualidades* 4, p. 48.

Pralain, J. (1969) *Oléagineux* 24, p. 395.

Praloran, J.C. (1968) *Fruits, France* 23(2), p. 107.

Prasanna, H.A. *et al.* (1969) *J. Food Sci. Technol. India* 6, p. 187.

Prevot, P. and Duchesne, J. (1955) *Oléagineux* 10, p. 117.

Price, W.C. *et al.* (1967) *FAO Plant Proct. Bull.* 15, p. 105.

Purseglove, J.W. (1968) *Trop. Sci.* 10, p. 191.

Purvis, C. (1956a) *West Afr. Inst. for Oil Palm Res. Jour.* 1(4), p. 60.

——— (1956b) *West Afr. Inst. for Oil Palm Res. Jour.* 1(1), p. 60.

Py, C. (1968a) *Fruits, France* 23, p. 139.

——— (1968b) *Fruits, France* 23, p. 3.

——— (1970) *Fruits, France* 25, p. 199.

——— *et al.* (1968) *Fruits, France* 23, p. 403.

——— and Guyot, A. (1970) *Fruits, France* 25, p. 253.

Quarterly-Papafio, E. (1964) *Cocoa Growers Bull.* 3, p. 12.

Radjino, A.J. (1969) *J. Rubber Res. Inst.* (Malaya) 21(1), p. 56.

Raghavan-Nair, K.N. (1970) *Planters' Chron.* 65, pp. 457, 477, 481.

Raheja, P.C. (1951) *Ind. J. Agr. Sci.* 21, p. 203.

Rajaratnam, J.A. (1971) Thesis. U. of Singapore.

——— and Williams, C.N. (1970) *The Planter* (Malaya) 46, p. 339.

Ramachandran, P. (1969) In: *Progress in Oil Palm,*

p. 192.

Ramaiah, P.K. and Vasudeva, N. (1969) *Turrialba, Rev. Interam. Ciencias Agr.* 19, p. 455.

Ramiah, K. (1954) *FAO Agri. Dev. Paper,* No. 45.

Ramirez, O.D. (1966) *J. Agri. Univ. Puerto Rico* 50, p. 231.

———— (1970) *Rev. Café Puerto Rico* 25, p. 29.

Ranger, J.B. *et al.* (1969) *Proc. Ann. Congr. S. Afr. Sugar Tech. Assn.* 43, p. 30.

Ratnam, T.C. (1968) *Ind. Farming* 17, p. 21.

Rayner, R.W. (1956) *Mon. Bull. Coffee Board* (Kenya) 21, pp. 65, 75.

Rees, A.R. (1958) *Nature* (Lond.) 182, p. 735.

———— (1959a) *West Afr. Inst. for Oil Palm Res. J.* 3(9), p. 76.

———— (1959b) *West Afr. Inst. for Oil Palm Res. J.* 3(9), p. 83.

———— (1962) *Ann. Bot.* 26, p. 569.

———— (1963) *Ann. Bot.* 27, p. 325.

———— and Tinker, P.B.H. (1963) *Plant & Soil* 19, p. 19.

Renard, J. (1970) *Oléagineux* 25, p. 581.

Reuther, W. and Rios-Castano, D. (1970) *Fruits, France* 25(5), p. 313.

Reyne, A. (1948) *De Landbouw in den Indischen Archipel.* IIA, p. 453.

Reynhardt, J.P.K. and Heerden, V.L. van (1968) *Farming in S. Afr.* 44, p. 57.

———— and Blommestein, J.A. van (1966) *Agr. Res. Dep. Agr. Techn. Serv. Rep. S. Afr.* 1, p. 112.

———— *et al.* (1967) *Farming in S. Afr.* 43, pp. 13, 15, 39.

———— (1970) *Agr. Res. Dep. Agr. Techn. Serv. S. Afr.* 1, p. 129.

Reys, C de *et al.* (1966) *Philippine Agriculturist* 50(4), p. 334.

Richards, A.V. (1966) *Tea Quart.* (Ceylon) 37, p. 154.

Rice Production Manual, Compiled by Rice Information Co-op. Effort (RICE) U. of Philippines and IRRI.

Rios-Castano, D. *et al.* (1968) *Proc. Trop. Region Amer. Soc. Hort. Sci.* 12(16), p. 154.

Ripley, E.A. (1967) *E. Afr. Agr. Forestry J.* 33, p. 67.

Rishbeth, J. (1957) *Ann. Bot.* 21, p. 215.

Rixon, A.J. and Sherman, G.D. (1962) *Soil Sci.* 94, p. 19.

Robertson, J.S. (1959) *West Afr. Inst. for Oil Palm Res. J.* 2(8), p. 310.

Robin, J. and Champion, J. (1962) *Fruits, France*

17, p. 93.

Robinson, J.B.D. (1964) In: *A Handbook of Arabica Coffee in Tanganyika,* Tanganyika Coffee Board.

———— (1969) *Exp. Agr.* 5, p. 311.

Robson, B.R. (1971) *J. Tea Boards Kenya, Uganda & Tanganyika* 12, pp. 33, 35.

Rocha, H.M. and Medieros, A.G. (1968) *Ciencias Agr.* 13, p. 10.

Rodha, K. and Lal, S.B. (1969) *FAO Techn. Working Party on Coconuts* PL: CNP/68/42, p. 1.

Rogers, D.J. (1963) *Bull. Torrey Bot. Club* 90, p. 43.

———— and Milner, (1963) *Econ. Bot.* 17, p. 261.

Rognon, F. (1971) *Oléagineux* 26, p. 307.

Rojas, C.V.J. and Zambrano, S.G. (1969) *Technologia* (Colombia) 11(58), p. 23.

Romney, D.A. (1968) In: *Techn. Meeting on Herbicides,* Coconut Industry Board, p. 11.

Romney, D.H. (1968) In: *Techn. Meeting on Herbicides for Tree Crops* (Jamaica), p. 5.

Rondon, A. *et al.* (1970) *Agron. Trop.* (Venezuela) 20, p. 13.

Roperos, N.I. and Bangoy, V.O. (1967) *Philippine J. Plt. Ind.*

Roy, B.N. *et al.* (1970) *Allahabad Farmer* 44(5), p. 275.

Roy, R.N. *et al.* (1970) *Two and a Bud* 17, p. 49, p. 74.

Rubber Res. Inst. (Malaya) *Ann. Reports* 1960, 1961.

Ruer, P. (1967) *Oléagineux* 22, p. 535.

———— (1969) *Oléagineux* 24, p. 327.

———— and Rosselli del Tureo, M. (1967) *Oléagineux* 22, p. 231.

Ruinard, J. (1966) *Neth. J. Agr. Sci.* 14, p. 263.

Saccas, A.M. and Charpentier, J. (1969) *Café Cacao Thé* 13, pp. 131, 221.

Said, M. and Inayatullah (1965) *Agr. Pakistan* 16(4), p. 45.

Sajoso, (1965) *Circ. Agr. Res. Sta.* (Manokwari), Agr. Ser. 10, p. 1.

Sakar, S.K. (1970) *Two and a Bud* 17, p. 65.

Salazar, C.G. (1966) *J. Agr. U. Puerto Rico* 50(4), p. 316.

Sale, P.J.M. (1965) *Ann. Rep. Cacao Res.,* p. 30.

———— (1966) *Ann. Rep. Cacao Res.,* p. 33.

———— (1967a) *Ann. Rep. Cocoa Res.,* p. 36.

———— (1967b) *Ann. Rep. Cocoa Res.,* p. 45.

Salgado, M.L.M. and Abeywardena, V.C. (1964) *Ceylon Coconut Quart.* 15, p. 95.

Salibe, A.A. (1966) *UN Dev. Prog. FAO* 2237,

I–III, p. 1.

———— and Cortez, P.E. (1966) *FAO Plant Proct. Bull.* 14, p. 141.

Salter, P.J. and Goode, J.E. (1967) *Crop Response to Water at Different Stages of Growth. Comm. Ag. Bureau Res. Rev.* No. 2.

Samosorn, S. and Ratana, G. (1972) *Tech. Rep. Rubb. Res. Centre* (Thailand) 27, p. 1.

Sampayan, T.S. (1966) *Philippine J. Plt. Ind.* 31, p. 193.

Samson, J.A. (1966) *Surenaamse Landbouw* 14(4), p. 143.

Samuels, G. (1969) *Foliar Diagnosis for Sugar Cane,* Puerto Rico.

Sanchez Nieva, F. *et al.* (1970) *J. Agr. U. Puerto Rico* 54, p. 195.

Sanderson, G.W. and Sivapalan, K. (1966) *Tea Quart.* (Ceylon) 37, p. 140.

Sartorius, G.B. (1929) *J. Agr. Res.* p. 195.

Santhirasegaram, K. (1967) *Ceylon Coconut Planters' Rev.* 1, p. 12.

———— (1967b) *Ceylon Coconut Quart.* 18, p. 38.

Satchuthananthavale, V. and Halangoda, L. (1971) *Quart. J. Rubb. Res. Inst. Ceylon* 48, p. 82.

Sato, K. (1969) *Mem. Number Okitsu Branch, Hort. Res. Sta.* 46, p. 1543.

Sato, I. (1957) *Tohoku Univ. Res. Inst. Bull.* 8, p. 173.

Sato, S. (1955) *J. Agr. Met.* (Japan) 10, p. 84.

———— (1956) *J. Agr. Met.* (Japan) 12, p. 24.

Satyabalan, K. *et al.* (1968a) *Ind. J. Agr. Sci.* 38, p. 155.

———— *et al.* (1968b) *Ind. J. Agr. Sci.* 38, p. 720.

———— *et al.* (1968c) *Trop. Agr.* (Trinidad) 45, p. 69.

———— *et al.* (1970) *Ind. J. Agr. Sci.* 40, p. 1088.

Savin, G. (1966) *Pesquisa Agropec. Brasileira* 1, p. 391.

———— and Christoi, R. (1965) *Oléagineux* 20, p. 585.

Scaife, A. (1966) *Ukiriguru Prog. Rep.* (Tanzania) 3, p. 1.

———— (1968) *East Afr. Agric. Forest J.* 33, p. 231.

Schaefers, G.A. (1969) *J. Agr. Univ. Puerto Rico* 53, p. 1.

Schatz, A.S. and Drescher, R.W. (1971) *Ser. Techn. Estac. Exp. Agropec. Concordia, INTA* 40, p. 1.

Scheiber, E. (1969) *Turrialba Rev. Interam. Ciencias Agr.* 19, p. 340.

Schoorel, A.F. (1949) *Archif. voor de thee culture* 16, p. 127.

Schreure, J. (1971) *Neth. J. Plt. Pathol.* 77, p. 113.

Schwarz, R.E. (1970) *Plt. Disease Reporter,* USDA 54, p. 1003.

Schweizer, J. (1939) *De Bergcult.* 13, p. 74.

———— (1949) *De Bergcult.* 18, p. 314 (cited by Dijkman, 1951).

———— (1936) *Hydratur und assimilations bestimmungen bei Feldver suche in den Tropen.* Handelingen le Ned. Ind. Natuurw. Congress (cited by Dijkman, 1951).

Sealy, J.B. (1958) *A Revision of the Genus Camellia. Royal Hort. Soc.*

Seberny, J.A. and Hall, E.G. (1970) *Agr. Gaz.* (NSW) 81, p. 565.

Seeyave, J. (1966) *Rev. Agr. Sucr. Ile. Maurice* 45, p. 299.

Sekhar, B.C. and Subramaniam, S. (1972) *3rd Int. Symp. on Sub-trop. & Trop. Hort.,* Bangalore. *Field crops of Ceylon,* Colombo, Lake House Investments.

Senewiratne, S.T. and Appadurai, R.R. (1966) *Field crops of Ceylon,* Colombo.

Senanayake, Y.A.D. and Samaranayake, P. (1970) *Quart. J. Rubb. Res. Inst. Ceylon* 46, p. 61.

Seow, K.K. and Wee, Y.C. (1970) *Malaysian Agr. J.* 47, p. 499.

Sequera, T.P.E. (1965) *Agronomica* (Venezuela) 2, p. 24.

Seth, A.K. and Abu Bakar bin Baba (1970) *The Planter* 46, p. 68.

Shanmuganathan, N. (1969) *Tea Quart.* (Ceylon) 40, p. 164.

Shanmugham, K.S. (1966) *Indian Farming* 16, p. 47.

Shanta, P. *et al.* (1964) *Ind. Coconut J.* 18, p. 25.

Sharma, A.K. and Roy, A.R. (1972) *Ind. J. Agr. Sci.* 42, p. 493.

Sharma, H.C. and Bajpai, P.N. (1969) *Ind. J. Sci. and Industr.* 3, p. 9.

Sharma, P.C. (1968) *Two and a Bud* 15, p. 96.

Sharples, A. (1936) *Diseases and Pest of the Rubber Tree,* London, Macmillan.

Shaukat, Ali, M.A. and Ehsen, A. (1971) *Sci. & Industr.* 8, p. 347.

Shaw, H.R. (1937) *Hawaiian Plant. Rec.* 41, p. 199 (from Dillewijn, *loc. cit.*).

Shaw, M.E. (1963) *Fertilité* 19, p. 13.

Shaw, R.D. (1971) *Oléagineux* 26, p. 155.

Sheldrick, R.D. (1965) In: *The Oil Palm,* London, Trop. Prod. Inst.

———— (1969) In: *Progress in Oil Palm,* p. 220.

Shimoya, C. (1967) *Rev. Ceres* 14, p. 13.

Shorrocks, V.M. (1964) *Mineral Deficiencies in Hevea and Associated Cover Crops*, Rubber Res. Inst., Malaysia.

———— (1965a) *J. Rubber Res. Inst.* (Malaya) 19, p. 32.

———— (1965b) *J. Rubber Res. Inst.* (Malaya) 19, p. 48.

Sideris, C.P. and Krause, P.H. (1955) *Amer. J. Bot.* 42, p. 707.

Siebert, R.J. (1947) *Ann. Missouri Bot. Garden* 34, p. 261 (cited by Dijkman, 1951).

Silva, J. Ribeiro da, and Freire, E.S. (1968) *Bragantia* 27, p. 357.

Silva, L.F. da and Cabala, R.F.P. (1966) *Techn. Working Party on Cocoa Prod.* (Rome). (Ca) 66/20, p. 1.

Silva, M.A.T. (1967) *Ceylon Coconut Planters' Rev.* 5, p. 25.

———— (1968) *Ceylon Coconut Planters' Rev.* 5, p. 108.

———— and Abeywardena, V. (1970) *Ceylon Coconut Planters' Rev.* 6, p. 59.

Silva, R.L. (1967) *Tea Quart.* (Ceylon) 38, p. 340.

———— (1968) *Tea Quart.* (Ceylon) 39, p. 42.

———— (1967) *Tea Quart.* (Ceylon) 38, p. 336.

———— and Fernando, S.R.A. (1968) *Tea Quart.* (Ceylon) 39, p. 87.

Silva, S. *et al.* (1969) *Rev. Café Puerto Rico* 24, p. 16.

Silverstre, P. (1967) In: *Proc. 1st. Internat. Symp. on Trop. Root Crop* (Trinidad) 1, p. 84.

Simmonds, N.W. (1966) *Bananas*, 2nd. ed. London, Longmans.

Simmonds, N.W. (1953) *J. Exp. Bot.* 4, p. 87.

Singh, R.B. and Shanker, G. (1968) *Allahabad Farmer* 42, p. 1.

Siregar, M. and Basuki, R. (1972) *Menara Perkebunan* 40, p. 115.

Sivaraman-Nair, P.C. and Mukherjee, S.K. (1970a) *Ind. J. Agr. Sci.* 40, p. 379.

———— and Mukherjee, S.K. (1970b) *Ind. J. Agr. Sci.* 40, p. 461.

Sivasubramaniam, S. and Talibudeen, O. (1972) *Tea Quart.* (Ceylon) 43, p. 4.

Smit, E.H.D. (1967) *Agron. Trop.* (Venezuela) 17, p. 163.

Smith, F.E. and Soria, V.J. (1968) *Cacao Inst. Interam. Ciencias Agr.* 13, p. 14.

Smith, G.W. (1966) *J. Appl. Ecol.* 3, p. 117.

Smith, R.W. (1964) *Emp. J. Exp. Agr.* 32, p. 249.

———— (1964/5, 1965/6) *Reps. Res. Dep. Coconut Ind. Board* (Jamaica), pp. 9, 35; pp. 19, 45.

———— (1968) In: *Cocoa and Coconuts in Malaya*, Inc. Soc. Planters, p. 87.

———— (1969a) *FAO. Working Party Rep.* No. 7, p. 1.

———— (1969b) *Exp. Agr.* 5, p. 133.

———— (1970) *Oléagineux* 25, p. 593.

———— and Romney, D.H. (1969) *Farmer* (Jamaica) 74, p. 411.

Smith, P.F. (1968) *Proc. Soil Crop Sci. Soc. Florida* 28, p. 97.

Smyth, A.H. (1966a) *Cocoa Growers Bull.* 6, p. 7.

Smyth, A.J. (1966b) *Soils Bull. FAO* 5, pp. 1, 4.

Snoeck, J. (1971) In: *Symp. sur lé dés herbage dés cultures tropicales*, Antibes.

Snoep, W. (1932) *Berg. Cultures* 6, p. 687 (cited by Salter & Goode, *loc. cit.*).

Soerjobroto Widjanarko (1967) *Menara Perkebunan* 36, p. 22.

Somaatmadja, D. and Ali, D. (1969) *FAO Working Party on Coconut* (Rome), No. 12, p. 1.

Soria, V.J. (1970) *Cocoa Growers Bull.* 15, p. 13.

———— and Esquivel, O. (1968) *Cacao Inst. Interam. Ciencias Agr.* 13, p. 11.

Soule, J. *et al.* (1969) *Market Res. Rep. USDA*, p. 1.

Southern, P.J. (1967) *Papua New Guinea Agr. J.* 19, pp. 18, 38.

Southorn, W.A. (1969) *J. Rubber Res. Inst.* (Malaya) 21, p. 494.

———— and Gomez, J.B. (1968) Unpublished (cited by Southorn, 1969).

Sparnaaij, L.D. *et al.* (1964) *Nigerian Inst. for Oil Palm Res. Jour.* 4, p. 111.

———— and Gunn, J.S. (1959) *West Afr. Inst. for Oil Palm Res. Jour.* 2, p. 281.

Spector, W.S. (1956) *Handbook of Biological Data*, Philadelphia and London.

Spencer, W.F. (1966) *Soil Sci.* 5, p. 296.

Srinivasan, C.S. (1969) *Conicafé* 20, p. 20.

Srivastava, R.P. and Arya, S.K. (1969) *Allahabad Farmer* 43, p. 241.

Stafford, L.M. (1970) *Aust. J. Exp. Agr. & Anim. Husbd.* 10, p. 644.

Stannard, M.C. *et al.* (1970) *Agr. Gaz.* NSW. 81(5), p. 258.

Stevenson, G.C. (1965) *Genetics and Breeding of Sugar Cane*, London, Longmans.

Stone B.C. *et al.* (1973) Biotropica 5, p. 102.

Story, G.E. and Halliwell, R.S. (1969) *Plant Disease Reporter* USDA 53(9), p. 757.

Strasburger, E. (1926) *Jarb. Niss. Bot.* 43, p. 580.

Stylianou, Y. and Orphanos, P.I. (1970) *Tech. Bull. Cyprus Agr. Res. Inst.* 6, p. 3.

Su, N.R. (1965) *Soils and Ferts.* (Taiwan), p. 104.

—————— (1965a) *Soils and Ferts.* (Taiwan), p. 49.

Subramaniam, T.R. *et al.* (1969) *J. Ind. Soc. Soil Sci.* 17, p. 33.

Sumbak, J.H. (1968) *Oléagineux* 23, p. 579.

—————— (1970) *Papua & New Guinea Agr. J.* 22, p. 1.

Summerville, W.A.T. (1944) *Queensl. J. Agric. Sci.* 1, p. 1.

Sun, V.C. and Chow, N.P. (1949) *Rep. Taiwan Sug. Exp. Sta.* 4, p. 1.

Swezey, J.A. and Wadsworth, H.A. (1946) *Hawaiian Plt. Rec.* 44, p. 49.

Tailliez, B. (1971) In: *Symp sur lé dés herbage dés cultures tropicales*, Antibes.

—————— and Valverde, G. (1970) *Oléagineux* 25, p. 335.

—————— and Valverde, G. (1971) *Oléagineux* 26, p. 753.

Tanaka, A. (1964a) In: *Mineral Nutrition of the Rice Plant*, p. 37, Johns Hopkins Press.

—————— (1964b) In: *Mineral Nutrition of the Rice Plant*, p. 419, Johns Hopkins Press.

—————— (1966) *Plant and Soil* 25, p. 201.

—————— *et al.* (1964) *IRRI Tech. Bull.* 3.

—————— Kawano, K. and Yamaguchi, J. (1966) *Int. Rice Res. Inst. Tech. Bull.* No. 7.

Tang, T.L. (1971) *Malaysian Agr. J.* 48, p. 57.

Tanimoto, T. and Nickell, L.G. (1965) *Proc. 12th ISSCT Congr.* (Puerto Rico), p. 893.

Tanner, J.W. and Jones, C.E. (1965) In: *Genes to Genus*, Illinois, Internat. Min. & Chem. Corpn.

Tasugi, H. and Yoshida, K. (1958) *Min. of Education Publn.* (Japan) 48, p. 31.

Tay, T.H. (1968) *Malaysian Agr. J.* 46, p. 458.

—————— *et al.* (1969) *Malaysian Agr. J.* 47, p. 175.

Taysum, D.H. (1961) *Nature* (London) 191, p. 1319.

Teaotia, S.R. and Singh, R.N. (1967) *Indian Agriculturist* 11, p. 45.

Teisson, C. (1970) *Fruits, France* 25, p. 210.

Teixeira, S.L. *et al.* (1971) *Rev. Ceres* 18, p. 406.

Templeton, J.K. (1968) *J. Rubber Res. Inst.* (Malaya) 20, p. 136.

—————— (1969) *J. Rubber Res. Inst.* (Malaya) 21, p. 259.

Tengwall, T.A. and Zyl, van der C.E. (1924) *Meded. Proefst.* (cited by Dillewijn, *loc. cit.*).

Teoh, K.S. (1972) *Planters Confr.*, Rubber Res.

Inst. (Malaysia) Paper No. 4.

Thorold, C.A. (1955) *J. Ecol.* 43, p. 219.

Thomas, M.D. and Hill, G.R. (1949) In: *Photosynthesis in Plants*, Eds. J. Frank and W.E Loomis, USA.

Thuljaram, J. Rao (1951) *Proc. Int. Soc. Sug. Cane Tech.* 7, (from Dillewijn, *loc. cit.*).

Thung Tjiang Pek (1966) *Menara Perkebunan* 35, p. 49.

Tidbury, G.E. (1937) *East Afr. Agr. J.* 3, p. 119.

Tisseau, M. (1969) *Fruits, France* 24, p. 241.

Tjioe Sien Bie (1965) *Warta Bulanan Balai Penyelidekan Perusahaan2 Gula* 12, p. 348.

Tolhurst, J.A.H. (1970) *Tea Jour. Tea Boards of E. Afr.* 11, p. 16.

—————— (1971) *Tea Jour. Tea Boards of E. Afr.* 11, p. 32.

Tollenaar, D. (1967) *Cocoa Growers Bull.* 8, p. 15.

Topper, B.F. (1957) *Cocoa Confr.* (Lond.), p. 49.

Torii, H. (1966) *Jap. Agr. Res. Quart.* 1, p. 20.

Torres, M.R. and Rios-Castano, D. (1968) *Proc. Trop. Region Amer. Soc. Hort. Sci.* 12, p. 107.

—————— and Rios-Castano, D. (1968a) *Agricultura Trop.* 24, p. 107.

Toxopeus, H. (1969) *Cocoa Growers Bull.* 13, p. 14.

—————— and Okoloko, G.E. (1967) *Nigerian Agr. J.* 4, p. 45.

—————— and Wessell, M. (1970) *Neth. J. Agr. Sci.* 18, p. 132.

Tree, E.F. (1966) *Qld. Agr. J.* 92, p. 546.

Trelease, S.F. (1923) *Philippines J. Sci.* 23, p. 85.

Tremlett, R.K. (1952) *Rep. Dep. Agr.* (Uganda) 49–50, p. 52.

Triana, B.J. (1957) *Centro Nac. Invest. Cafe Chinchina, Colombia* 8, p. 5.

Tsunoda, S. (1962) *Jap. J. Breeding* 12, p. 49.

—————— (1964) In: *Mineral Nutrition of the Rice Plant*, p. 401, Johns Hopkins Press.

—————— (1965) In: *Mineral Nutrition of the Rice Plant*, John Hopkins Press.

Tubbs, F.R. (1936) *Jour. of Pomology* 14, p. 317.

Turner, D.W. (1970) *Aust. J. Exp. Agr. Anim. Husbd.* 10, p. 231.

—————— (1971) *Trop. Agr.* (Trinidad) 48, p.283.

—————— (1972a & b) *Aust. J. Exp. Agr. Anim. Husbd.* 12, pp. 209, 216.

—————— and Bull, J.H. (1970) *Agr. Gaz.* (NSW) 81, p. 365.

Turner, P.D. (1960) *Trans. Brit. Mycol. Soc.* 43, p. 665.

—————— (1967) *Exp. Agr.* 3, p. 129.

Turner, P.D. and Gillbanks, R.A. (1974) *Oil Palm Cultivation and Management*: Inc. Soc. Planters, Malaysia.

Turpin, J.W. *et al.* (1970) *Agr. Gaz. NSW.* 81, p. 50.

Uchida, K. (1963) *Gendai-Nogyo* 42, p. 44 (cited by Yoshio, M. *loc. cit.*).

Urquhart, D.H. (1961) *Cocoa*, Longmans.

Vakis, N. *et al.* (1970) *Proc. Trop. Region Amer. Soc. Hort. Sci.* 14, p. 89.

Varisai Muhammed, S. *et al.* (1971) *Madras Agr. J.* 58, p. 517.

Velsen, R.J. van and Edward, I.L. (1970) *Papua & New Guinea Agr. J.* 21, p. 106.

Venkata Ram, C.S. (1967) *Planters' Chron.* 62, p. 243.

————— (1968) *Planters' Chron.* 63, p. 111.

Venkataramani, K.S. (1968) *Planters' Chron.* 63, pp. 54, 212.

————— and Venkata Ram, C.S. (1968) *Planters' Chron.* 63, p. 156.

Veno, I. and Nishiura, M. (1969) *Bull. Hort. Res. Sta. Ser. B.* 9, p. 11.

Ventocilla, J.A. *et al.* (1969) *Comun. Tech. CEPLAC* 24, p. 1.

Vergara, B.S. (1963) Paper presented to *Thursday Seminar IRRI Philippines*, (from IRRI, Rice Prod. Manual).

————— and Jennings, P.R. (1963) *Project Report, IRRI* (from IRRI Rice Prod. Manual).

Verhagen, A.M.W., Wilson, J.H. and Britten, E.J. (1963) *Ann. Bot.* 27, p. 726.

Verliere, G. (1967) *Confr. Int. Res. Agron. Cacaoyeres* (Abidjan, 1965), p. 24.

Verma, A.N. *et al.* (1970) *Allahabad Farmer* 44, p. 139.

Verma, H.P. (1965) *Ind. Sug. J.* 9, p. 219.

Vernon, A.J. (1967) *Trop. Agr.* (Trinidad) 44, p. 223.

————— (1971) *Cocoa Growers Bull.* 16, p. 9.

Vijaendraswamy, R. (1971) *Ind. Coffee* 35, p. 115.

Vilardebo, A. *et al.* (1972) *Fruits, France* 27, p. 777.

Villezar, W.J. and Oyardo, E.O. (1968) *Philippine J. Plt. Ind.* 33, p. 49.

Vincente-Chandler, J. *et al.* (1969) *J. Agr. Univ. Puerto Rico* 53, p. 124.

Vine, P.A. and Mitchell, H.W. (1969) *Kenya Coffee* 34, p. 24.

Vink, N. *et al.* (1971) *Econ. Rurale Libanaise* 36, p. 62.

Vishveshwara, S. and Govindarajan, A.G. (1970) *Ind. Coffee* 34, p. 71.

Visser, T. (1969) *Neth. J. Agr. Sci.* 17, p. 234.

Voelcker, O.J. (1955) *Cocoa Confr.* (Lond.) (cited by Murray, 1964).

Vollema, J.S. (1949) *Arch. v.d. Rubbercult in Ned. Indie.* 26, p. 257.

————— and Dijkman, M.J. (1939) *Arch. v.d. Rubbercult in Ned. Indie.* 23, p. 47.

Vossen, H.A.M. van der (1969) *Ghana J. Agr. Sci.* 2, p. 113.

Wadsworth, H.A. (1949) *Index to Irrigation Investigations in Hawaii*, U. of Hawaii.

Waidyanatha, U.P. de S. and Pathiratne, L.S. (1971) *Quart. Rubb. Res. Inst.* (Ceylon) 48, p. 47.

Waldron, J.C., Glasziou, K.T. and Bull, T.A. (1967) *Aust. J. Biol. Sci.* 20, p. 1043.

Walker, W.M. and Melsted, S.W. (1971) *Trop. Agr.* (Trin.) 48, p. 237.

Walter, A. (1910) *The Sugar Industry of Mauritius: A Study in Correlation*, Lond. (from Dillewijn, *loc. cit.*).

Wardlaw, C.W. (1935) *Diseases of the Banana*, Lond.

————— (1961) *Banana Diseases*, Lond.

Watts, I.E.M. (1957) *Equatorial Weather*, N.Y., Pitman Publishing Corpn.

Weathers, L.G. *et al.* (1969) *Agric. Tech.* (Chile) 29, p. 166.

Webster, B.N. (1953) *Tea Quarterly* 24, p. 26.

Wee, Y.C. (1969) *Malaysian Agr. J.* 47, p. 164.

————— and Ng, J.C. (1968) *Malayan Agr. J.* 46, p. 469.

Weller, D.M. (1937) *Repts. Assoc. Hawaiian Sug. Tech.* 6, p. 73.

Wellman, F.L. (1961) *Coffee*, London, Leonard Hill.

Wen, L.Y. and Ponnamperuma, F.N. (1966) *Plant and Soil* 25, p. 347.

Wessell, M. (1970) *Cocoa Growers Bull.* 15, p. 22.

————— (1967–8) *Ann. Rep. Cocoa Inst.* (Nigeria), p. 38.

————— *et al.* (1967) *Confr. Int. Res. Agron. Cacaoyeres* (Abidjan, 1965).

Westgarth, D.R. and Buttery, B.R. (1965) *J. Rubber Res. Inst.* (Malaya) 19, p. 62.

Weststeijn, G. (1967) *Mem. Cocoa Res. Inst.* (Nigeria) 16, p. 1.

Wettasinghe, D.T. (1968) *Tea Quart.* (Ceylon) 39, p. 119.

Whitehead, R.A. (1965a) *Econ. Bot.* 19, p. 267.

————— (1965b) *Trop. Agr.* (Trinidad) 42, p. 369.

——— (1966a) *Trop. Agr.* (Trinidad) 43, p. 277.

——— (1966b) 'Coconut Germplasm in the Pacific.' *M.O.D. Res. Publ.* No 16.

——— *et al.* (1966) *Oléagineux* 21, p. 153.

——— and Smith, R.W. (1968) *Trop. Agr.* (Trinidad) 45, p. 127.

Whitlam, A.E. (1966) *Tea Quart.* (Ceylon) 37, p. 237.

Wickremasuriya, C.A. (1968) *Ceylon Coconut Quart.* 19, p. 152.

Wier, C.C. (1966) *Bull. Citrus Res.* 5, p. 1.

——— (1968) *Proc. Trop. Region Amer. Soc. Hort. Sci.* 12, p. 138.

——— (1969a) *Techn. Bull. Citrus Res. U. of W. Indies* 2, p. 1.

——— (1969b) *Plant and Soil* 30, p. 405.

——— (1970) *Caribbean Farming* 2, p. 15.

Willatt, S.T. (1971) *Agr. Metrology* 8, p. 341.

Willatt, S.T.A. (1970, 1971) *Trop. Agr.* (Trinidad) 47, p. 243; 48, p. 271.

Williams, A.H., Wallace, T. and Marsh, R.W. (1953) *Science and Fruit*, University of Bristol.

Williams, C.N. (1971) *Exp. Agr.* 7, p. 49.

——— (1972) *Exp. Agr.* 8, p. 15.

——— (1974) *Exp. Agr.* 10, p. 9.

——— and Ghazali, S.M. (1969) *Exp. Agr.* 5, p. 183.

——— and Hsu, Y.C. (1970) *Oil Palm Cultivation in Malaya*, University of Malaya Press.

——— and Joseph, K.T. (1970) *Climate, Soil and Crop Production in the Humid Tropics*, Oxford U. Press.

——— and Rajaratnam, J.A. (1971) *World Crops* 23, p. 182.

——— and Thomas, R.L. (1970) *Ann. Bot.* 34, p. 957.

Williams, E.N.D. (1971) In: *Water and the Tea Plant*, Tea Res. Inst. E. Afr.

Williams, R.F. and Williams, C.N. (1968) *Aust. J. Biol. Sci.* 21, p. 835.

Wills, J.M. (1951) *Queensl. Agric. J.* 72, p. 147, p. 223.

Wilson, K.C. (1967/8) *Tea Jour. Tea Boards of E. Afr.* 8, p. 19; 9, p. 23.

Winter, E.J. (1969) *Tea Jour. Tea Boards of E. Afr.* 10, p. 21.

Wohl, J. and James W.O. (1942) *New Phythologist* 41, p. 230.

Wong, T.K. and Varghese, G. (1966) *Exp. Agr.* 2, p. 305.

Wood, B.J. (1969) *The Planter* (Malaya) 45, p. 510.

Wood, G.A.R. (1963) *Cocoa Growers Bull.* 1, p. 6.

——— (1964) *Cocoa Growers Bull.* 2, p. 16.

Woodhead, T. (1969) *Agr. Meterol.* 6, p. 195.

Wormer, T.M. (1965) *Ann. Bot.* 29, p. 523.

——— and Gituanja, J. (1970a) *Kenya Coffee* 35, p. 270.

——— and Gituanja, J. (1970b) *Exp. Agr.* 6, p. 157.

Wright, W. (1942) *Indian Tea Assoc.*, Memorandum No. 15.

Wycherley, P.R. (1963) *J. Trop. Geog.* 17, p. 143.

Wycherley, P.R. *et al.* (1964) *16th Int. Hort. Congr.* (1962), Brussels.

Yahampath, C. (1968) *Quart. J. Rubb. Res. Inst.* (Ceylon) 44, p. 27.

Yamane, I. and Sato, K. (1961) *Tohoku Univ. Inst. Agr. Res. Rept.* 12, p. 73.

Yanase, Y. (1970) *Jap. Agr. Res. Quart.* 5, p. 21.

Yoon, P.K. (1967) *Planters Bull. Rubber Res. Inst.* (Malaya) No. 92, p. 240.

——— (1971) *Planters Confr. Rubber Res. Inst.* (Malaysia) Papers No. 9 & 10.

——— (1972a) *Planters Confr. Rubber Res. Inst.* (Malaysia) Proceedings, p. 143.

——— (1972b) *Technique of Crown Budding*, Rubber Res. Inst. Malaysia.

——— (1972c) *Planters Confr. Rubber Res. Inst.* (Malaysia) Paper No. 5, plus supplement.

Yoshi, H. (1941) *Kyusu Imp. Univ. Sci. Fakultato Terkultura Bull.* 9, p. 277.

Yoshio Murata (1966) *FAO Newsletter* 15(2).

——— (1966) *IRRI Newsletter* 15, p. 20.

Yuji Kawamura (1971) In: *3rd Asian Pacific Weed Confr.*, Malaysia.

Zelcer, A. *et al.* (1971) *Israel J. Agr. Res.* 21, p. 137.

Zelitch, I. (1967) In: *Harvesting the Sun*, Eds. San Pietro F.A. Greer & T.J. N.Y. Army, Acad. Press.

Zevallos, A.C. and Alvim, P. de T. (1966) *Cacau Atualidades* 3, p. 2.

Zeven, A.C. (1967) *Agr. Res. Repts.* No. 689.

Zuniga, L.C. (1965) *Philippine J. Plt. Industry* 29, p. 19.

——— (1969) *Philippine J. Plt. Industry* 34, p. 9.

INDEX